Aquaculture in the Ecosystem

Marianne Holmer • Kenny Black
Carlos M. Duarte • Nuria Marbà
Ioannis Karakassis
Editors

Aquaculture in the Ecosystem

Springer

Marianne Holmer
Institute of Biology
University of Southern Denmark
Campusvej 55, 5230 Odense M
Denmark
holmer@biology.sdu.dk

Kenny Black
Scottish Association for Marine Science
Oban, Argyll
Scotland, PA37 1QA
kenny.black@sams.ac.uk

Carlos M. Duarte
Institut Mediterrani d'Estudis
Avançats (CSIC-UIB)
Miquel Marquès 21
07190 Esporles (Illes Balears)
Spain
carlos.duarte@uib.es

Nuria Marbà
Institut Mediterrani d'Estudis
Avançats (CSIC-UIB)
Miquel Marquès 21
07190 Esporles (Illes Balears)
Spain
nuria.marba@uib.es

Ioannis Karakassis
Marine Ecology Laboratory
Biology Department
University of Crete
PO Box 2208
Heraklion 714 09 Crete
Greece
karakassis@biology.uoc.gr

ISBN-13: 978-90-481-7732-5

e-ISBN-13: 978-1-4020-6810-2

Softcover reprint of the hardcover 1st edition 2008

Printed on acid-free paper

9 8 7 6 5 4 3 2 1

springer.com

Foreword

Aquaculture in the Ecosystem – An Introduction

The growth of Aquaculture and its future role as a food supplier to human society has environmental, social and economic limitations, affecting marine ecosystems and socio-economic scales from local to global. These are close links with human health requirements and societal needs for various goods and services provided by marine ecosystems. This book shows this broad spectrum of dependencies of the future growth of aquaculture and highlights both relevant problems and expectations.

Compensating for stagnant wild capture fisheries and the increasing demand for marine products, marine aquaculture is one of the fastest growing industries in the world, comparable to the computer technology industry (Chapters 9 and 10). The demand for marine products is controlled by a complexity of factors in our society, not least the increasing human population and the increasing global affluence that allows the consumer to buy higher priced marine products such as salmon, tuna and shellfish (Chapter 9). The populations of several of these top-carnivore species are seriously compromised and it will be impossible in the future to maintain wild captures at the level of consumer demand. In less affluent areas including SE Asia and Africa, aquaculture for both domestic consumption and export has major nutritional and economic benefits. The production of fish in aquaculture is thus expected to increase under the assumption that the bottlenecks for expansion can be overcome (Chapter 10). This book discusses a range of bottlenecks, not only the environmental, but also technological, social and economic constrains.

Aquaculture is an ancient activity enduring over millennia. Cultivation in historic times was primarily for domestic use but, at the beginning of the 20th century, larger farms started to appear, such as rainbow trout farms in fresh water ponds in Northern Europe (FAO 2006). Since then the number of species domesticated for aquaculture production has increased exponentially now exceeding the number of species domesticated on land (Duarte et al. 2007). There is a large potential for further species in aquaculture as only about 450 species are currently cultured out of about 3,000 aquatic species used for human consumption. Characteristically, the first initiatives in aquaculture were simple, low technology systems with limited demands for maintenance and low operating costs. These aquaculture systems were dependent on high water quality which was often easy to achieve because of their low intensity. It was not until greater intensification of aquaculture in the 1970s, increasing the pressure on the environment significantly, that it became urgent to monitor and regulate aquaculture (Chapter 2). The current expansion rate in world aquaculture production of 3.5–4.6% yr^{-1} can only be sustained if the major pressures exerted on the environment and dependence on natural resources, such as feed and brood stocks (Chapter 10), are reduced.

With regard to regulation and monitoring at present time, the Water Framework Directive (WFD) is being implemented all over Europe and will become important for the regulation of aquaculture and other human activities in the coastal zone (Chapter 1). Chapter 1 clarifies present understanding of eutrophication and provides an insight into water quality models on as they are expected to be used under the WFD, providing examples from Scotland different scenarios for the future regulation of marine aquaculture in the coastal zones. Aquaculture producing countries outside Europe regulate aquaculture activities through a number of different laws and conventions, often with several laws enforced on different aspects of the production cycle (Chapter 2). In Norway, which is one of the top five producers in the world and where the production of salmon in net cages in the coastal zone is an important contributor to the national economy, the monitoring of environmental impacts of the industry has been developed since the beginning of the industry 30 years ago and is now a classified program according to national standards implemented throughout the country (Chapter 2). As an example of a more recent developed program, the monitoring in Malta is presented (Chapter 2). During the 1990s, the Mediterranean experienced an exponential growth in the production of sea bream and sea bass in net cages and, as the environmental conditions in the Mediterranean are unique (e.g. widespread oligotrophy), some of the environmental pressures differ considerably from those in Northern Europe. One example is the prevalence of seagrass meadows of the species *Posidonia oceanica* as a benthic ecosystem along Mediterranean coasts. As this is a sensitive ecosystem, facing general declines in the coastal zone (Marbá et al. 2005), it is important to monitor this ecosystem in fish farm surroundings to avoid accelerating declines (Chapter 2). Tuna farming (or ranching) is a major activity in Malta as well as in several other Mediterranean countries and, although it is debated whether this industry is "real" aquaculture or should be considered as a fattening industry instead, the environmental impacts differ from sea bream and sea bass aquaculture due to the use of wet feed (fresh/frozen fish) instead of dry feed pellets.

A new development in aquaculture monitoring and regulation, which will play an important role for future development, is in considering aquaculture as an integrated part of the marine ecosystem. This means that aquaculture should be managed together with a number of other industries and other users of the marine ecosystem (Chapter 3), but also that the production is a part of ecosystem and has to be managed at different scales, not only the water column and sediment floor in the vicinity of the net cages, but also at larger scales in the coastal zones (Chapter 1). One example of scale can be found in Chapter 5, which addresses the issue of introductions of alien species into coastal zones caused by aquaculture operations. This is particularly important since it is well known that aquaculture is the second most important vector for species introductions after maritime transport. Also the attraction of wild fish to net cages adds constraints to the ecosystem structure and function, in particular in areas such as the Mediterranean, where wild fish are abundant around cages and may be more available to fisheries (Chapter 3). Although the presence of wild fishes at the farms can minimize the environmental impacts, e.g. through reducing inputs of organic matter to the seafloor, there are risks such as transfer of diseases to wild populations (Chapter 3). A related issue is the genetic pollution of wild stocks through either inadvertent (as in farm escapes) or deliberate (as in stocking/ ranching) introduction of cultured species into the wild (Chapter 4). Genetic impacts have been extensively studied for salmon in Northern Europe, where there are problems with interbreeding, and are now under consideration for other cultured species such as sea bream and sea bass in the Mediterranean and for other species in the tropics (Chapter 4). Chapter 4 discusses the possible future solutions to the genetic interactions between farmed and wild fish.

One major constrain to aquaculture growth is the availability of fish meal and fish oil for production of carnivore fish (Chapters 6 and 10). There is currently a major research effort in optimizing feed through substituting fish meal and oil with vegetable flour and oil. As there is substantial scientific evidence of human health benefits from consumption of marine products, primarily due to the omega-3 fatty acids, the aims of the current research is to maintain the composition of the cultured fish product while reducing dependence on fishery-derived feedstocks (Chapter 6). There are also other future options for solving the bottle neck of feed availability, which involve not only breakthroughs in feed technology but also changing the way humanity interacts with the oceans (Chapter 10). Such breakthroughs could be through use of marine plants for feed or moving production from carnivore to herbivore species.

Aquaculture is expected to develop along two main lines, either in net cages at sea or on land-based facilities (Chapter 10). To keep up with the production needs the size of the farms will expand and net cage farms will move from coastal sites to open-ocean locations. Land-based farms have the advantage of reuse of the water and treatment facilities, but are at the present constrained by high energy costs.
In addition to technological constrains there are several other bottlenecks, which are less predictable. These are related to attitudinal issues (Chapters 8 and 10) and to the economic development of the industry (Chapter 9). Aquaculture production has for instance become of active interest to a number of non-governmental organizations (NGOs) around the world, which is discussed in Chapter 7. NGO concerns about aquaculture are not solely in its growth or where the product is consumed. Rather, their interest is in the on-the-ground environmental or social impacts that threaten or undermine the NGO's ability to deliver on their overall missions of conservation or social welfare. Public and consumer attitudes and legislation, related to, e.g., ethics, environment and health can play important roles, such as observed with the threatened bird flu pandemic, where suddenly almost every consumer stopped eating chicken. This did affect the sales of salmon from aquaculture positively, whereas the news on high dioxin levels in cultured salmon resulted in a

major, if transitory, reduction in the consumption of fish. One possible way to comply with public attitudes and to impose legislation is through resolution of externalities through monetary valuation of the interactions between aquaculture and the environment and vice versa (Chapter 8). Externalities can be used for policy formulation, e.g., through introduction of environmental taxes and make the producer aware of the environmental costs.

Changes in the market may significantly affect the development of the aquaculture industry, as production only takes place if there are economic benefits to the producer. Chapter 9 analyses the past development in the economics of the industry and from this analysis predicts future trends. It is predicted that production will move towards a few high-volume species supplemented with a large number of small-volume species for local markets. High-volume species have the advantage of predictability and can be sold in the large and global supermarket chains, where weekly sales can be promoted founded on the stability of delivery. High-volume productions are characterized by relatively low production costs. On the other hand, the small-volume species can be sold at a higher price at local markets depending on season and demand.

Aquaculture has increased tremendously in the last decades and is predicted to continue this increase. The aim of this book is to provide a scientific forecast of the development with a focus on the environmental, technological, social and economic constraints that need to be resolved to ensure sustainable development of the industry and allow the industry to be able to feed healthy seafood products to the future generations.

References

Duarte CM, Marbà N, Holmer M. (2007) Rapid domestication of marine species. Science 316:382–383

FAO (2006) FAO-STAT (UN Food and Agriculture Organization, Rome) http://www.fao.org/fi/

Marbà N, Duarte CM, Díaz-Almela E, Terrados J, Álvarez E, Martinez R, Santiago R, Gacia E, Grau AM (2005) Direct evidence of imbalanced seagrass (*Posidonia oceanica*) shoot population dynamics in the Spanish Mediterranean. Estuaries 28:53–62

Contents

Chapter 1
Fish Farm Wastes in the Ecosystem

Paul Tett

Abstract Fish farms release dissolved and particulate waste into the ecosystem and the most important impacts on the water column and the sediments are described at different scales (A, B, C zones). An overview of the ethical and legal frameworks for management of aquaculture is given, introducing the ecosystem approach to regulation through the DPSIR (Driver-Pressure-State-Impact-Response) approach and EQSs (Environmental Quality Standards). The Scottish loch Creran is used as a case study due to the existence of long term monitoring and the presence of aquaculture in the loch. Finally the prospects for management of aquaculture within the European Water Framework Directive is discussed, and it is predicted that the implementation may either result in limited changes (e.g., same practice but outphasing of environmental hazards) or major changes (e.g., ecosystem approach to aquaculture through polycultures) to Scottish regulation.

Keywords Eutrophication, water framework directive

1.1 Introduction

This chapter is about the interactions between fish-farming and its environment, and how these interactions might be managed in the best interests of ecological sustainability. Despite humanity's generally bad record in this respect, there is evidence that we can learn how to live with, as well as in, Nature (Diamond 2005). There is an increasing will to do this, made concrete within the European Union by the Water Framework and other Directives, and an increasing body of scientific knowledge that can be used for management. I aim to give overviews of both the relevant science and an ethical and legal framework for management. This framework grows out of

---

School of Life Sciences, Napier University, 10 Colinton Road, Edinburgh EH10 5DT, Scotland.
Tel. (+44) 0131-455-2526; E-mail: p.tett@napier.ac.uk

M. Holmer et al. (eds.), *Aquaculture in the Ecosystem.*

the "ecosystem approach", which is grounded not only in the scientific theory of ecosystems but also in views about how we might or should try sustain our species' existence on spaceship Earth. Unlike the planetary-scale problem of global warming, the fish farm–environment interaction is more tractable both to management and to discussion within the space of this chapter: it largely takes place on space and time scales that are easy to see. Nevertheless, the general principles are the same, and if we cannot deal with the impacts of fish-farming – and I think we can – we are unlikely to be able to deal with the bigger matters.

Because I am writing for regulators, policy makers, human health and nutrition community, and coastal zone managers, as well as post graduate students in the field of aquaculture, I include in this chapter some accounts of ecological principles and attempt to explain them without assuming any prior ecological knowledge. And so I start by explaining why there are concerns about the environmental impact of marine aquaculture.

1.2 Humans and Pollution

Once upon a time there was (or may have been) an Edenic age in which small bands of Eves and Adams and their children wandered through a unspoilt Mediterranean landscape of small woods and pastures, trapping wild animals and tending wayside gardens where grew the plants that later became fully domesticated (Mithen 2003). These small bands stopped for the night or perhaps for a few weeks before moving on, and, like all humans, they pissed and shat and threw away uneaten bones or fruit. As human population density, and agricultural skills, increased, the settlements grew larger and less temporary: but never long-lasting, because human wastes polluted water supplies, and wood cutting and agriculture damaged local ecosystems. So villages rose and decayed, and populations moved on, or died from disease and malnourishment, until humans began to learn how to regulate their waste.

It became possible to live in cities, giving rise to another period of population increase and environmental pollution. Classical Rome dealt with waste by piping it down a "cloaca maxima" into the Tiber, where it was flushed out to sea; but elsewhere, Roman mining of metals such as copper and silver created toxic zones where the soils were rich in heavy metals and streams ran red with acid water. By the late 19th century most large European cities had recreated Roman sanitation, and by the late 20th century most European countries were trying to decrease pollution by industrial poisons. But at the same time, the growing populations of these cities required, and provided markets for, huge quantities of food, which increasingly tended to be produced by semi-industrial methods.

Some of this food came initially from the exploitation of populations of wild fish: but the supply of this apparently free resource was often unpredictable because the fish had to be caught far from land and in all weathers, and their imperfect management led to overfishing. In consequence, aquaculture has grown to provide a replacement source of marine protein, albeit sometimes by converting small fish into larger ones. And, just as was the case during the early development of human societies, this farming initially generated large amounts of waste, which accumulated in an environment hitherto thought to be pristine.

The metabolism of fin-fish is not dissimilar to that of humans, and, like people, fish produce solid and dissolved wastes. Waste food and faeces voided into the water tend to sink to the seabed. Many farmed fish are carnivores, and so must be fed a protein rich diet, which they use inefficiently compared with the herbivores and omnivores that are farmed on land. Consequently, they excrete dissolved compounds of nitrogen (especially, ammonia) and phosphorus (especially, phosphate) by way, mainly, of their gills. These processes are natural; the problems due to these wastes arise from intensive or semi-intensive farming, which takes in food from an extensive region but concentrates the waste in a much smaller area around a farm.

As an example, a farm stocked with 200,000 young salmon, and harvesting about a thousand tonnes of fish towards the end of a 2-year production cycle, uses about 1,200 t of feed made from 3,600 to 5,900 t of wild fish (according to conversion ratios in (Black 2001)). The food supply represents a share of the primary organic production of hundreds of square kilometres of sea. During the second year of the cycle the farm releases an amount of nitrogen, phosphorus, and faecal matter similar to that in the untreated sewage from several tens of thousands of humans. But whereas these people would inhabit at least a few square kilometres even in the most densely settled European cities, typical netpen farms of this size cover only a fraction of a square kilometre. Furthermore, whereas the most human and industrial wastes are now, in cities in the developed world, collected and treated before discharge, farm waste enters directly into the sea.

Although such wastes are in themselves natural, and so harmful only in excess, some mariculture results in the production of a second category of wastes. These are the man-made chemicals used to treat fish for disease, to make them grow faster, or to prevent seaweeds, seasquirts and barnacles from growing on fish cages. Speed-reducing *fouling* by these organisms has long been a problem for ships, and the success of the British Navy during the Napoleonic wars was partly due to the use of copper plating to prevent fouling of their wooden hulls (Rogers 2004). Copper is expensive, however, and can cause problems due to electrolytic corrosion, and there was a search for other compounds that could be applied to hulls in paint. The invention of the antifouling compound *tributyl tin, or TBT,* seemed to be a break-through. After several decades of use, however, it was found to be harmful to marine invertebrates, causing female dogwhelks to grow penises and farmed oysters to become mis-shapen (Readman 2005). It is now banned from use by fish farms and all small craft that anchor in coastal waters.

Thus, nutrients, organic matter and toxic pollutants have the potential to do harm to marine organisms. Their actual impact depends, however, on the environment into which these wastes are released. The next section looks at the properties of one type of environment much used for aquaculture, and uses this example of a water body to explain the idea of *an ecosystem.*

1.3 The Ecosystem in Loch Creran

The west coast of Scotland is cleft in many places with long arms of the sea. Called *loch* in Scots Gaelic (with the final *ch* a soft sound made in the back of the mouth), most are technically *fjords*: river valleys internally deepened by glaciers during the Ice Age and then flooded with salt water as the level of the ocean rose when the main ice sheets melted. For several millennia, these sheltered sea-lochs have provided highways and food sources to the people who lived in this otherwise unproductive and mountainous region. Now they are both a tourist attraction and a site for fish-farming, especially Atlantic salmon and mussels.

Halfway up this coast, the large fjord of the Firth of Lorne runs north-eastwards, along the line of the Great Glen fault that separates two ancient tectonic plates and continues to shake us locals with mini-earthquakes about once a decade. Big fjords often have little fjords, made by tributary glaciers, and the Firth of Lorne is no exception: loch Spelve, on the island of Mull, and on the mainland side, lochs Eil, Linnhe, Leven, Creran, Etive, Feochan and Craignish. All these have the characteristic feature of a fjord: a narrow and shallow entrance, with at least one deeper and wider basin inside. My friend Anton Edwards once wrote that although there is no such thing as a typical sea-loch, if you make lists of the Scottish saltwater lochs ranked in terms of their physical attributes, such as greatest depth, or freshwater inflow from the rivers discharging to their heads, then Creran comes close to the middle of most lists.

Seem from the top of a nearby hill, Creran looks like a lake: the winding channel that connects it to the Firth of Lorne is hidden behind a wooded hill (Fig. 1.1).

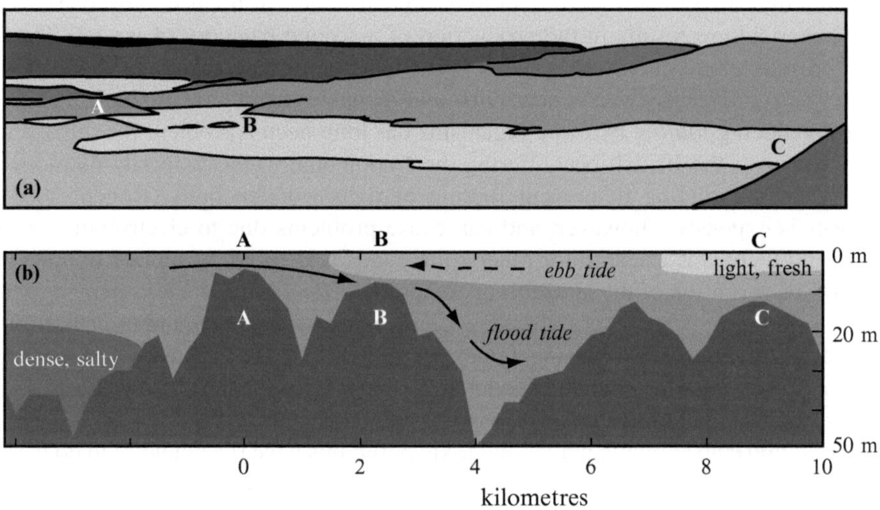

Fig. 1.1 A Scottish site for aquaculture: (a) sketch of loch Creran, looking west towards the larger fjord of the Firth of Lorne; (b) section, showing density and deduced circulation

But through this channel come pouring millions of cubic metres of salt water on each rising tide, and a slightly greater volume leaves on the ebb tide, swirling past small islands where seals lie and black birds perch on the lookout for fish. The outflow volume is greater because it must include the water added by rivers: in normal circumstances only a few percent of the tidal flow, but with a major effect on the circulation within the loch. Fresh water is less dense than salt water, and, where it mixes with seawater forms a lighter superficial layer that floats seawards, while the heavier saltwater, brought in by the tide, penetrates underneath.

This circulation renews water and oxygen within the loch, and creates good conditions for the growth of the fish and seabed animals that feed the seals and birds. On the seabed, there were once-abundant beds of the European oyster, and there still are extensive reefs made from the calcareous tubes of serpulid worms. Both oysters and serpulid worms are members of the **benthos**. Some benthic animals feed on organic matter within seabed mud, but the oysters and serpulids get food by filtering suspended particles. The most nutritious of these are the tiny floating algae of the phytoplankton, too small to be seen, as individuals, by the naked human eye. These micro-algae are well known as the "grass of the sea", the main marine source of organic food made by photosynthesis. When my colleagues and I studied it (Tett et al. 1985; Tett and Wallis 1978), Creran was typically rich in a variety of phytoplankters, especially those belonging to the group known as *diatoms*, which absorb dissolved silica from sea-water and use it to make glassy cases for their cells. The circulation of water through the loch provided a continuing source of compounds of nitrogen, phosphorus and silicon; and the layering created by the freshwater input allows phytoplankters to remain in a superficial layer that is well-lit by sunlight for much of the year.

Phytoplankton is not the only source of organic food in Creran: seaweeds are also important *primary producers*, and there is a further input of dead organic matter from rivers (Cronin and Tyler 1980; Tyler 1984). But I have described enough to make my point: that loch Creran is an *ecosystem*, a term invented by Roy Clapham in 1930, published by Arthur Tansley (1935) and defined by Eugene Odum (1959) as

> *any area of nature that includes living organisms and nonliving substances interacting to produce an exchange of materials between the living and nonliving parts…*

Formally, the *nonliving substances* form the *environment* and the *living organisms* form the (biotic) *community*; but a ecosystem is not simply *environment* plus *community* but also the interactions between and amongst them; it is both *structure* and *function* – the food web and how it works.

Thus, the interactions in loch Creran include the biogeochemical fluxes of organic matter and nutrients amongst the biota and between them and their surroundings; the effects of the serpulid reefs in stabilizing the seabed in Creran; the transport of animal as well as micro-algal plankton by currents; the addition of oxygen by algal photosynthesis and air–sea exchange, and its consumption by the respiration of all the animals and bacteria living in the waters of the loch or on or in its seabed. By analogy with human health, we can say that an ecosystem is healthy when all its parts are in good order and also when the interactions are in balance with the needs of the biota. This is a topic to which I'll return later – but for now, please note a significant difference between the health of a human – for whom the environment is something outside of the body and which is seen as a factor conducive to good or bad health, depending on whether air or water is clean or polluted – and the health of an ecosystem – which includes the state of the non-living part. Suppose we add a fish farm – either fin fish or shellfish – to an ecosystem such as Creran. Should we view the farm as bolted on to the outside of the ecosystem – potentially able to perturb it through waste products and liable to harm if some of this waste, for example, decays and consumes oxygen – or as an addition to the loch's ecosystem, participating in the *exchange of materials*? And what about the humans who operate the farm and truck in fishmeal caught in distant seas?

1.4 Aquacultural Pressures and Potential Impacts on Ecosystems

Any fish farm is a site of concentrated food production. Shellfish such as mussels take their food from the water flowing past them, and so one of their impacts on the ecosystem is the removal of the phytoplankton that forms much of this food. Depending on the extent of water movements, a mussel farm may harvest planktonic primary production from a wide area of sea – an area much greater than the extent of the mussel farm itself.

In contrast, the feed given to farmed salmon is largely made from other fish, caught in a different part of the ocean, but again harvesting the primary production of much wider area of sea than the extent of the fish farm. Think of both types of farm as the drain at the end of a bath, a vortex through which must flow large quantities of material. Both mussels and salmon draw oxygen from the water to support their metabolism of this food, and, because of the vortex effect, can potentially cause oxygen depletion – which would be fatal for the fish and shellfish. The way to avoid this is to site a farm in a region of strong water flow – which will also carry away the potentially toxic ammonia released by the animals' metabolism, and any other harmful dissolved substances such as those involved in ridding salmon of sea-lice or preventing fouling on nets.

However, although *the answer to pollution is dispersion and dilution*, the dilution of fish farm wastes has to be sufficient for undesirable ecological consequences to be avoided. It is, unfortunately, possible to site a farm in a region of flow sufficiently strong to avoid oxygen depletion or ammonia build-up around the farm, but insufficiently flushed to avoid the accumulation of wastes on a larger scale. Bearing this in mind, let us look at three types of potential ecological disturbance associated with fish-farming. Figure 1.2 exemplifies these in a fjord, but most can occur anywhere in the sea.

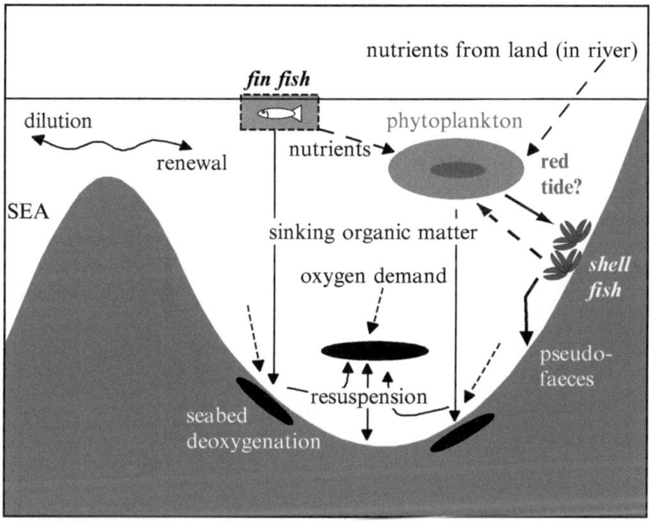

Fig. 1.2 Effects of aquaculture in a fjord

The first type of disturbance is a result of fall of fish faeces, uneaten food, and similar, towards the seabed. Water currents and eddies disperse these particles, and their "footprint" on the seabed depends on water depth and turbulence. In small amounts this organic matter provides food for benthic animals and demersal fish, but when it accumulates on the seabed, it can block the supply of oxygen to burrowing animals and can drive an increase in oxygen consumption by micro-organisms. It may be that all oxygen is removed from the water between sediment particles, leading to the replacement of aerobic bacteria (which release carbon dioxide as a product of metabolism) by anaerobic bacteria, whose by-products are methane, sulphur, and poisonous hydrogen sulphide. The effects of increasing organic input on the benthic fauna in fjords was systematically described by Pearson and Rosenberg (1976, 1978) in relation to the waste from wood pulp processing, and although fish-farm waste is more labile and nutrient-rich, it seems to have much the same effect – shown in simplified form in Fig. 1.3(a).

The second kind of potential disturbance is *eutrophication*, defined by OSPAR (2003) as

> *the enrichment of water by nutrients causing an accelerated growth of algae and higher forms of plant life to produce* ***an undesirable disturbance to the balance of organisms present in the water and to the quality of the water concerned****…*

These nutrients are the dissolved compounds of nitrogen and phosphorus – especially nitrate, ammonium and phosphate – which are necessary for the growth of photosynthetic organisms. Eutrophication thus defined is different from the effects of the organic matter needed by animals and by non-photosynthetic

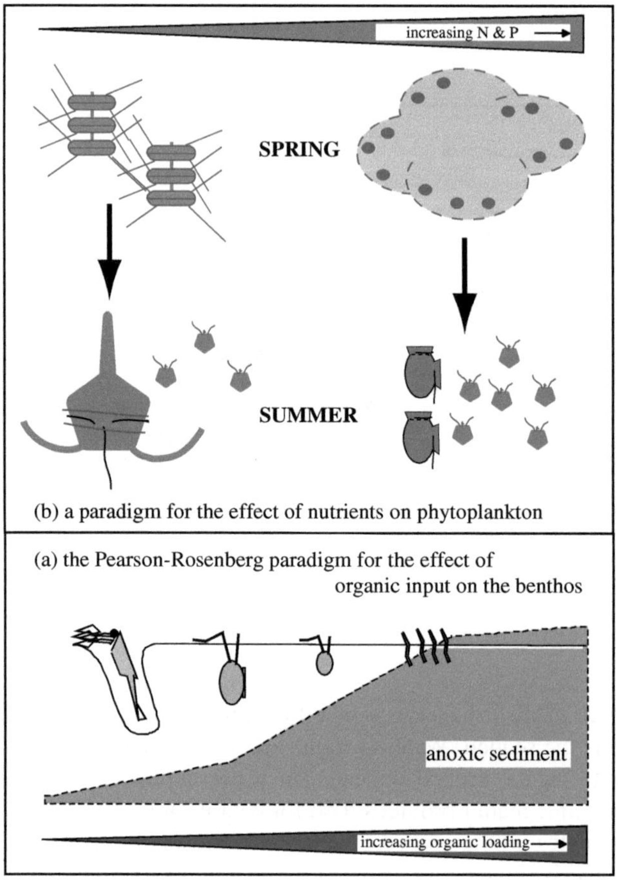

Fig. 1.3 Paradigms for disturbance: (a) Pearson–Rosenberg paradigm Pearson & Rosenberg (1976, 1978), for effects of organic waste, increasing in amount from left to right, leading initially to the loss of water-pumping animals (bio-irrigators) and finally to complete replacement of oxygen-requiring organisms by anaerobes; (b) an attempt, inspired by Margalef (1978) to schematize the phytoplankton response to anthropogenic nutrient enrichment of temperate waters; the diatom-(dino)flagellate seasonal succession is shown giving way to gelatinous colonial algae in the spring and to toxic dinoflagellates and small flagellates during summer

micro-organisms. The key distinction is that the growth stimulated by the mineral nutrients is accompanied by the photosynthetic release of oxygen, whereas growth on preformed organic matter consumes oxygen. Of course, the first may lead to the second, recycling the nutrient elements nitrogen and phosphorus back into their mineral forms, and consuming the oxygen released during photosynthesis. The problems associated with eutrophication typically come about when the coupling between the first and second parts of this natural cycle is weakened because of excess primary production and the formation, in the absence of sufficient grazing by planktonic or benthic consumers, of excess phytoplankton or seaweed biomass.

Thus, the harmful consequences that may result from nutrient enrichment include increasing frequencies and intensities of *Harmful Algal Blooms (HABs)*, including Red Tides, nuisance blooms causing foaming, toxic blooms that can kill farmed fish, and increased occurrences of incidents of shellfish-vectored toxins, such as those causing paralytic shellfish poisoning (Anderson and Garrison 1997). If blooms sink into deeper water, the decay of their biomass can cause oxygen depletion. Increased amounts of phytoplankton attenuate light more strongly, with the consequence that the growth of seaweeds and seagrasses may be retarded. Opportunistic green or brown seaweeds spread over seagrass meadows or over the slower-growing brown fucoid and laminarian seaweeds that are the natural flora of temperate seashores and the shallow sublittoral. Although green seaweed growth can be stimulated close to cages, eutrophication is a phenomenon that is more typical of water bodies, such as lochs or coastal seas, as a whole. It is thus distinct from the local impacts of particulate waste, although the *change in the balance of* pelagic *organisms* associated with eutrophication (Fig. 1.3(b)) can be likened to the changes caused by organic input to the benthos (Fig. 1.3(a)).

The third type of potential disturbance is that from chemicals that are used to prevent or treat fish illnesses or parasitical infections, to improve fish growth, or to prevent fouling of nets or farm structure. Let us look at two groups of such chemicals, starting with the compounds azamethiphos and emamictin benzoate, used to rid farmed salmon of parasitic sea-lice.

These lice are crustaceans that burrow under the scales of the fish, causing sores that irritate the salmon and offer a route for infection by pathogenic micro-organisms. Young lice are planktonic, and so can infect other farmed or wild salmon. For all these reasons, fish-farmers in Scotland are required to treat their fish to keep lice infestation to a minimum. The two chemicals are arthropocides – that is, they are intended to kill lice, which are members of the arthropod phylum, but not salmon, which are vertebrates.

The problem is that many members of the plankton are also arthropods, the group that includes insects, spiders and crustaceans. To be precise, the sea-lice are copepod crustaceans, as their planktonic larvae show, and so chemicals that kill sea-lice are also at risk of killing planktonic copepods and thus of damaging an important link in marine food webs. Azamethiphos, which is applied externally, is a greater hazard than emamectin, which is given to salmon in their food and reaches the lice by way of the fish bloodstream. However, some emamectin reaches the sediment in fish faeces and uneaten food, and here it may harm benthic crustaceans. Both the chemicals are degraded by light and oxygen, and can also be removed by adsorption on particles; and these processes augment dilution and dispersion in bringing concentrations below levels at which harm might result.

Whereas azamethiphos and emamectin are solely of human manufacture, and hence were never present in ecosystems before humans introduced them, the story about antifouling compounds is more complex (Readman 2005). These compounds are used to prevent the growth of bacterial slime and seaweed sporelings on nets and supporting structures. TBT, which did this effectively, was entirely synthetic, but is now banned. Modern paints and steeping liquids use compounds of copper, and sometimes zinc, which dissolve slowly in seawater, releasing ions of copper and zinc. It is these ions that are harmful to micro-organisms that might settle and grow on the netting or cage. Paradoxically, copper and zinc are needed in small amounts by living creatures, being essential for some biochemical reactions, and are toxic only at higher concentrations. So the challenge for the designers of anti-fouling materials is to ensure that they release sufficient copper etc to kill bacteria and algal spores close to the surfaces they are intended to protect, but without dissolving too quickly, which would increase the risk of wider harm and would require more frequent treatments.

Consequently, some manufacturers add "booster biocides" to augment the anti-fouling action. These include the synthetic chemical, copper pyrithione. However, research suggests that when zinc is present, the pyrithione part can swop from copper to zinc, resulting in zinc pyrithione. This compound, used in anti-dandruff shampoos and as a fungicidal additive for plastics, has been found to be highly toxic to copepods as well as planktonic micro-algae (Hjorth et al. 2006; Maraldo and Dahllöf 2004).

The last part of this story is that farmed fish need copper, and so it is added to their food, perhaps in unnecessarily large amounts that the fish excrete into the water or by way of their faeces; because of the latter, the seabed beneath fish cages may contain high levels of copper, which dissolves to increase the concentration of copper ions in the sediment pore waters, and which may diffuse back into the water column.

1.5 DPSIR and EQS

The ***DPSIR*** system breaks the ecosystem effects of pollutants into 5 steps. In this acronym, *D* stands for ***driver***, *P* for ***pressure***, *S* for ***state***, *I* for ***impact***, and *R* for response. The ***state*** is that of the ecosystem under consideration; the ***pressures*** are those generated by human activity whose change provides the ***drivers***. Thus the growth of salmon-farming is the ***driver*** that has led to increasing loading of Scottish fjords with farm waste, with consequential ***pressures*** on the fjordic ecosystems from organic matter, mineral nutrients, and chemicals. A build-up of particulate waste beneath a fish cage, with consequent death of larger sea-bed animals, exemplifies a highly visible ***impact***, and the ***response*** to this impact has been for society to impose more stringent conditions on the location and management of fish farms.

Environmental Quality Standards (EQS) have been used to set limits to pressures. The Water Framework Directive, which we will come to later, defines a standard as:

> *the concentration of a particular pollutant or group of pollutants in water, sediment or biota which should not be exceeded in order to protect human health and the environment.*

As an example, the current Scottish EQS for azamethiphos is 40 ng/L (SEPA 1997, 1998). In laboratory studies, 50% of lobster larvae exposed to an azamethiphos concentration of 500 ng/L died within 4 days. The EQS was set below this value in order to avoid any harm to free-living marine animals, taking into account the natural decay of the chemical when released into the water.

In the case of such toxic pollutants there is an obvious relationship between pressure and impact, and the aim is to avoid any such impact. In the case of pollutants such as nutrients, which cause problems only when in excess, the setting of EQS is more difficult. The aim, of course, is to avoid the undesirable disturbances associated with eutrophication or the smothering of seabed communities by particulate waste from fish farms. The European Urban Waste Water Treatment Directive (UWWTD) of 1991 concerns the prevention of pollution by discharges of sewage, but the causes of such pollution are the same wastes as those from fish farms: organic waste, biological oxygen demand, and compounds of nitrogen and phosphorus; and some aspects of the UK response to the UWWTD can be applied just as well to fish farms as to urban waste water outflows. (There are differences, of course: human waste is treated before discharge; fish waste is not.) The United Kingdom set up a "Comprehensive Studies Task Team" to define standards and evaluative procedures for UK estuaries and coastal waters. The team (CSTT 1997) suggested that:

> *Hypernutrification exists when winter values of nutrient concentrations, outwith any area of local effect, significantly exceed 12 mmol DAIN m^{-3} in the presence of at least 0.2 mmol DAIP m^{-3}... Hypernutrification should not, however, be seen as a problem in itself. It causes harmful effects only if a substantial proportion of these nutrients is converted into planktonic algae or seaweed.*
>
> *A region is potentially eutrophic only if the relative rate of light-controlled phytoplankton growth is greater than the relative water exchange rate plus the relative loss rate of phytoplankton by grazing; and the predicted summer maximum chlorophyll is greater than 10 mg chl m^{-3}... A region is eutrophic is observed chlorophyll concentrations regularly exceed 10 mg m^{-3} during summer.*

The acronym DAIN refers to "dissolved available inorganic nitrogen", a useful and precise way of mentioning those compounds of the element that are useful to phytoplankton and seaweeds – what I have named earlier as nitrate and ammonia. DAIP refers to "dissolved available inorganic phosphorus", for which the shorter abbreviation DIP or "dissolved inorganic phosphate" will do as well.

These CSTT proposals suggest that, in the case of nutrients, it is difficult to set simple EQS, because the impact resulting from a given pressure depends on conditions in the water body receiving the discharge. Sensitivity to pressure is the topic of the next section.

1.6 Ecohydrodynamics and Sensitivity to Pressures

Although laboratory experiments can, for example, measure the concentration of copper or zinc pyrithione that kills 50% of phytoplankton (Maraldo and Dahllöf 2004) or the amount of DAIN that must be added to generate a phytoplankton biomass in excess of the CSTT threshold of 10mg chlorophyll m^{-3} (Edwards et al. 2003), the uncontrolled variability of conditions in the sea means that it is much harder to predict the impact of waste. For example, the food and faeces sinking from a small salmon farm in sheltered shallow waters might rapidly blanket the seabed beneath the farm, causing conditions to fall below those tolerable, whereas a larger farm moored in more turbulent and deeper waters might have no visible effect on the seabed, because the waste is dispersed by turbulence and spread over a wide area. However, the larger farm's waste has a greater potential to contribute to the widespread build-up of chronically harmful levels. Whereas the smaller farm may suffer from nutrient-stimulated seaweed growth on its cages, the water body containing the larger farm may suffer eutrophication because nutrients remain high for sufficiently long, and over sufficient extent, for phytoplankton to benefit from them.

Such considerations lead to two key ideas: first, that the sensitivity to waste of the waters or sea bed at a particular farm site, depend on ecohydrodynamic conditions at and around that site; second, that the impact of a particular environmental pressure depends on the spatial and temporal scale on which that pressure is applied. *Scales* are considered in the next section. *Sensitivity* can be roughly defined as the ratio of *impact* to *pressure*, and *ecohydrodynamics* refers to the physical conditions at a site and in a water body, and the chemical and biological conditions that would naturally occur under such conditions. An *ecohydrodynamic typology* provides a mean of classifying water bodies on the basis of such conditions. Tett et al. (2007) proposed a typology based on four key factors: lateral exchange; vertical mixing; illumination conditions; and the type and abundance of grazers.

The first distinction in the typology is that between open waters and partly enclosed coastal and transitional waters, called *Regions of Restricted Exchange*, or *RREs*. In RREs, exchange of water with the open sea is an important environmental condition; Tett et al. (2003a) compared a number of European fjords and barrier-protected bays in which the proportion of water exchanged each day varies from 2.5% (in the Swedish Himmer fjord) to more than 200% (in the Portuguese Ria Formosa) of the RRE's volume at mid-tide. The exchange rate for Creran lies between 0.1 and 0.3 d^{-1}. Clearly, well-flushed RREs can accept a greater loading of dissolved waste per unit surface area than can a poorly flushed water body, so long as the outside sea contains a lower concentration of the polluting substance.

The availability of light for photosynthesis is an important factor. Light does not penetrate far into water, because it is scattered by particles and absorbed by water itself, by chlorophyll and accessory photosynthetic pigments in phytoplankton, and by the dissolved substances than can give water a yellow or brown colour. The ***euphotic zone*** includes the part of the water column in which there is sufficient light for the growth of plants, seaweeds, micro-algae and photosynthetic bacteria; its depth reaches up to a hundred metres in clear ocean waters, such as parts of the Mediterranean, but may be only 1 or 2 m in some very turbid coastal waters. The next group of distinction in the typology arises from the relationship between the euphotic zone, the seabed, water column layers, and natural and human supplies of nutrients. A key distinction is that between waters in which the seabed is within the euphotic zone, allowing seaweeds, seagrasses or micro-algae to flourish, and those where it lies deeper, so requiring phytoplankton to provide the primary production. In the first case, nutrient enrichment may lead to replacement of seagrasses or brown seaweeds by green seaweeds or epiphytic micro-algae, and there will be concern if an increase in phytoplankton results in less light reaching the seabed. In the second case, the seasonal pattern of phytoplankton growth, and the ecosystem's sensitivity to nutrient enrichment, depends on seasonal patterns of water layering.

In the second case, we need to distinguish between waters that are well-mixed in the vertical, due to strong stirring by tidal or other currents, or by wind or surface cooling, and waters that are layered in density as a result of surface heating or freshwater input. The term ***pycnocline*** is used by oceanographers to refer to a zone of strong vertical gradient in density (due to temperature or salinity) that separates mixed layers. Phytoplankters growing above such a pycnocline are better illuminated, on average, than those in deep mixed waters. On the other hand, the upper layer tends to become depleted in nutrients during the main season of phytoplankton growth, and this constrains micro-algal growth. Nutrients added to such an impoverished layer can have a striking effect by fertilizing phytoplankton when there are few planktonic animals to eat the micro-algae. Organic matter produced during these blooms can give rise, later to an increased risk of deoxygenation when uneaten material sinks, and decays, below a pycnocline.

At the latitude of Scotland, there is generally too little light for phytoplankton production during the winter, and the typical pattern in coastal seas is that of a spring bloom as the surface of the sea is warmed by the sun and forms a distinct layer. Within this well-illuminated surface layer, algae can rapidly convert winter nutrients into biomass. This is, typically, followed by a summer period of low biomass because of nutrient exhaustion, and sometimes by an autumn bloom as nutrients are remixed into the surface water. In the Mediterranean, in contrast, the main seasons of phytoplankton growth are the autumn and Winter; in summer the surface layer is typically intensely nutrient-depleted, but there may be a subsurface layer of high chlorophyll. As demonstrated by loch Creran (Tett and Wallis 1978), layering (Fig. 1.1) resulting from freshwater input can extend the season of phytoplankton growth, unless the freshwater supply is so great that it brings the salinity down below a level tolerated by marine phytoplankton or flushes the algae from the system.

A final part of ecohydrodynamics takes into account the type of grazers on the primary producers. This is important in relation to eutrophication, for a poor coupling between producers and consumers can allow nutrient enrichment to stimulate a large increase in producer biomass – red tides of dinoflagellates, or blooms of opportunistic green seaweeds, for examples. In shallow waters, removal of pelagic micro-algae by water-filtering benthic animals can be important, but in deeper systems the benthos is passive: its members simply eat what sinks from the euphotic zone. Thus the efficiency of coupling in these waters depends on the numbers of protozoan microplankters and copepod and other mesozooplankters seeking micro-algal food. Algal blooms may be more likely if the growth of these animals is stunted by toxic pollutants. Conversely, adding a shellfish farm to a water body can artificially increase grazing.

1.7 Scales

Now let us consider the scales on which aquaculture can impact on ecosystems. These depend on a combination of the nature of the pressure, the dispersion characteristics of the water at and near the farm site, and the response time for the impact. The CSTT (1994, 1997) proposed that 3 scales be considered, applying to what the team called *zones A, B, and C* (Fig. 1.4). The key defining feature is the residence time of neutrally buoyant particles within the zone: citrus fruits can serve as suitable, and easily seen, particles, and so I like to imagine a modern Nell Gwyn tipping her basket of oranges into the sea from a farm, so that we can ask where are most of the oranges after a few hours (zone A scale), a few days (zone B) or a few weeks (zone C).

The ***zone A*** scale is that the water volume and sediment area immediately influenced by a fish farm, and corresponds to the *mixing zone* at the end of a pipe

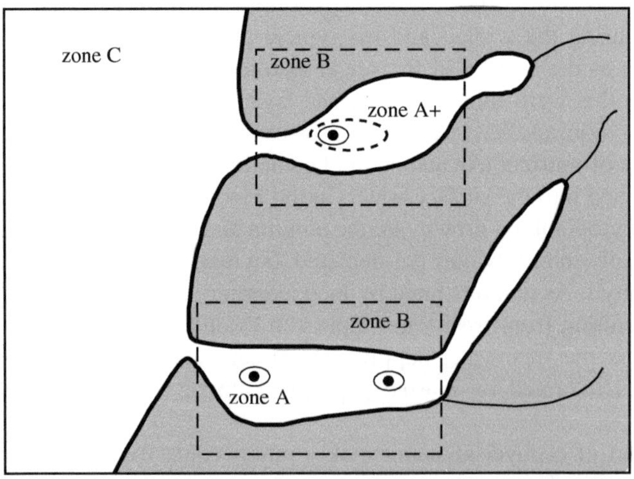

Fig. 1.4 Illustrated the 3 scales proposed by the UK Comprehensive Studies Task Team (CSTT). Zone A is the farm scale; it includes the part of the seabed that receives organic waste sinking from a farm and the part of the water column in which wastes and pollutants remain for a few hours. In tidally active waters, this water column zone is shown as A+. Zone B is the water body scale, and is exemplified by the main basin of loch Creran. Zone C is the regional scale

discharging waste into the sea, within which concentrations are allowed to exceed those specified by a *far-field* EQS. In general, it is easy to see benthic impact (Nickell et al. 2003) but difficult to detect pelagic impact on this scale, although it is sometimes possible to find a local increase in ammonia and a decrease in dissolved oxygen (Gowen and Bradbury 1987), and, in the case of shellfish farms, a local decrease in chlorophyll.

In the simple case of a fish farm in waters without tides or residual currents, the zone A scale is shown by the footprint of the cage on the sea, i.e., the area impacted by sinking waste, and a relatively small volume of water around the farm, the dimensions of which are set by the intensity of eddy diffusion. Under these unfavourable conditions the scale's dimensions are unlikely to exceed twice those of the farm. Now let us add a persistent current, which will transport the imaginary oranges in a downstream plume, broadening as it moves away from the farm. If the main flows are tidal, the oranges will move in an ellipse, returning after one complete tide to somewhere near their starting point, so that in this case, zone A for dissolved waste may be several kilometres long. We may take the (slightly over) 12 hours of a tidal cycle in NW European waters as the upper limit to the zone A timescale, and on this timescale it is impossible for added nutrients to impact on the plankton, although fast-acting chemical toxins may harm plankton before they are diluted by dispersion outside the zone. In order to apply this idea to non-tidal waters, such as those in the Mediterranean, we keep the half-day timescale and consider the limits of the zone in the water column as that reached by the oranges during this time. Unless the farm is sited in very energetic waters, the benthic footprint will likely be obvious, and smaller than the pelagic zone A.

The main basin of loch Creran provides an example of a stratified ***zone B*** scale water body and a region of restricted exchange. The residence time of water within this basin has been estimated as about a week (Tett 1986), although the contents of the surface layer leave the loch more quickly, within about 3 days, because of the freshwater driven, tidally enhanced, circulation described earlier. Such residence times are sufficient for nutrients to turn into planktonic algae before the latter are flushed out of the loch, and it is this, and the existence of stratification, that makes the loch potentially sensitive to the effects of nutrient enrichment. Extra growth of phytoplankton might be controlled by the grazing of the abundant seashore and seabed animals in Creran, and by the pelagic protozoans found in the water column. Except during times when benthic animals release their larvae into the water, the effect of crustacean zooplankton is small, because these animals tend to get flushed from Creran before they can complete their life cycles within the loch.

The Firth of Lorne, with which loch Creran exchanges, is a much larger body of water. The residence time of this water is not well known, but it is probably in the order of weeks or longer – sufficiently long for nutrients to become phytoplankton and then be grazed and recycled. Thus it is an example of a ***zone C*** scale water body, and provides the *boundary conditions* for loch Creran – that is to say, the water that enters Creran from the Firth already contains a certain amount of nutrients and phytoplankton, depending on the season, and enrichment or grazing within the loch will add to, or subtract from, these incoming concentrations. Thus it may be as important to control nutrient levels of the Firth of Lorne as it is to restrict enrichment within loch Creran. Indeed, we know from the results of a mathematical model that only during the summer, when nutrients are scarce in the Firth of Lorne, does farm input make an important contribution to Creran DAIN and phosphate (Laurent et al. 2006).

Fortunately, the waters of the Firth of Lorne are in a largely pristine condition, their moderate nutrient concentrations being set mainly by natural processes in the sea to the west of Scotland. Fish farms may, of course, become sufficiently to increase nutrients even on this larger scale. The region called the Minch, between the Scottish mainland and the island chain of the Outer Hebrides, has a sea area of about 10,000 km^2. The production of 64,000 t of salmon may have increased the concentration of DAIN and DIP in summer 1999 by a few percent (Tett and Edwards 2002), a scarcely measureable amount. Nevertheless, concerns about the effect of a greater enrichment may set an upper limit to the size of the industry here.

The Mediterranean Sea, being oligotrophic, might be considered at greater risk from enrichment, in that it takes only a little anthropogenic nutrient to double the naturally lowconcentration in each cubic metre of seawater. However, the Mediterranean is large; recent calculations suggest that input from fish farms will increase the total nutrient stock of the sea by at most 1%, whereas total human-driven inputs might double it (Karakassis et al. 2005).

1.8 Regulation of Pollution and Conservation of Species

At the core of the DPSIR scheme are the links between ***pressures***, ***states*** and ***impacts***. As humans became aware that the sea was neither an infinite garbage can for wastes nor an inexhaustible source of fish (McIntyre 1995), our societies began to legislate either to *prevent pollution* of the environment – corresponding to the regulation of ***pressure*** – or to *protect* certain animals or plants – corresponding to the prevention of ***impacts*** on these organisms. This was initially a piecemeal approach, which I will illustrate for the case of Scotland with two United Kingdom Laws – the *Control of Pollution Act (COPA)* of 1974, and the *Wildlife and Conservation Act* of 1981 – as these have used by the Scottish Environment Protection Agency (SEPA) to minimize the environmental impact of salmon-farming and to maintain water quality for shellfish.

My account greatly simplifies the complexities of a legal framework used to apply these UK laws in the separate, and different, jurisdictions of each part of the Kingdom. In most cases the generalities of the Acts of the UK Parliament (and, since 1999, also of the Scottish Parliament) are interpreted by detailed "Regulations" which are also commonly used to implement European Directives. Since the UK's accession to the European Community (as it was then called) on 1 January 1973, it has acquired (Graham 2002),

> *legal commitments to meet individual directive requirements that, in general, are transposed into UK law by means of regulations or other forms of secondary legislation issued as statutory instruments. A regulation identifies the competent regulatory authority and the actions required of it in order to achieve the directive's requirements. ... It is primarily regulations and directions passed by the UK or Scottish Parliament, which impose obligations on SEPA, as the competent authority, to deliver the objectives and standards so transposed from an EC Directive.*

The "Control of Pollution" Act (COPA) of 1974 marked the beginning of marine pollution control in the UK, although it took a decade to implement fully. The main regulatory tool is the "consent to discharge" from a "point source" such as a waste pipe or a fish farm. According to its web site (SEPA 2005a),

> SEPA has a duty to control discharges to surface waters and groundwaters *[in Scotland]*, including tidal waters out to the three-mile limit. SEPA does this by issuing a legally-binding consent to discharge under the Control of Pollution Act 1974..... Where consent is granted this will include specific conditions to limit the effects that the discharge may have upon the receiving environment. Monitoring will be carried out by the discharger and SEPA to ensure that the impacts of the discharge remain within acceptable levels.

Thus, anyone wishing to establish or extend a salmon farm in these waters must, amongst other legal requirements, make an application for a consent to discharge the waste from the farm. Then (SEPA 2005b),

> *SEPA will impose consent limitations on the maximum permitted fish biomass which may be held at any time. This is designed to minimise accumulation of organic wastes on the sea bed to prevent anoxic and polluted sediments and associated deleterious effects on the normal benthic fauna outwith the allowable zone of effects. In certain instances to protect important wild salmonid stocks, SEPA will limit the biomass to that which can be treated at the site using an authorised sea lice medicine* [without exceeding environmental quality standards for these medicines]. ... *SEPA will* [also] *limit consented biomass to ensure that the receiving water will not be* [at risk of eutrophication].

An *allowable zone of effects*, or *AZE* is a small region beneath fish cages where some impact is allowed. SEPA accepts

> *that a certain area immediately below and around the cages may experience carbon accretion to a level which may change the community structure of sediment fauna. Within this AZE quality standards ensure a minimum number of sediment re-workers will be available to breakdown wastes and prevent total anoxia developing.*

Two salmon farming sites have been consented in loch Creran, each of 1,500 t maximum biomass; however, only one site is available at a time, because each site is required to lie fallow for two years between use, in order to allow recovery of the benthos in the AZE.

The "Wildlife & Conservation" Act of 1981 has been used to implement the European "Habitats" Directives of 1992/1997 and the "Birds" Directive of 1979. It protects wild birds, and certain other animals, and plants that have been officially listed, together with designated sites. UK regional conservation agencies, exemplified by Scottish National Heritage (SNH), work under this law. The agency's web site (SNH 2006b) explains that

Special Areas of Conservation (SACs) are areas designated under the European Directive commonly known as the 'Habitats' Directive. Together with Special Protection Areas, which are designated under the Wild Birds Directive for wild birds and their habitats, SACs form the Natura 2000 network of sites. SNH acts as the advisor to Government in proposing selected sites for Ministerial approval as possible SACs. SNH then consults with... owners and occupiers of land, local authorities and other interested parties … [and] *negotiates the longer term management of these sites. Following consultation, SNH forwards all responses to Scottish Ministers who then make a decision about whether to submit the site to the European Commission as a candidate SAC. … sites which are adopted by the Commission become Sites of Community Importance (SCIs), after which they can be finally designated as Special Areas of Conservation by national governments. All candidate SACs in Scotland were approved by the European Commission as SCIs on 7 December 2004. Scottish Ministers then formally designated all these sites as Special Areas of Conservation on 17 March 2005.*

Under Regulation 33(2) of the ***Habitats Regulations*** *once a marine area becomes a designated SAC (European marine site), SNH is obliged to advise other relevant authorities as to a) the conservation objectives for that site, and b) any operations which may cause deterioration of natural habitats or the habitats of species, or disturbance of species, for which the site has been designated.*

Loch Creran have been designated as a SAC because of the

biogenic reefs of the calcareous tubeworm Serpula vermicularis, *which occur in shallow water around the periphery of the loch. The species has a world-wide distribution but the development of reefs is extremely rare: Loch Creran is the only known site in the UK to contain living S.* vermicularis *reefs and there are no known occurrences of similarly abundant reefs in Europe. Biogenic reefs of the horse mussel* Modiolus modiolus *occur in the upper basin of the loch. M. modiolus reefs are an important element of Scotland's marine biodiversity, and are considered to be habitats of high conservation value.*

SNH's advice about Creran (SNH 2006a) includes the following comments:

Finfish farming has the potential to cause deterioration of reef habitats and communities through changes in water quality, smothering from waste material, physical disturbance (in the case of rocky reefs), and physical damage (in the case of more fragile biogenic reefs) from mooring systems. There is also potential for accidental introduction of new non-native species and increasing the spread of existing non-native plants and animals…

[Shellfish farming] *has the potential to cause deterioration of the reef habitats and communities through physical damage (e.g. installation of mooring blocks and continued scouring by riser chains) and changes in community structure caused by smothering from pseudo-faeces (undigested waste products) and debris (including dead shells) falling from the farm. There is also potential for accidental introduction of new non-native species and increasing the spread within the UK of existing non-native plants and animals... through importation and translocation of shellfish stocks.*

[In both cases,] *invasive species have the potential to cause deterioration of the qualifying interest by altering community structure and quality. The … environmental effects* [associated with aquaculture] *are usually localised but the reduced water exchange within Loch Creran may exacerbate these effects and cumulative impacts should be considered.*

It was also noted that domestic and commercial effluents (whether treated or untreated) have

> *the potential to cause deterioration of reef habitats and communities. This would be through the effects of pollution and/or nutrient enrichment, which may cause subsequent changes in community structure* [of the reef].

Some of this advice has to be taken into account when permission for fish farms or other new developments is given by the planning departments of local government: it would certainly prevent farms being sited over reefs, or where their particulate wastes might accumulate on the reefs. An Environmental Statement, submitted as required by the Environmental Impact Assessment (Fish Farming in Marine Waters) Regulations 1999, should bring to light potential impacts of this sort. SEPA has a role to play, both at this stage and during the operation of the farm, as it is (Graham 2002):

> *a 'relevant authority' for European marine sites in Scotland, which are any SACs or SPAs that extend below the mean low water mark of spring tides. SEPA must, as a relevant authority, participate with other relevant authorities in drawing up a single management scheme for any European marine site where any relevant authority considers that one is necessary.*

Shellfish farming is much less strictly regulated, because it is not seen as producing a point source discharge. Instead, the industry is protected by the Shellfish Waters Directive of 1979, and much of loch Creran has been designated as a *Shellfish Growing Water* (SEPA 2004) under this Directive and the *Surface Waters (Shellfish) (Classification) (Scotland) Regulations 1997*. It is thus subject to monitoring by SEPA to ensure compliance with the standards set for metals and organohalogens in the water column and shellfish, faecal coliform bacteria in the shellfish, and dissolved oxygen. The aim is to protect the shellfish from environmental pressures and not to protect the rest of the ecosystem from the shellfish.

In summary, although some of the legislation discussed in this section takes account of links between pressures and impacts, the legal emphasis has been on polluting substances and their effects on particular commercial organisms or rare habitats; there is little of the general concern with the state of aquatic ecosystems that lies at the heart of the ***ecosystem approach***, the topic of the next section.

1.9 The Ecosystem Approach

The ***ecosystem approach*** can be seen, empirically, as a strategy for joined up management of the natural world, and scientifically, as arising from a modern understanding of community ecology and the interconnected processes within ecosystems. A web page of the UK Joint Nature Conservancy Council (JNCC 2004) provides a summary of the empirical view.

> *The phrase 'ecosystem approach' was first coined in the early 80s, but found formal acceptance at the Earth Summit in Rio in 1992 where it became an underpinning concept of the Convention on Biological Diversity, and was later described as: 'a strategy for the integrated management of land, water and living resources that promotes conservation and sustainable use in an equitable way.'*

> *Ecosystem-based management is currently a highly topical issue and is being widely discussed in the context of fisheries management. Introduction of the new Common Fisheries Policy (CFP) in January 2003 focused on this approach as the way forward to a sustainable fishing industry. Marine fisheries are one of the remaining examples of human endeavour involving the direct exploitation of wild animal populations. Fisheries are dependent on the productivity of the ecosystem, and fisheries have an effect on, and are affected by, the supporting ecosystem of the target species. It, therefore, follows that prudent and responsible fisheries management should take account of the profound interactions between fisheries and their supporting ecosystem.*
>
> [However,] *ecosystem-based management is not about managing or manipulating ecosystem processes, something that is clearly beyond our abilities. Rather, ecosystem-based management is concerned with ensuring that fishery management decisions do not adversely affect the ecosystem function and productivity, so that harvesting of target stocks (and resultant economic benefits) is sustainable in the long-term. Traditional systems of management, which have tended to focus on individual stocks or species, have not achieved this objective and consequently the economic activity that the ecosystem supports has become compromised.*

To my mind, this account falls short in several ways. First, it tends to suggest that the purpose of ecosystems is merely to produce food, or other services, for humans: it may be prudent to take account of the dynamic interconnections, but they are not valuable in themselves. Second, it is quite evidently not beyond human abilities to manipulate ecosystem processes. The matter at issue is, of course, to manage ecosystems wisely – at least in our own interests, but also, I believe, in the interests of all the creatures within them, and perhaps also in the interests of ecosystems as "emergent systems" whose properties are greater than the sums of their parts. My view is that an "environmental ethic" is also practical: we can only ensure sustainability if we treat all organisms and natural systems as having "interests worthy of consideration" (Johnson 1991).

My standpoint is close to that summarized by Miller's (2006) account of one millennia-old strand of Chinese thought, that of Daoism.

> *Daoists view morality in medical terms: goodness consists of the optimal health of a system comprised of various interdependent subsystems. This medical concept of virtue can... be useful in constructing an ecological ethics, one that recognizes that humans cannot act for their own good without considering the overall health of the ecosystems in which they are embedded.... the ideal state is achieved through embodying the complex transformative power of nature rather than denying it.*

Such emphasis on "connectedness" does have its own intellectual pitfalls, exemplified by the false "science" of astrology, based on the notion of connection between the human microcosm and the astronomical macrocosm. Nevertheless, I think that most of our present-day ecological science is well grounded in Enlightenment rationality and scientific methods, and the idea of ecosystem health is, at the very least, useful for devising monitoring programmes. I will return to this idea later.

It may be that the western, utilitarian approach, grows from our biblical heritage. In Genesis 1:26 it is written:

> *And God said, Let us make man in our image, after our likeness: and let them have dominion over the fish of the sea, and over the fowl of the air, and over the cattle, and over all the earth, and over every creeping thing that creepeth upon the earth.*

Hence, humans have souls as well as sentience and so are qualitatively different from other living things, and indeed may be seen as inhabiting Earth only briefly whilst on their way to heaven or hell; they are distinct from the rest of Nature as well as entitled, perhaps even required, to look after it as well as use it. As I wrote above, managing ecosystem processes is clearly within our ability: humans have been doing it for millennia. The problem is that we have often done it badly and unintentionally. Thus, although we might look at present-day environmental problems in China and wonder what has happened to the Daoist ideal, I prefer the idea that we are embedded in the ecosystem, and will sink or swim with the rest of Nature, rather than the idea that a better world awaits us somewhere else.

I should not claim that there are clear-cut distinctions between the religious traditions. The relationship between Daoism and science is complex (Ronan and Needham 1978). The Christian tradition has included St Francis of Assisi and the romantic poet, Coleridge, who write the *Rime of the Ancient Mariner* in 1798. These lines, taken from near the end, sum up his philosophy, which seems to place humans on the same plane as the rest of creation:

He prayeth well, who loveth well
Both man and bird and beast.
He prayeth best, who loveth best
All things both great and small;
For the dear God who loveth us
He made and loveth all.

I like to think that if St Francis had had a microscope, he would have loved nematode worms as much as birds, and, indeed, the whole of the magnificent "tree of life" that is being revealed by nucleic acid sequencing studies. Whether or not one accepts the theologies of Coleridge or the saint, the idea that we humans are made of the same stuff as the rest of creation is one to cherish, I believe, both for its own sake and because it may help prevent *Homo sapiens* from going extinct.

And that is as much of a sermon as I wish to offer in this chapter. Now to return to more mundane considerations of how such an ethic can be turned into regulatory and management practices.

1.10 The Water Framework Directive

As already mentioned, there are hints of an integrated approach to ecosystems in earlier laws, but it is in the "***Water Framework Directive***", or ***WFD*** that the approach begins to be clearly visible. The WFD is formally entitled *DIRECTIVE 2000/60/EC OF THE EUROPEAN PARLIAMENT AND OF THE COUNCIL, of 23 October 2000, establishing a framework for Community action in the field of water policy*, and Article 1 states that:

The purpose of this Directive is to establish a framework for the protection of inland surface waters, transitional waters, coastal waters and groundwater which:

(a) prevents further deterioration and protects and enhances the status of aquatic ecosystems …

(c) aims at enhanced protection and improvement of the aquatic environment, inter alia, through specific measures for the progressive reduction of discharges, emissions and losses of priority substances and the cessation or phasing-out of discharges, emissions and losses of the priority hazardous substances;.... and thereby contributes to:... the protection of territorial and marine waters, and... achieving the objectives of relevant international agreements, including those which aim to prevent and eliminate pollution of the marine environment,... with the ultimate aim of achieving concentrations in the marine environment near background values for naturally occurring substances and close to zero for man-made synthetic substances.

The first part that I have underlined refers to ***transitional waters*** (those *substantially influenced by river flow*, hence, typically, estuaries) and ***coastal waters*** (extending at least to 1 nautical mile from a coastal baseline, to 3 nautical miles in Scotland). These are the waters relevant to marine aquaculture as considered in this chapter. In addition, however, the Directive's protection of rivers, lakes, and their catchments, should improve the quality of discharges to estuaries and coastal waters, and so improve the background conditions here, to the advantage of aquaculture.

The third group of underlined words concerns the reduction of environmental pollution. In this respect, the WFD may be seen simply as intensifying earlier legislation, such as that of the UK's COPA or the Dangerous Substances Directive; but it goes beyond the use of experimental toxicology to set values for EQS. Notice the distinction between the "man-made synthetic" substances, and "naturally occurring" substances that are enhanced in wastes. The former are to be, ultimately, excluded from seawater; the latter should not be allowed to exceed "background" values by very much. The distinction can be made from the *Indicative list of the main pollutants* provided in Annex VIII of the WFD:

Man-made synthetics: *1. Organohalogen compounds and substances which may form such compounds in the aquatic environment. 2. Organophosphorous compounds. 3. Organotin compounds. 4. Substances and preparations, or the breakdown products of such, which have been proved to possess carcinogenic or mutagenic properties or properties which may affect steroidogenic, thyroid, reproduction or other endocrine-related functions in or via the aquatic environment. 5. Persistent hydrocarbons and persistent and bioaccumulable organic toxic substances. 6. Cyanides. 7. Metals and their compounds. 8. Arsenic and its compounds. 9. Biocides and plant protection products.*

Naturally-occurring substances: *10. Materials in suspension. 11. Substances which contribute to eutrophication (in particular, nitrates and phosphates). 12. Substances which have an unfavourable influence on the oxygen balance (and can be measured using parameters such as BOD, COD, etc.).*

Of course, it may be necessary to be a little more subtle than I have been. For example, some copper compounds occur naturally in seawater, whereas others, such as copper pyrithione, are synthetic.

The second underlining, referring to *the status of aquatic ecosystems*, highlights the ecosystem approach. In fact, the WFD implements the approach in two main ways: through the management of river basins (and their coastal waters) as a whole, including the joint consideration of point and diffuse sources of nutrients; and through the ecological component of quality status. ***Quality status*** is defined in article 2 in the following terms:

17. "Surface water status" is the general expression of the status of a body of surface water, determined by the poorer of its ecological status and its chemical status.

18. "Good surface water status" means the status achieved by a surface water body when both its ecological status and its chemical status are at least "good"

21. "Ecological status" is an expression of the quality of the structure and functioning of aquatic ecosystems associated with surface waters, classified in accordance with Annex V.

24. "Good surface water chemical status" means the chemical status required to meet the environmental objectives for surface waters established in Article 4(1)(a)....

As before I have underlined the key point and novelty: the focus on ecosystem structure and function. Details are given in Annex V, which is the longest single part of the Directive and (in my view) provides its beating heart. The Annex defines *ecological status* as consisting of *biological elements, physico-chemical elements supporting the biological elements* and *hydromorphological elements supporting the biological elements*. The *biological quality elements* for transitional and coastal waters are: *phytoplankton; macroalgae and angiosperms; benthic invertebrate fauna;* and *fish fauna* (in transitional waters only). Table 1.1 presents some general

Table 1.1 Some definitions of quality, from the Water Framework Directive: (a) Annex V section 1.2. Normative definitions of ecological status classifications: Table 1.2. General definition for rivers, lakes, transitional waters and coastal waters

Status	General definition
High	There are no, or only very minor, anthropogenic alterations to the values of the physico-chemical and hydromorphological quality elements for the surface water body type from those normally associated with that type under undisturbed conditions. The values of the biological quality elements for the surface water body reflect those normally associated with that type under undisturbed conditions, and show no, or only very minor, evidence of distortion. These are the type-specific conditions and communities.
Good	The values of the biological quality elements for the surface water body type show low levels of distortion resulting from human activity, but deviate only slightly from those normally associated with the surface water body type under undisturbed conditions.
Moderate	The values of the biological quality elements for the surface water body type deviate moderately from those normally associated with the surface water body type under undisturbed conditions. The values show moderate signs of distortion resulting from human activity and are significantly more disturbed than under conditions of good status.
Poor	Waters showing evidence of major alterations to the values of the biological quality elements for the surface water body type and in which the relevant biological communities deviate substantially from those normally associated with the surface water body type under undisturbed conditions, shall be classified as poor.
Bad	Waters showing evidence of severe alterations to the values of the biological quality elements for the surface water body type and in which large portions of the relevant biological communities normally associated with the surface water body type under undisturbed conditions are absent, shall be classified as bad.

The *biological quality elements* for transitional and coastal waters are: *phytoplankton*; *macroalgae and angiosperms*; *benthic invertebrate fauna*; and *fish fauna* (in transitional waters only).

Table 1.1 Some definitions of quality, from the Water Framework Directive, continued: (b) Annex V, section 1.2.4. Example of standards for *physico-chemical quality elements*, in *coastal waters*

Status	General conditions	Specific synthetic pollutants	Specific non-synthetic pollutants
High	The physico-chemical elements correspond totally or nearly totally to undisturbed conditions. Nutrient concentrations remain within the range normally associated with undisturbed conditions. Temperature, oxygen balance and transparency do not show signs of anthropogenic disturbance and remain within the ranges normally associated with undisturbed conditions.	Concentrations close to zero and at least below the limits of detection of the most advanced analytical techniques in general use.	Concentrations remain within the range normally associated with undisturbed conditions (background levels…).
Good	Temperature, oxygenation conditions and transparency do not reach levels outside the ranges established so as to ensure the functioning of the ecosystem and the achievement of the values specified above for the biological quality elements. Nutrient concentrations do not exceed the levels established so as to ensure the functioning of the ecosystem and the achievement of the values specified above for the biological quality elements.	Concentrations not in excess of the standards set in accordance with the procedure detailed in section 1.2.6[but not required to be below background levels] without prejudice to Directive 91/414/EC and Directive 98/8/EC. (<EQS)	
Moderate	Conditions consistent with the achievement of the values specified… for the biological quality elements [at moderate status].		

definitions of ecological and physico-chemical status. In essence, the ecological status of a water body is high when its phytoplankton, seaweeds or seagrasses, and benthos, all appear to be in a natural condition.

Figure 1.5 is a flow diagram to show how a regulator might apply the Directive. The starting point is the definition of water bodies and the identification of the type to which each belongs; then the present quality status of each water body identified in relation to a "type-specific reference condition" as *high* (the same as a water body in the reference condition), *good* (acceptable), *moderate*, *poor*, or *bad*. If the status is worse than *good*, ***Programmes of Measures*** must be implemented in order to improve the quality status, and monitoring programmes put in place to check on this. The WFD is also an integrating framework, bringing together provisions from earlier directives, including the UWWTD and the Habitats Directive. Any special requirements of these directives are dealt with by the concept of *Protected Areas* within which additional management might be

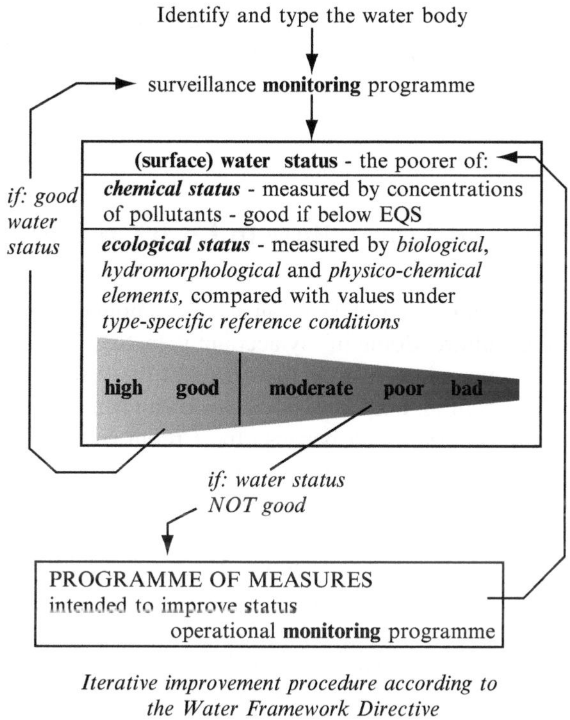

Fig. 1.5 Flow diagram for the operation of the Water Framework Directive – showing the relationship between the objective of maintaining good status, programmes of measures, and monitoring

needed. The timetable incorporated into the WFD requires one complete cycle of evaluation and management to be completed by the end of 2015, and some of the steps have already been carried out.

In Scotland, our (regional) parliament, which met in 1999 for the first time since 1707, used newly devolved powers to pass the *Water Environment and Water Services (Scotland) Act (2003)*, summarized as a law that, amongst other objectives, *make provision for protection of the water environment, including provision for implementing European Parliament and Council Directive 2000/60/EC*. This law gives the Scottish Environment Protection Agency (SEPA) the responsibility of drawing up the *River Basin Management Plans* required by the Directive to report *pressures* on water bodies, the existing quality status, protected areas, monitoring plans, and programmes of measures. Local authorities, analogous to municipalities or counties in other parts of Europe, must liaise with SEPA and take account of the WFD, and of programmes of measures, when giving permission for new building works. It is expected that some of the management measures will result from consent, because the WFD explicitly requires "stakeholder involvement", and that some will be enforced by SEPA using its "consent to discharge" powers under COPA, strengthened and modified in the *Water Environment (Controlled Activities) (Scotland) Regulations 2005* which provide for registration and licensing of discharges.

1.11 The WFD and Aquaculture in Loch Creran

In this section I am going to use Loch Creran as an example of how the WFD might come to bear on aquaculture. Some of my account is factual, and draws on material published by SEPA concerning its implementation of the WFD in Scotland. Some, however, must be conjectural, both because River Basin Management Plans are not due for publication until 2009, and because the Creran river basin, including the loch and adjacent coastal water, is only one of many such basins on the west coast of Scotland, and detailed plans by water body are initially only available for the "protected areas" within each. Nevertheless, I will roughly stick to the format set out for plans in WFD Annex VII, and will include: (i) a description and typing, including identification of reference conditions; (ii) a summary of significant pressures and impacts; (iii) a list of protected areas; (iv) a description of monitoring networks; (v) a list of specific environmental objectives; and, (vi) a summary of the "programmes of measures" required to achieve these objectives. All, of course, with the focus on aquaculture.

The first step in the application of the WFD in Scotland was the identification in 2003 of River Basin Districts, defined in the Directive as: *the area of land and sea, made up of one or more neighbouring river basins together with their associated groundwaters and coastal waters, which is... the main unit for management of river basins*. Most waters in Scotland, including loch Creran, fall into a single "Scotland" RBD. In contrast with many parts of continental Europe, where RBDs correspond to the catchments and coastal waters of single large rivers, the Scotland RBD includes many rivers, especially on the west coast, where rainfall is heavy and short steep rivers discharge into sea-lochs. Hence the Creran river basin and associated coastal water is but a small part of the Scotland RBD, and receives no specific description in the account so far published of the environmental features of the Scotland district. So the reader can turn to the description of loch Creran earlier in this chapter.

Completion of part (i) requires identification of the type of water body exemplified by loch Creran, so that reference conditions can be specified. Annex II of the Directive sets out the principles for two (alternative) typologies, and the UK, in collaboration with the ROI, has implemented these principles as a set of types for the coastal and transitional waters around our islands (UKTAG 2003). Creran can thus be identified as a coastal water of type 12, a "deep sea-loch" in the "Atlantic Ocean" ecoregion of Annex XI of the WFD. It is a coastal water because its depth- and time- averaged salinity is above 30 and hence close to that of seawater (in contrast to transitional waters, in which the mean salinity is less than 30), and a "deep fjord" because its greatest depth exceeds 30m. The adjacent loch Etive, receiving much more freshwater, has been identified as a 'transitional sea-loch' (transitional water type 5). Although there are some sea trout farms in Etive, the low mean and strongly fluctuating salinities make it less good for farming the salmon than the higher mean salinity of the coastal water fjords. In contrast, Etive has a good reputation for mussels, because the intermittently low salinities reduce fouling.

Table 1.2 illustrates the UK/ROI typology for the three fjordic types found in Scotland, and gives details both of the physical conditions that define the type, (UKTAG 2003) and of the proposed reference conditions (UKTAG 2004).

Protected areas in Creran include the serpulid reefs which are a SAC, and the *Shellfish Growing Waters* that occupy the main basin of the loch. SEPA's published description of the shellfish waters (SEPA 2004) gives the following for "land use and potential diffuse pollution sources":

> *The predominant land use is coniferous forestry but there is some extensive livestock agriculture on the north and far western shores. The main freshwater inflow is the River Creran, draining both forest and moorland. Loch Creran is remote from centres of population and is popular with visitors, particularly in the summer months.*

Point-source discharges include those from about 50 private houses, and the consented, major discharges from a fish farm and a fish processing factory. Laurent et al. (2006) used a mathematical model to show that the nutrients from the fish farm could make a significant contribution during summer, when the concentrations in the inflow from the Firth of Lorne are low. Nickell et al. (2003) found high organic loading and oxygen demand immediately beneath the farm, falling off rapidly at 60 m distance and returning to normal background levels for shallow coastal waters at 2 km from the farm. Only in the sediment immediately beneath the farm was the benthic community composition grossly perturbed.

Creran's waters are monitored for shellfish purposes from two sites, one near the mouth and one near the head of the main basin. In addition the river Creran is sometimes sampled for nutrients above its discharge into the upper basin: concentrations are typically low, as might be expected in runoff from granitic rocks and unimproved acidic grassland.

SEPA (2004) reports that:

> *In 2002, all samples from both monitoring sites met all shellfish waters imperative and guideline environmental quality standards. Biannual sampling is continuing for metals and organochlorines in waters along with monthly sampling for T, Sal, DO and pH at South Creagan and North Shian. Mussels will be sampled annually for organohalogens and metals at North Shian. This site will also be monitored quarterly for faecal coliforms in mussels and in addition, collection of mussels for TBT and PAH analysis will begin in 2004 as part of a SEPA Environmental Improvement Plan.... SEPA will continue to pursue a policy of no new discharges of sewage effluent to designated waters, to avoid incremental increase in microbiological loading. In the event that discharges to the designated waters cannot be avoided, they will be subject to appropriate treatment to ensure compliance with the [Shellfish Waters] Directive's standards.... All farms in catchment area will be inspected according to the Scottish Executive's... Plan to reduce point source farm discharges into inland and coastal waters. SEPA intend to initiate an Environmental Improvement Plan of agricultural inspections and improvement requirements, designed to reduce diffuse pollution.*

Table 1.2 The three types of fjord in the UK and ROI set of types for WFD coastal and transitional waters (UKTAG 2003). With descriptions of the reference conditions for some physico-chemical and biological elements (UKTAG 2004)

UK & ROI type	TW5	CW11	CW12
Name	Transitional Sea Lochs	Sea Lochs (Shallow)	Sea Lochs (Deep)
physical conditions	Polyhaline, mesotidal, sheltered	Euhaline, mesotidal, sheltered	Euhaline, mesotidal, sheltered
European water type	TW (not distinguished)	CW-NEA6 shallow fjordic type	CW-NEA7 deep fjordic type
Additional physical conditions		Low current; residence time – days	low current; residence time – weeks to months
Type-specific reference conditions			
Physico-chemical conditions	**Nutrient** concentrations will be elevated above Atlantic Shelf Concentrations by a factor dependent on local geological, hydrological and natural input regimes and characterised by a conservative dilution regime.	**Nutrient** concentrations in undisturbed conditions will be a function (± x %) of Atlantic Shelf nutrient concentrations. The winter concentration of nitrates and phosphates correspond totally or nearly totally to regional undisturbed conditions.	
	Dissolved oxygen annual range is 80–100%.	Maximum **temperature** range is 0–24.5°C, but typically between 5 and 15°C (mean 10°C), depending on latitude.	
	Maximum **temperature** range 0.8–21°C, typically between 5 and 15°C (mean 10°C) depending on latitude and type-specifics.	**Dissolved oxygen** annual range is 80–120%, with a mean of 100%. Range 2–9 mg l^{-1} (temperature & salinity dependent).	
	Transparency: Light availability is likely to be naturally lower for TWs than for CWs. It is a natural function of physical processes, estuary size, phytoplankton blooms and other organic and inorganic components.	**Transparency:** Light availability will range from clear to highly turbid, depending on type-specific conditions. This may be a natural function of phytoplankton blooms and other organic and inorganic components.	

Phytoplankton	TWs prone to higher levels of production compared with CWs, though light availability, salinity and hydrological effects may naturally temper this. Patterns of seasonal growth and succession are similar to coastal dynamics but demonstrate greater variability, in peak duration and composition. Nuisance/toxic species are at persistently low levels compared with local background levels. Peaks in chlorophyll a, used as a proxy for phytoplankton bloom biomass, are infrequent and inter-bloom periods are low compared with background levels.	Species richness high. Normal patterns of seasonal growth, biomass & succession, i.e., diatom dominated spring bloom and low summer biomass. Diatoms persist throughout growth-period. Increasing numbers of dinoflagellates from late spring. Transition from heterotrophic to autotrophic dinoflagellates from summer to autumn. Autumnal bloom dominated by diatoms or autotrophic dinoflagellates. Nuisance/toxic species at persistently low levels compared with local background levels. Peaks in chlorophyll infrequent & inter-bloom periods low compared with local background.	
Benthic macro-invertebrates	The littoral habitats of the upper reaches of sea lochs are typically comprised of coarse sediment (shingle, gravels and coarse sand). The habitats are subject to variable and reduced salinity conditions, and are typically species-poor and characterised by oligochaete worms.	Shallow sea lochs typically support species characteristic of mixed sediments. Due to the quite variable nature of the sediment type, a widely variable array of communities may be found, including those characterised by bivalves, anemones and file shells. Where physically very stable muds extend from the extreme lower shore to about 15 m depth. the biotopes can be more specific; anemones, brittlestars (e.g., *Amphiura*, *Ophiura*), the opisthobranch gastropod *Philine* and synaptid holothurians being characteristic of shallow muds. Where small stones and shells are abundant on the sediment surface, these can provide a substratum for hydroids, ascidians and other epifauna to attach	Typically, undisturbed circalittoral fine mud is found in deep sea lochs. These habitats are heavily bioturbated by megafaunal burrowers, such as *Nephrops norvegicus*. The infauna may contain populations of polychaetes such as, *Pholoe, Glycera, Nephtys, Pectinaria* and *Terebellides*, bivalves such as, *Nucula, Corbula* and *Thyasira,* and echinoderms such as, *Amphiura* and *Brissopsis*. The gastropod *Turritella* may also be present in large numbers. Epibenthos such as *Asterias rubens*, *Pagurus bernhardus and Liocarcinus depurator* may also be present. Habitats may occur characterised by conspicuous populations of sea pens such as *Funiculina quadrangularis*, *Virgularia mirabilis* and *Pennatula phosphorea*.

(continued)

Table 1.2 (continued)

UK & ROI type	TW5	CW11	CW12
	Lower shore of tide-swept areas, a mixed substratum (mainly cobbles and pebbles on muddy sediments) with dense aggregations of the mussel, *Mytilus edulis* may be found. In high densities the mussels bind the substratum and provide a habitat for many species more commonly found on rocky shores; the mussels are usually encrusted with barnacles and whelks and small crabs are common amongst the mussels. Areas of sediment may contain polychaetes such as the genus *Arenicola* and *Lanice*, the bivalve *Cerastoderma* and other infaunal species. Subtidally, variable salinity cohesive muddy sediment can be dominated by the polychaete *Aphelochaeta marioni* and the oligochaetes *Tubificoides*. The polychaetes *Polydora, Cossura longocirrata* and *Melinna palmata* may also occur in high numbers. The cirratulid polychaete *Caulleriella zetlandica* may be present.		
Examples: hydrography, phytoplankton	Loch Etive (Wood et al., 1973)	Loch Ardbhar (Gowen et al., 1983)	Loch Creran (Tett et al., 1981; Tett & Wallis, 1978)

Definitions: tidal range: micro-tidal < 1 m, meso-tidal 1–5 m, and macro-tidal > 5 m; euhaline: salinity greater than 30.0; polyhaline = 18 or 20 to 30; deep = greater than 30 m. *This except last from WFD CIS Guidance Document No. 5 - which gives shallow <30, intermediate 30–50, deep >50 (COAST, 2003)*

Sources: text copied, with a few abridgements, from UKTAG (2003) for typology and UKTAG (2004) for reference conditions.

Examples are of reference conditions, in the author's opinion and at the time the reported studies were done.

Much of the substance of these plans will no doubt become part of the "programme of measures" and the sampling networks required to monitor their effect. However, there is not much here relevant to the impact of the salmon farm on Creran, and so I must make an informed guess as to what the "environmental objectives" set for the loch will be. Although there are likely to be some changes in SEPA's regulation of benthic AZEs, I am assuming that changes brought about by the implementation of the WFD will increase emphasis on the phytoplankton biological quality element. The "reference condition" for this element in the CW12 type was given in Table 1.1, and Table 1.3 shows how Annex V of the WFD distinguishes the top three quality states of this element for coastal waters.

The main concern here appears to be that nutrient enrichment will lead to signs of eutrophication such as disturbance to the balance of organisms, increased phytoplankton biomass and bloom frequency, and decreased water transparency. Although the definition of "moderate status" does not mention the "undesirable disturbance" that is diagnostic of eutrophication, even moderate disturbances will require remediation, and the resulting "measures" may include more severe constraints on fish-farming if it can be shown to be contributing substantially to nutrient loads.

In the case of loch Creran we have an unexpected finding. When my research student, Céline Laurent, began to sample Creran in 2003, we had expected, on the basis of a simple mathematical model, that there would be a small increase in

Table 1.3 Definitions of *high, good* and *moderate* phytoplankton biological quality in coastal waters, from the WFD Annex V

High status	Good status	Moderate status
The composition and abundance of phytoplanktonic taxa are consistent with undisturbed conditions.	The composition and abundance of phytoplanktonic taxa show slight signs of disturbance.	The composition and abundance of planktonic taxa show signs of moderate disturbance.
The average phytoplankton biomass is consistent with the type-specific physicochemical conditions and is not such as to significantly alter the type-specific transparency conditions.	There are slight changes in biomass compared to type-specific conditions. Such changes do not indicate any accelerated growth of algae resulting in undesirable disturbance to the balance of organisms present in the water body or to the quality of the water.	Algal biomass is substantially outside the range associated with type-specific conditions, and is such as to impact upon other biological quality elements.
Planktonic blooms occur at a frequency and intensity which is consistent with the type specific physicochemical conditions.	A slight increase in the frequency and intensity of the type-specific planktonic blooms may occur.	A moderate increase in the frequency and intensity of planktonic blooms may occur. Persistent blooms may occur during summer months.

Quoted from the Water Framework Directive, Annex V, Table. I have underlined the sentence that refers to the definition of eutrophication.

amount of phytoplankton compared with the period 1970–1976, because of the additional nutrient input by the salmon farm now present in the loch. Instead, we found a decrease in average concentrations of chlorophyll (Laurent et al. 2006). The cause of this has yet to be explained. Are farmed mussels eating more phytoplankton? Is the loch chemically polluted by antifouling compounds? Are new chemicals in use on land surrounding the loch? Have its waters become more turbid? The WFD calls for *investigative monitoring* where:

> *surveillance monitoring indicates that the objectives set out in Article 4 for a body of water are not likely to be achieved and operational monitoring has not already been established, in order to ascertain the causes of a water body or water bodies failing to achieve the environmental objectives.*

However, since the emphasis by the WFD, as shown in Table 1.3, is on accelerated growth of algae, it is not clear that there is a failure to achieve the Directive's objectives. The question may turn on whether there has been *an undesirable disturbance to the balance of organisms*, which takes us to the topic of ecosystem health.

1.12 Ecosystem Health

The WFD's *type specific reference conditions* can be interpreted to imply that there is, or was, an ideal, "natural", or "pristine" state, and that any change from this state is a deterioration. But there is a practical problem in identifying reference conditions, given that *high* status corresponds to a state in which

> *no, or only very minor, anthropogenic alterations to the values of the physico-chemical and hydromorphological quality elements... from those normally associated with that type under undisturbed conditions. The values of the biological quality elements... reflect those normally associated with that type under undisturbed conditions, and show no, or only very minor, evidence of distortion.*

The practical problem is that there are few sites in Europe that are completely free of such disturbance and there is, indeed, some uncertainty about separating developing human influence from natural changes since the ending of the last glaciation. Given that it is unfeasible to seek completely pristine conditions, a realistic aim might be to describe the way things might have been before the industrial revolution in the 19th century. This is exemplified by a modeling study of nutrient discharge from the river Seine, and its effects on the trophic status of the Seine estuary (Cugier et al. 2005).

I think of this interpretation of WFD as seeking a return to a past Eden or golden age, or at least a tolerable approximation thereto. But we live in the here and now, and it may be better to seek a definition of reference conditions that takes account of this. So let us consider the alternative idea that the ideal state for an ecosystem is that of good health, irrespective of whether this state is natural or the result of human management. According to Costanza (1992), a healthy ecosystem, like a healthy human body, is a system that functions well and is able to resist or recover from disturbance. Ecosystems, which have the emergent property of homeostasis, are most healthy when their self-regulatory ability is fully functioning, and ecologists argue that this requires an appropriate balance of organisms performing different functions within the ecosystem. When the balance is disturbed to the extent that the ecosystem is no longer able to self-regulate properly, and is in danger of collapse or becoming something else, then it is unhealthy. This view envisages an internal rather than an external reference for good status. An unhealthy ecosystem is quite obviously not good for the organisms that form part of it, nor for sustainable human use, and it is clearly undesirable for humans to bring about such disturbances to the balance of organisms.

Mageau et al. (1995) propose that ecosystem health has quantifiable components of ***vigour***, ***organization***, ***resistance*** to disturbance, and ***resilience***. Tett et al. (2007) explore ways in which these components might be monitored in marine ecosystems, focusing on the relationship between ***organization*** and ***vigour*** that is shown diagrammatically in Fig. 1.6. The terms can be illustrated by considering the impact of fish farm organic waste on the benthic community underneath a salmon farm at the start of a 2-year cycle.

Initially the benthic community contains a mixture of species and the full range of "guilds" of functional types, such as burrowers and filter feeders. The first result of extra organic input is that existing animals are better fed, and so grow and multiply better. Initially, then, the *vigour* of the community, as measured by the flow of energy through it, increases. As extra organic matter continues to arrive, however, the burrows of animals that pump aerating water through the sediment become blocked, and these animals either die or move away. Oxygen levels within sediment pore water begin to decrease, creating conditions in which fewer species of animals can survive: those which do survive, typically small, specialized worms, have plenty of food and grow numerous. Under very high levels of organic input, all animal life is impossible, and bacteria capable of surviving in oxygen free conditions multiply, consuming all available oxygen and then turning to other compounds that they can use to oxidize organic matter. They may, for example, use the sulphate ions in seawater for this purpose, excreting either sulphur (which makes a white layer on the seabed) or the gas, hydrogen sulphide, which is poisonous to most multicellular animals including fish and humans. There may be a high flow of energy through the seabed, but little of it is put to good purpose within the ecosystem – at least, judged from the standpoint of multicellular animals, so that *vigour* is much decreased. Certainly *organization*, measured by the taxonomic and functional variety of the benthos, has much decreased.

The *resistance* of the benthic community to the pressure of increased organic input is shown by the community's initial increase in *vigour* with load; it is when the burrowers are overwhelmed that this resistance begins to be exceeded and *organization* begins to decline markedly – a state of affairs captured by the cartoon of the Pearson–Rosenberg paradigm in Fig. 1.3(a).

Now, let us assume that, as required by regulation in Scotland, the impacted benthic zone is confined to a small *Allowable Zone of Effect*, and that after 2 years the farm is moved to a new site. Experience has shown that the benthic community

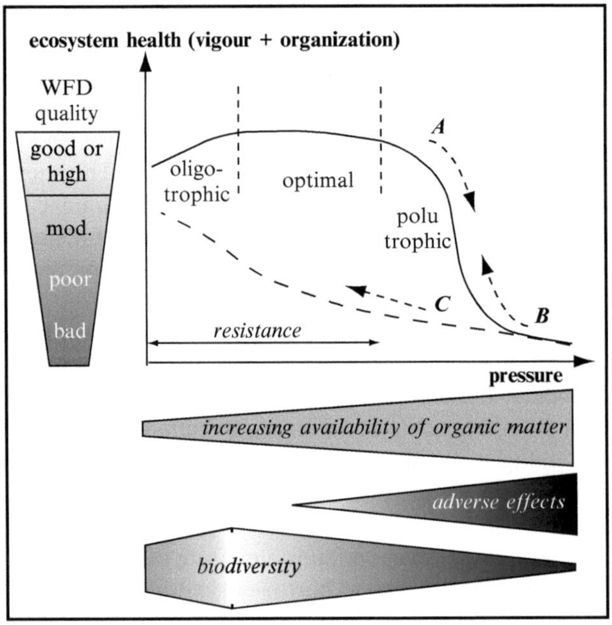

Fig. 1.6 Ecosystem health: changes with pressure. This complex diagram shows one variant of the current ecological paradigm for the behaviour of ecosystems under pressure. It also attempts to relate health to WFD quality. It is based, with modifications, on Tett et al. (2007). Read in the direction shown by arrow A, the main curve shows the response of an oligotrophic (low-production) ecosystem to increasing supply of organic matter, due either to additional inputs from outside, or to nutrient-stimulated primary production. Small increases can add to the vigour and structure of the ecosystem, but larger amounts tend to overwhelm assimilative capacity, so that harmful effects become dominant and the ecosystem state collapses. This is, of course, bad, but a crucial question is whether reducing the pressure leads to ecosystem recovery along curve B, or the persistent change in ecosystem state shown by curve C. Ecosystem resistance denotes the system's self-regulatory property (a function of health) that maintains structure. The diagram uses the term polutrophic (from the classical Greek for "excess nourishment") for the state which is often called eutrophic in contradiction of that word's etymology (from classical Greek for "good feeding"). The WFD would identify the zero-pressure (reference) state as high, and what is here called optimal (because it contains maximum biomass, structure and biodiversity) as, at best, good. If the ecological theory shown here is correct, the line separating good from moderate should be drawn at the point where the ecosystem approaches the edge of the "cliff", after which (from A onwards) its state decays rapidly as pressure increases

recovers rapidly – that is, it shows a high level of *resilience* – because the AZE is surrounded by plenty of healthy benthos to reseed the impacted area with larvae and migration within the sediment. So, on a zone A scale, disturbance to benthic health is of little serious concern so long as confined to one or a few AZEs which comprise only a small fraction of the seabed of a water body such as a sea-loch. But what could happen on the zone B scale?

Suppose that the whole of a sea-loch was given over to mussel farming, so that mussels took by far the largest share of the available primary production and their pseudo-faeces began to buildup a loose layer of mainly inorganic particles on and near the sea bed. Under such conditions the loch would become a "sink" for phytoplankton, consuming more than it produced, and the depletion of food, and benthic changes, might mean that the biomass of plankton and benthos decreased and that some species were eliminated. The increase in turbidity due to the resuspended matter might prevent seaweeds from growing where the sea bed was below the low tide mark. Or, suppose the amount of salmon farmed in Creran was greatly increased, so that nutrient enrichment caused eutrophication, with many algal blooms and increased downwards flow of organic matter, causing many locally impacted areas on the seabed, and depletion of oxygen in the deeper parts of the loch. Again, the effect might be to degrade the *structure* of the benthic community and suppress the natural extent of seaweed primary production. In all these cases the loch's ecosystem would have suffered an undesirable disturbance that can be described in terms of the ecosystem's *resistance* being overwhelmed, so that ecosystem state plunges over the "cliff" in Fig. 1.6, bringing about a state of ill health in which self-regulation is poor.

In such a state there would be little *resilience* within the loch's ecosystem to bring about recovery if mariculture was removed. However, this hypothetically impacted water body is, fortunately, part of a larger world, and it is reasonable to assume that the import of plankton, seaweed spores, and the pelagic larvae of benthic animals, would eventually restore the ecosystem. Nevertheless, we know little about the processes of ecosystem reassembly, and it is possible that the resulting system would be unlike that which was made unwell. Nor do we know how long it would take to restore a zone B ecosystem: very likely, much longer than the 1–2 years required to restore an impacted AZE beneath a fish cage. And finally, as the reader may already have spotted, pressures on zone C scales that lead to a weakening of the health of ecosystems over large parts of coastal seas, will damage the recovery prospects for any zone B scale waterbody for which the zone C scale ought to offer the reseeding potential.

1.13 Phytoplankton Community Index

Ecosystem *vigour* is easy to understand: it refers to the intensity of life, including its production and consumption of organic matter, its turnover of nutrient elements, and its ability to restore a good state after local disturbance. *Organization* is more complicated. If we were dealing with a coral reef, its organization would include the physical structure of the reef, together with the diversity of the organisms living there and there food web interrelationships. A similar account could be written for the benthic community, as shown in Fig. 3(a) where increasing organic loading results in organizational degradation. But what about the plankton? Plankters are passive riders on water motion, and their population abundances can change rapidly.

Can the plankton be said to have *organization?* This section attempts to answer that question for the phytoplankton, by introducing a monitoring tool called the *Phytoplankton Community Index*, or PCI.

An ecosystem is made up of many parts: in the case of loch Creran, of water, mud, dissolved substances and populations of many species of animals, algae and bacteria. These components are continuously changing: water is exchanged with the sea, benthic animals reproduce, seeding the water with planktonic larvae; the balance amongst the populations of species of phytoplankters changes with the season. So, like humans whose every atom is said to be replaced every seven years, ecosystems remain identifiable while subject to flux. There is a way to describe the essence of such changing systems in terms of *state variables*. In the case of Creran, these variables might include the volume and salinity of water in the loch, the concentrations of nutrients and oxygen in the water, and the abundance of species of benthic animals and of phytoplankton. System theory states that a system is in the same state whenever all state variables have the same value. This may seem obvious, or perhaps even tautologous, but it allows us to find ways of describing ecosystems so as to discover whether they are indeed in the same state, which is a precursor to deducing whether the state is "good" (from the perspective of the WFD), or "healthy", or whether it has changed in a way that would be regarded as an "impact" (from the standpoint of the DPSIR terminology). Now let us zoom in to consider phytoplankton alone (but as a component of an ecosystem).

What state variables can we define to capture the essence of phytoplankton in ecosystems? Ecologists have for a long time been interested in species diversity and questions about number of species and the relative abundance of each species (Tett and Barton 1995). However, the list of species of phytoplankters in a typical water sample may be as long as several hundred, and in most cases we know little about what particular species "do" in the pelagic ecosystem. An alternative is to consider that there are a number of functional rôles to be played by pelagic photosynthesizers and that all these rôles must be properly played for proper functioning (and hence health) of the ecosystem. The functions include the cycling of nutrients, and this suggests that there is a distinction between the glassy-walled diatoms, which use and cycle silicon as well as nitrogen and phosphorus, and most other phytoplankters, which do not use silica. Another distinction might be between small phytoplankters, which are suitable food for pelagic protozoa, and larger phytoplankters, which offer a tasty mouthful for copepods and pelagic crustaceans. To cut a potentially long story (Tett et al. 2003b) short, we may view the phytoplankton as being made up of populations of a handful, or double handful, of *life forms*, and a healthy "balance of organisms" being a balance of these life forms able to carry out all the functions that the ecosystem requires of the phytoplankton, and without which it will degrade into an unhealthy state.

This is not the place to list life forms. Indeed, we probably do not know enough about phytoplankton ecology to make a single undisputed list. For the sake of illustration, let us take just two life forms: *pelagic diatoms* (PD, so called to distinguish them from the thick-walled diatoms that normally grow on the seabed but which can be lifted into the phytoplankton by turbulence); and *medium-sized autotrophic dinoflagellates* (MAD, a name that emphasizes the need to distinguish photosynthesizing dinoflagellates from their relatives that live by eating other micro-organisms, and which excludes certain large-bodied dinoflagellates characteristic of summer in deep, temperature-layered, waters). Sampling loch Creran at a particular time, counting the phytoplankters in these samples using a microscope, and assigning the relevant counts to these life forms, gives a pair of values: an abundance of the PD and an abundance of the MAD. Next, draw a pair of axes: one for the abundance of the PD and another, at right-angles, for the abundance of the MAD. (For reasons of statistical methods, we actually use the logarithm of abundance.) Onto the resulting Cartesian co-ordinate system, plot the point specified by the abundances of PD and MAD on the date of sampling. It is this point that defines the state of the ecosystem – or at least of its phytoplankton components – on that day.

This, however, is not enough. It is a characteristic of phytoplankton in temperate seas that the absolute and relative abundance of phytoplankter life forms changes with the seasons, and we must take account of this. So we continue to take samples from Creran and to plot additional points until we have several years worth of data displayed on the PD-MAD axes Now we can see that a graph linking the points makes loops on the "PD-MAD" surface, and we can define the "state" of the loch Creran ecosystem as being the area on the PD-MAD diagram that is occupied by all these points. I have made such a diagram in Fig. 1.7, using data obtained from 1979, 1980 and 1981. If we assume that the loch was in a natural and healthy state during these years, and if Creran is typical of its WFD coastal water type, we can argue that this diagram defines a "type-specific reference condition" for the balance of organisms in the phytoplankton. This is the approach taken by the PCI-LF: a "state-space" diagram of this sort is made for a reference condition; an envelope is drawn about this reference condition; and the PCI value is measured by plotting new data onto the same diagram and counting the proportion of new points that fall outside the reference condition envelope (Tett 2006).

We are in course of doing this with more recent data from loch Creran, and some early results are shown in part (d) of the diagram. What is to be expected in a case when a fish farm added sufficient nutrients to disturb the balance of organisms? Because the nutrients would be compounds of nitrogen and phosphorus, but not silicon, they would favor dinoflagellates rather than diatoms; and hence new points should be found more towards the top (the MAD axis) than the right-hand side (the PD axis) of the diagram. This is to some extent what we seem to be finding, although it also seems that diatom abundance has decreased. That decrease might be the proximate cause of the decrease of chlorophyll in loch Creran that was mentioned previously.

I have taken this detour into details of how to assess change in phytoplankton in order to penetrate a little deeper into some of the theory underlying the "ecosystem approach" and to show how such theory may contribute to the practical matter of assessing ecological quality. There are two ways in which a PCI might be used. If it is to be used to quantify the health of the phytoplankton, ecologists need better knowledge of the relationship between organization and vigour in pelagic communities. To be more concrete, which parts of the PD-MAD surface represent a healthy

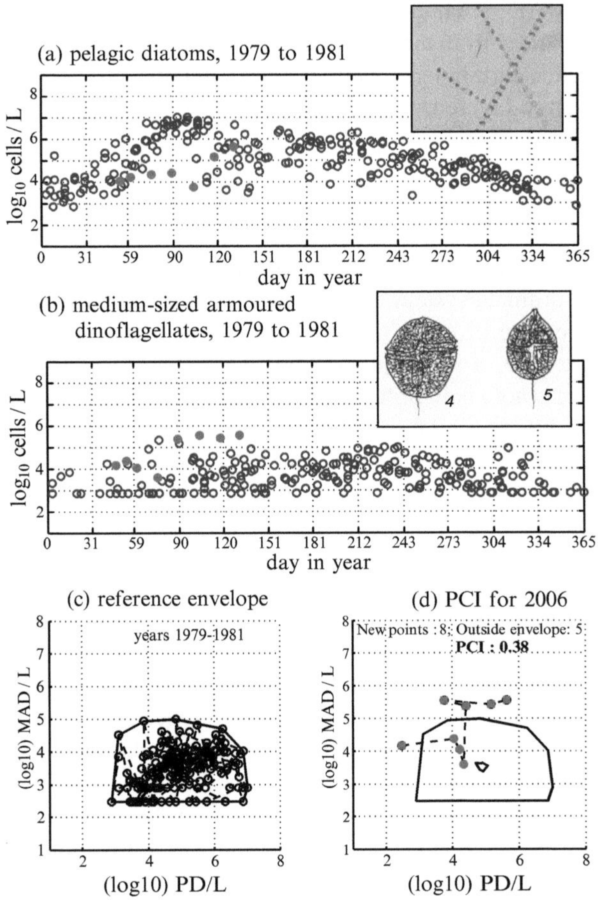

Fig. 1.7 Evaluation of a Phytoplankton Community Index illustrated with data from loch Creran. Part (a) shows the seasonal cycle of pelagic diatoms (PD), illustrated by the common species *Skeletonema costatum*, and was obtained by plotting abundances in all phytoplankton samples taken in the loch during 1979, 1980 and 1981. The vertical scale is logarithmic, in order to show, clearly, the wide range in abundance. The horizontal scale gives days in the year, with day 1 being 1st January. The data from 1979–1981 is plotted as open circles; the small set of filled circles shows observations made during 2006. Part (b) shows a similar graph for the medium-sized autotrophic dinoflagellates (MAD), illustrated by drawings of a typical species of the genera *Gonyaulax* and *Scrippsiella*. In part (c) the 1979–1981 data from (a) and (b) have been plotted onto a surface whose axes are the abundances of the PD and the MAD, and an envelope has been drawn around the points to define reference conditions. In part (d) the envelope has been redrawn, and points from 2006 plotted onto the surface. The PCI is the proportion of new points that remain inside the reference envelope

state and which do not? In the absence of adequate knowledge, the WFD Annex V assessment strategy serves to provide an empirical appraisal of change away from a reference state which is by definition healthy. If a sufficiently large proportion of points fall outside the reference envelope, then the PCI can be used to indicate a change in quality from *high* or *good* to *moderate* or worse. This is the second use, but even it needs agreement about critical values of the PCI – at the *good/moderate* boundary, above all.

1.14 Assimilative Capacity

Given regulation according to the WFD and the need to maintain ecosystem health and sustainable human use, how many finfish or shellfish can be farmed within a water body? The size of a sustainable aquaculture is said to be the ***carrying capacity*** of the water body for the stock concerned; I approach it here from the alternative perspective of the ***assimilative capacity*** of the water body for the wastes of – or, more generally, the pressures generated by – fish-farming and other human activities. What, for example, is a water body's ability to absorb anthropogenic DAIN without significant adverse effect on the health of the ecosystem?

Figure 1.8 shows some of the principles involved in the estimation of assimilative capacity. The horizontal axis represents increasing pressure from anthropogenic activity. This pressure could be quantified as the number of fish farms in a water body or the number of humans who would produce waste equal to the total input to the water body from all sources, but it is better to relate the waste input to the receiving system. Thus, suitable indicators of pressure would be the annual rate of organic matter arriving on each square metre of seabed in the AZE below a farm, or the daily total of nutrients input to a zone B water body, divided by the volume of the water that is replaced each day from the adjacent sea.

The vertical axis is something that measures impact on the ecosystem – that is, the change in state from a reference condition as defined for the WFD or a decrease in the health components *organization* and *vigour*. Examples of such benthic indicators include the AZTI Marine Biotic Index (AMBI) (Borja et al. 2003) and the Infaunal Trophic Index (ITI) (Word 1990). These assess the balance of the several kinds of large benthic animal needed to maintain a healthy ecosystem in the mud. Examples for the water column include the excess of chlorophyll concentration over that in a reference condition, and the PCI described above.

There is a scale issue: the pressure variable on the x-axis and the impact variable on the y-axis must relate to the same scale: A, B or C as defined previously. Given that, the next part of the task is to find a relationship between the two axes, as shown by the diagonal line in the diagram. A simple relationship might be that of linear regression, so that $y = a + b.x$, where a and b are constants. As suggested by the curve in Fig. 1.6, the true relationship in Fig. 1.8 is unlikely to be simple, but this is not a problem so long as it can be expressed by a mathematical equation, or by a table in which values of impact, y, can be looked up for values of pressure, x.

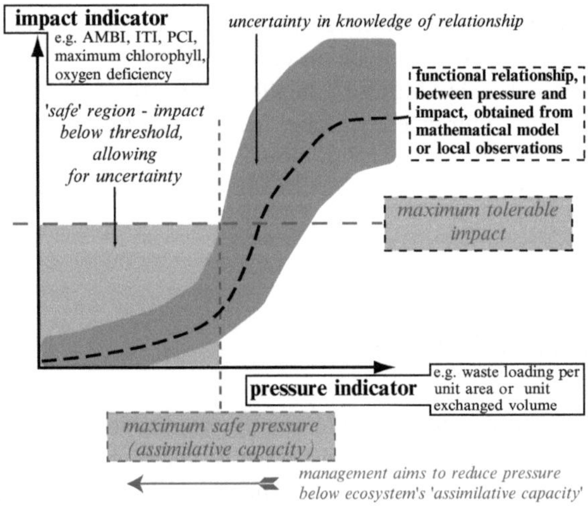

Fig. 1.8 The estimation of assimilative capacity. Note that the impact indicators: "maximum chlorophyll" and "oxygen deficiency" directly indicate impact; the AMBI, ITI and PCI are constructed so that their values are high for high ecological status, and thus, strictly, it is their inverse that is an indicator of impact

The relationship can be gotten in two ways: by observing *y* for many values of *x*; or by developing a mathematical model that predicts *y* from *x*. In either case, there will be some uncertainty in any prediction of *y*, and this is shown by the grey area that surrounds the relationship line.

Regulators need to set thresholds for the impact indicators, exemplified for the water column and the zone B scale by the CSTT's threshold of 10 mg chlorophyll m^{-3} in summer. The greatest tolerable pressure is that which takes the *y*-axis variable up to, but not past, the threshold. This pressure is the *assimilative capacity* of that particular water body for the waste responsible for the pressure, and it is a regulator's or planner's task to consent discharges only up to this capacity, taking account of any natural contributions towards it. It may also be a farm manager's task (as a condition of a consent to discharge) to ensure that the zone A pressures, and the contribution to zone B scale pressures, are controlled so that the impacts do not exceed those allowed.

Note that an impact threshold is not an EQS, which is explicitly defined by the WFD in terms of concentrations of pollutants. An impact threshold can, also, be stated implicitly, by way of an *Ecological Quality Objective or* **EcoQO**, exemplified for the zone A benthos by the Scottish regulator's requirement that there must be at least 2 taxa of Polychaeta worms alive within a fish-farm AZE. Painting et al. (2005) consider the utility of several larger-scale EcoQOs intended to prevent eutrophication.

Account needs to be taken of the uncertainty in the *y*–*x* relationship. If the greatest allowable pressure were read at the point on the *x*-axis of Fig. 1.8 where the upper limit of the uncertainty crosses the impact threshold on the *y*-axis, the risk of an undesirable impact would be minimized. This is giving the benefit of the doubt to the ecosystem. Alternatively, the benefit could be given to the producers of waste by using the lower limit to the uncertainty, which will maximize the tolerable pressure. Or, the most probable *y*–*x* relationship (the thick dashed line) could be used, ignoring the risk of some excessive impacts and denying fish-farmers (and other waste producers) the benefit of some possible assimilative capacity. Carstensen (2007) has discussed this topic in the context of identifying to which WFD quality class a water body belongs.

Of course, wastes and pollutants enter ecosystems from many sources. In the case of nutrients, the anthropogenic sources include diffuse agricultural inputs and urban waste water discharges as well as aquaculture. Thus, the ability to estimate the maximum safe loading from all these sources gives regulators the new task of sharing a water body's nutrient assimilative capacity amongst its human users – some of whom have, historically, taken it as an inexhaustible gift of nature rather than something that they might have to share or pay for.

1.15 Sustainability and the Ecosystem Approach to Aquaculture

This chapter has mentioned the use of the terms *Environmental Quality Standard* and *Ecological Quality Objective*. Initially, EQSs were made to prevent pollution by harmful substances, and the term is used in the Water Framework Directive in exactly this way. The concept of specific EcoQOs came later, and the term is not always used explicitly: for example, the Water Framework Directive refers to general "environmental objectives" in its main text and to "quality status" in Annex V. Nevertheless, it is useful to see the statement of precise EcoQOs as a key device for maintaining the status of components of ecosystem quality or health or to prevent undesirable impacts. EQS's and EcoQOs can be enforced only if each is associated with an indicator than can be monitored. In the case of the highly toxic substances which are dangerous at any level, the scheme set out in Fig. 1.8 can be bypassed: the ecosystem has no assimilative capacity for these substances, and the Water Framework Directive aims to stop their release into the aquatic environment. In all other cases Fig. 1.8 summarizes the task facing aquacultural managers, regulators and scientists.

The figure is an outcome from ECASA, a European Commission Framework 6 project, concerned with the *Ecosystem Approach to Sustainable Aquaculture*. ECASA's main product (Box 1.1) is a "virtual toolbox" giving details of the models and indicators that can be used to apply the approach of the previous section. These tools do not in themselves guarantee sustainable aquaculture, because this requires economic efficiency and attention to the needs of local societies in addition to a

Box 1.1 Models for assimilative capacity in the ECASA project

ECASA stands for *Ecosystem Approach for Sustainable Aquaculture*. The project, which ran from December 2004 through November 2007, was part of the European Community's 6th framework program and funded by a contract from the Fisheries Directorate-General of the Commission of the European Communities. Its aims included:

Assessing the applicability (efficiency, cost effectiveness, robustness, practicality, feasibility, accuracy, precision, etc) of selected ***indicators*** *and developing operational* ***tools****, e.g.,* ***models****, establishing the functional relationship between environment and aquaculture activities.*

The models studied during ECASA include the following categories:

1. Models for the biology of a type of farmed organism, including mussels and other shellfish, and salmon and sea bream. Some of these models can be used for best management of the animals.
2. Zone A models for local impact of fish-farms, especially models able to predict the pattern made on the sea-bed by sinking organic waste and its effect on the sediment and benthos.
3. Zone B models for the water-body scale impact of finfish-farming, including effects on chlorophyll, transparency and deep-water oxygen that are associated with eutophication, together with basin-scale models for shellfish production.

Further details may be obtained from the ECASA web site at http://www.ecasa.org.uk, and details about the models can be found in the ECASA "toolbox", at http://www.ecasa.org.uk/toolbox.

concern for ecosystem health, but they can help to manage sites for sustainability by ensuring that conditions remain within the "safe" area in Fig. 1.8.

An indicator of sustainability is categorically different from indicators of pressures and impacts. The latter are like thermometers: their readings help describe the weather at a particular time, or show whether a human is well or sick from fever. A sustainability indicator must take account of time and the overall *state* of the ecosystem. More precisely, if the symbol Y_i refers to values of a particular impact indicator, $\{Y_i\}$ means the set of all relevant impact indicators, and $f(\{Y_i\})$ specifies a function of the values of each member of this set, such as the function that converts monitoring results into a WFD water quality status value. Then a sustainability indicator is $df(\{Y_i\})/dt$, and a generalized EcoQO for sustainability requires that: $d\,f(\{Y_i\})/dt \leq 0$ for a given site, water body or regional sea.

To understand this, imagine a set of diagrams, similar to Fig. 1.8, for each of the CSTT scales. To make things more concrete, let us consider the zone B scale water body that is loch Creran. The known environmental pressures on this scale are from nutrients, oxygen-demanding organic matter, and antifouling and anti-lice chemicals, mostly from several sources. For each, one or more a pressure-impact diagrams can be drawn. Models, such as those examined during ECASA, can be used to guide management of the pressures. Monitoring can be carried out to establish whether impacts within Creran remain within the "safe" region in each pressure-impact diagram. If they do, then human use of loch Creran is sustainable, and we can expect to go on using it in the same way in future as we have in the past.

Of course, this judgement is not eternal. Changes such as those due to global warming, for example, might increase or decrease the loch's assimilative capacity. In a warmer world, in which water can dissolve less oxygen, the oxygen demand of decaying waste might be more critical than the nutrients released by that decay. So the situation must be kept under review.

It is also possible that the ecosystem approach might be used not only to maintain the health of loch Creran but also to increase the efficiency with which humans can take goods (such as mussels) and services (such as nutrient assimilation) from it. Suppose, for example, that the critical pressure-impact diagram is the one that relates eutrophication impact to nutrient loading. This could constrain finfish-farming by means of setting a limit to the amount of nutrients that the farm could put into the water. In this case, farming shellfish or seaweeds in the loch might be a way of removing some of these nutrients and hence effectively increasing the nutrient assimilative capacity of Creran.

My other theme in this chapter has been the potential impact of the Water Framework Directive on fish farming. The Directive aims to *establish a framework* that *protects and enhances the status of aquatic ecosystems*, and that is – excepting my reservation about the difference between ecosystem health and ecosystem quality status – just what is needed to maintain health and ensure ecological sustainability. How much difference will the WFD make to aquaculture? As in many places in this chapter, I focus on Scotland, the only part of Europe where I am familiar with law and regulatory practice as well as the ecological impact of fish-farming. For Scotland there are two simple and apparently opposed answers: not much; and, a lot.

First, the "not much change" answer. Because the WFD builds on and synthesizes previous directives, and is implemented in Scotland using regulatory methods that are already well developed, the changes in regulation are likely to be gradual and, perhaps, will impact most on the most old-fashioned aspect – that of pollution by synthetic compounds. In my view, fish farmers should expect in the long run to do without these, which has implications for the management of fouling, sea-lice and diseases. It probably means farming fish at lower densities in more highly dispersive environments. However, some fish farmers are already exploring this, and those that do so are able to get a premium on their fish, both out of consumers' concern for animal welfare and environmental health, and because (in my view) fish thus farmed, taste better.

Second, the "big difference" answer. This is based on the argument that the WFD implements the ecosystem approach. If farmers and regulators become real converts to the ecosystem approach, their world view will change. Farmers will go from reluctantly conforming to AZE regulations to willingly embracing their part in maintaining ecosystem sustainability on the zone B scale, perhaps with collaboration between finfish aquaculture (which adds nutrients) and shellfish aquaculture (which benefits from increased amounts of phytoplankton). Whether such a change can take place in the existing economic environment is a matter for other chapters in this volume.

References

Anderson DM, Garrison DL (1997) The ecology and oceanography of harmful algal blooms: preface. Limnology and Oceanography 42: 1007–1009

Black KD (2001) Mariculture, environmental, economic and social impacts of. In: Steele J, Thorpe S, Turekian K (eds.), Encyclopedia of ocean sciences. Academic Press, London, pp. 1578–1584

Borja A, Muxika I, Franco J (2003) The application of a Marine Biotic Index to different impact sources affecting soft-bottom benthic communities along European coasts. Marine Pollution Bulletin 46: 835–845

Carstensen J (2007) Statistical principles for ecological status classification of water framework directive monitoring data. Marine Pollution Bulletin 55: 3–15

COAST (2003) Common Implementation Strategy for the Water Framework Directive (2000/60/EC). Guidance Document No 5, Transitional and Coastal Waters – Typology, Reference Conditions and Classification Systems. Produced by Working Group 2.4 – COAST. Office for Official Publications of the European Communities, Luxembourg.

Costanza R (1992) Towards an operational definition of health. In: Costanza R, Norton B, Haskell BD (eds.), Ecosystem health – New goals for environmental management. Inland Press, Washington, DC, pp. 239–256

Cronin JR, Tyler ID (1980) Organic carbon in a Scottish sea loch. In: Albaiges J (ed.), Analytical techniques in environmental chemistry. Pergamon Press, Oxford, pp. 419–426

CSTT (1994) Comprehensive studies for the purposes of Article 6 of DIR 91/271 EEC, the Urban Waste Water Treatment Directive. Published for the Comprehensive Studies Task Team of Group Coordinating Sea Disposal Monitoring by the Forth River Purification Board, Edinburgh

CSTT (1997) Comprehensive studies for the purposes of Article 6 & 8.5 of DIR 91/271 EEC, the Urban Waste Water Treatment Directive, 2nd edn. Published for the Comprehensive Studies Task Team of Group Coordinating Sea Disposal Monitoring by the Department of the Environment for Northern Ireland, the Environment Agency, the Scottish Environmental Protection Agency and the Water Services Association, Edinburgh

Cugier P, Billen G, Guillaud JF, Garnier J, Ménesguen A (2005) Modelling the eutrophication of the Seine Bight (France) under historical, present and future riverine nutrient loading. Journal of Hydrology 304: 381–396

Diamond J (2005) Collapse: How societies choose to fail or succeed. Penguin Books, London

Edwards VR, Tett P, Jones KJ (2003) Changes in the yield of chlorophyll a from dissolved available inorganic nitrogen after an enrichment event – applications for predicting eutrophication in coastal waters. Continental Shelf Research 23: 1771–1785

Gowen RJ, Bradbury NB (1987) The ecological impact of salmon farming in coastal waters: a review. Oceanography and Marine Biology, an Annual Review 25: 563–575

Gowen RJ, Tett P, Jones KJ (1983) The hydrography and phytoplankton ecology of Loch Ardbhair: a small sea-loch on the west coast of Scotland. Journal of Experimental Marine Biology and Ecology 71: 1–16.

Graham J (2002) European Community Law, Rep. No. DLM/COPA/LEG2. Scottish Environment Protection Agency, Stirling. http://www.sepa.org.uk/pdf/guidance/water/legislation/European_

Community_Law.pdfHjorth M, Dahllof I, Forbes VE (2006) Effects on the function of three trophic levels in marine plankton communities under stress from the antifouling compound zinc pyrithione. Aquatic toxicology 77: 105–115

Johnson LE (1991) A morally deep world. Cambridge University Press, Cambridge

JNCC (2004) The ecosystem based approach. Joint Nature Conservation Council, Peterborough, England. http://www.jncc.gov.uk/, web page accessed January 2006

Karakassis I, Pitta P, Krom MD (2005) Contribution of fish farming to the nutrient loading of the Mediterranean. Scientia Marina 69: 313–321

Laurent C, Tett P, Fernandes T, Gilpin L, Jones KJ (2006) A dynamic CSTT model for the effects of added nutrients in Loch Creran, a shallow fjord. Journal of Marine Systems 61: 149–164

Mageau MT, Costanza R, Ulanowicz RE (1995) The development and initial testing of a quantitative assessment of ecosystem health. Ecosystem Health 1: 201–213

Maraldo K, Dahllöf I (2004) Seasonal variations in the effect of zinc pyrithione and copper pyrithione on pelagic phytoplankton communities. Aquatic Toxicology 69: 189–198

Margalef R (1978) Life forms of phytoplankton as survival alternatives in an unstable environment. Oceanologica Acta 1: 493–509

Mithen S (2003) After the ice. Weidenfeld & Nicolson, London

McIntyre AD (1995) Human impact on the oceans: The 1990s and beyond. Marine Pollution Bulletin 31: 147–151

Miller J (2006) Daoism and Nature. In: Gottlieb R (ed.), Handbook of religion and ecology, Oxford University Press, Oxford

Nickell LA, Black KD, Hughes DJ, Overnell J, Brand T, Nickell TD, Breuer E, Harvey SM (2003) Bioturbation, sediment fluxes and benthic community structure around a salmon cage farm in Loch Creran, Scotland. Journal of Experimental Marine Biology and Ecology 285–286: 221–233

Odum EP (1959) Fundamentals of ecology, 2nd edn. Saunders, Philadelphia.

OSPAR (2003) 2003 Strategies of the OSPAR Commission for the Protection of the Marine Environment of the North-East Atlantic (Reference number: 2003–21). In: OSPAR Convention for the Protection of the Marine Environment of the North-East Atlantic: Ministerial Meeting of the Ospar Commission, Bremen: 25 June 2003, Vol. Annex 31 (Ref. B-4.2)

Painting SJ, Devlin MJ, Rogers SI, Mills DK, Parker ER, Rees HL (2005) Assessing the suitability of OSPAR EcoQOs for eutrophication vs ICES criteria for England and Wales. Marine Pollution Bulletin 50: 1569–1584

Pearson TH, Rosenberg R (1976) A comparative study of the effects on the marine environment of wastes from cellulose industries in Scotland and Sweden. Ambio 5: 77–79

Pearson TH, Rosenberg R (1978) Macrobenthic succession in relation to organic enrichment and pollution of the marine environment. Oceanography and Marine Biology Annual Review 16: 229–311

Readman JW (2005) Development, occurrence and regulation of antifouling paint biocides: historical review and future trends. Handbook of environmental chemistry, 5 part O

Rogers NAM (2004) The command of the ocean: a naval history of Britain 1649–1815. Allen Lane, London

Ronan CA, Needham J (1978) The shorter science & civilization in China, Vol. 1. Cambridge University Press, Cambridge

SEPA (1997) Policy No 17. Cage fish farms: sea lice treatment chemicals: risk assessment of azamethiphos Version 1, June 1997. Scottish Environment Protection Agency, Stirling

SEPA (1998) Sea lice treatment chemicals for cage fish farms: provisional environmental quality standards for azamethiphos. Policy No. 17 Version 2, November 1998 Scottish Environment Protection Agency, Stirling. http://www.sepa.org.uk/aquaculture/policies/17.pdf

SEPA (2004) Designated shellfish waters in Scotland: Site Data: Area 32: Loch Creran, Argyll. Scottish Environmental Protection Agency, Stirling http://www.sepa.org.uk/data/shellfish/sites.asp

SEPA (2005a) Introduction to aquaculture regulation. Scottish Enviroment Protection Agency, Stirling. http://www.sepa.org.uk/aquaculture/background/index.htm, accessed January 2007

SEPA (2005b) Section 5: Setting consent limits (version 1.3). Scottish Environment Protection Agency. http://www.sepa.org.uk/guidance/fishfarmmanual/manual.asp, accessed January 2007
SNH (2006a) Loch creran special area of conservation. Advice under Regulation 33(2) of The Conservation (Natural Habitats &c.) Regulations 1994 (as amended). Scottish Natural Heritage. URL: http://www.snh.org.uk/about/directives/ab-dir14.asp, accessed January 2007
SNH (2006b) Special areas of conservation. Scottish Natural Heritage, http://www.snh.org.uk/about/directives/ab-dir07.asp (and linked pages), accessed 4 January 2006
Tansley AG (1935) The use and abuse of vegetational concepts and terms. Ecology 16: 284–307
Tett P (1986) Physical exchange and the dynamics of phytoplankton in Scottish sea-lochs. In: Skreslet S (ed.), The role of freshwater outflow in coastal marine ecosystems. Springer, Berlin, pp. 205–218
Tett P (2006) Using the PCI-LF – a draft user guide. Napier University, Edinburgh. http://www.lifesciences.napier.ac.uk/research/Envbiofiles/PCI.htm
Tett P, Barton ED (1995) Why are there about 5000 species of phytoplankton in the sea? Journal of Plankton Research 17: 1693–1704
Tett P, Drysdale M, Shaw, J (1981). Phytoplankton in Loch Creran in 1979, and its effect on the rearing of oyster larvae, Rep. No. 52. Scottish Marine Biological Association, Oban.
Tett P, Edwards V (2002) Review of harmful algal blooms in Scottish coastal waters. Scottish Environmental Protection Agency, Stirling
Tett P, Gilpin L, Svendsen H, Erlandsson CP, Larsson U, Kratzer S, Fouilland E, Janzen C, Lee J-Y, Grenz C, Newton A, Ferreira JG, Fernandes T, Scory S (2003a) Eutrophication and some European waters of restricted exchange. Continental Shelf Research 23: 1635–1671
Tett P, Gowen R, Mills D, Fernandes T, Gilpin L, Huxham M, Kennington K, Read P, Service M, Wilkinson M, Malcolm S (2007) Defining and detecting undesirable disturbance in the context of Eutrophication. Marine Pollution Bulletin 53: 282–297
Tett P, Heaney SI, Droop MR (1985) The Redfield ratio and phytoplankton growth rate. Journal of the Marine Biological Association of the United Kingdom 65: 487–504
Tett P, Hydes D, Sanders R (2003b) Influence of nutrient biogeochemistry on the ecology of North-West European shelf seas. In: Schimmield G, Black K (eds.), Biogeochemistry of marine systems. Sheffield Academic Press, Sheffield, pp. 293–363
Tett P, Wallis A (1978) The general annual cycle of chlorophyll standing crop in Loch Creran. Journal of Ecology 66: 227–239
Tyler ID (1984) A carbon budget for Creran, a Scottish sea loch. Ph.D. thesis. University of Strathclyde
UKTAG (2003) Guidance on Typology for Coastal & Transitional Waters of the UK and Republic of Ireland (Final P2 v12, 14/4/04). UK Technical Advisory Group on the Water Framework Directive. Obtain from *Typology of Water Bodies*, http://www.wfduk.org/tag_guidance/Article_05/ Folder.2004–02–16.5312/view.
UKTAG (2004) (Draft) Type Specific Reference Condition Descriptions for Transitional and Coastal Waters for the UK (v2 02/08/04). UK Technical Advisory Group on the Water Framework Directive. Obtain from *Reference Conditions,* http://www.wfduk.org/tag_guidance/Article_05/Type specific reference conditions/view
Wood, BJB, Tett PB, Edwards, A (1973) An introduction to the phytoplankton, primary production and relevant hydrography of Loch Etive. Journal of Ecology 61: 569–585.
Word JQ (1990) The infaunal trophic index: a functional approach to benthic community analyses. Ph.D. thesis. University of Washington, 297 pp

Chapter 2
Monitoring of Environmental Impacts of Marine Aquaculture

Marianne Holmer[1,*], Pia Kupka Hansen[2], Ioannis Karakassis[3], Joseph A. Borg[4], and Patrick J. Schembri[4]

Abstract Marine aquaculture is regulated and monitored through international and national legislation that varies significantly between countries and regions around the world. Research is still needed to improve the monitoring programmes, in particular those related to the ecosystem approach at larger scales. Most monitoring programmes include examination of the benthic environment and some also of water quality, although impacts are difficult to detect due to rapid dilution. In the Mediterranean benthic monitoring may include use of the seagrass *Posidonia oceanica*, as this species is widespread and highly sensitive to aquaculture waste products. This chapter provides details of two monitoring programmes: (1) salmon farming in Norway and (2) sea bream/sea bass and tuna farming in Malta.

Keywords Ecosystem approach, Europe, case studies, Norway, Malta

2.1 Regulation and Monitoring of Marine Aquaculture

Marine aquaculture is a diverse production industry involving a variety of different species, production methods and husbandry. In this chapter we will primarily focus on finfish culturing and to some extent shellfish production. Environmental impacts of finfish culturing are widely documented (Hargrave 2005) and include a broad

[1]Institute of Biology, University of Southern Denmark, Campusvej 55, 5230 Odense M, Denmark. Tel + 4565502605; Fax + 4565930457; E-mail: holmer@biology.sdu.dk

[2]Institute of Marine Research, Box 1870 Nordnes, 5817 Bergen, Norway

[3]Marine Ecology Laboratory, Biology Department, University of Crete, PO Box 2208, Heraklion 714 09 Crete, Greece

[4]Department of Biology, University of Malta, Msida, Malta

* Corresponding author

range of impacts from aesthetic to direct pollution problems (Fig. 2.1, Pillay 2004). Fish production can generate considerable amounts of dissolved effluents, which potentially affect water quality in the vicinity of the farms, and due to rapid dilution, also at larger scales (km-scale). Due to the rapid dilution it has been difficult to document the effects of dissolved nutrients in farm vicinities, in particular in areas with relatively high nutrient concentrations such as in the Baltic Sea (Christensen et al. 2000). Other studies of fishery landings in the Mediterranean suggest that nutrients are rapidly transferred up the trophic chain enhancing secondary production (Machias et al. 2005), indicating that monitoring of nutrient losses should be done at different scales. Due to rapid settling of feed and faecal pellets in the vicinity of the farms, benthic impacts are much more widely documented (Holmer et al. 2005; Kalantzi and Karakassis, 2006). This input of organic rich material enhances the microbial processes in the sediments, often leading to anoxic conditions (Holmer and Kristensen 1992). This may have major effects on the benthic fauna and flora leading to lower fauna and flora densities under the cages or even defaunated sediments (Delgado et al. 1997; Karakassis et al. 2002). Other environmental impacts include release of chemicals, medicines and pesticides, which are used for treatment of the fish and the farm installations. Interactions with wild populations, spreading of disease and release of parasites from farms are also of environmental concern.

The environmental impacts of marine aquaculture within the European Union are regulated and managed, at a European level, through a variety of European Commission (EC) Directives and International Conventions. There are currently eight EC directives (Table 2.1) directly involved and an additional 50 + Directives, Decisions and Regulations, which have an indirect effect (Read et al. 2001). In addition, three International Conventions on marine pollution cover EU coastal waters (Table 2.2) and there are a further 30 + international agreements that have an indirect effect on the monitoring and regulation of marine aquaculture (Read et al. 2001). Within the European Union, the regulation of the aquaculture sector comes under the remit of the Common Fisheries Policy (CFP). The CFP states that Member States shall adopt provisions to comply with the objectives of regular monitoring of activities and technical controls. At EU level, environmental protection measures have been established at three levels: (1) general policy; (2) specific measures; and (3) regulations that control specific local conditions (Eleftheriou and Eleftheriou 2001).

The regulations controlling aquaculture vary between countries, but most countries use some form of Environmental Quality Objectives (EQOs) and Environmental Quality Standards (EQSs) (Table 2.3). Only a few countries apply a carrying capacity at the moment, but this has been suggested for future regulation within the Integrated Coastal Zone Management (ICZM) approach (see below). Most countries have specific demands for the location of the farms to avoid situating these near habitats of special interest (recreation, wild life, fishing zones) and near industries and sewage outfalls. Requirements on stocking density, feed type and sediment and water quality standards are also included in most regulations. A few countries regulate the production based on discharges, e.g., N and P release per kg

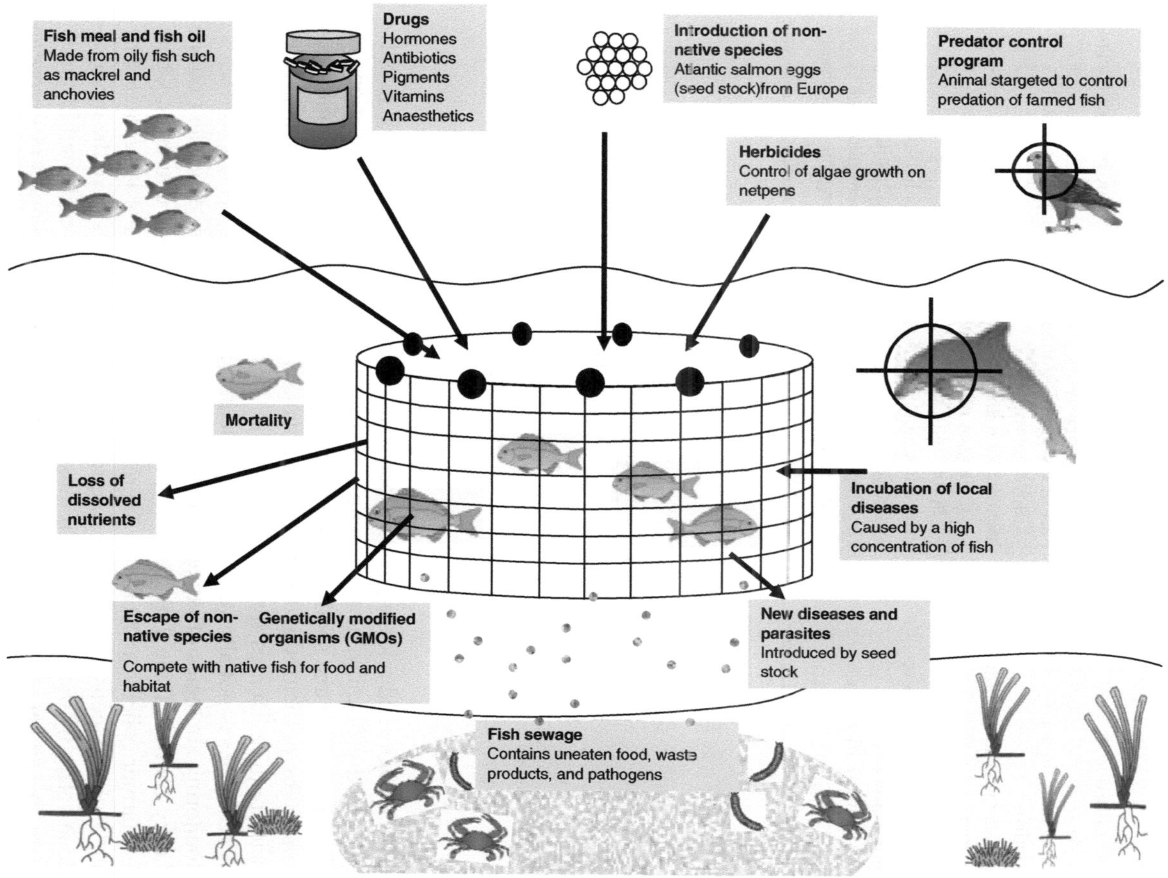

Fig. 2.1 Conceptual figure of aquaculture in the ecosystem (symbols from IAN Symbol Libraries)

Table 2.1 European Commission directives related to management of marine aquaculture

Directive	Target
Dangerous substances directive	Reduce pollution by list II substances
Quality of shellfish growing waters directive	Contribute to high quality of shellfish products through protection or improvement of shellfish flesh and shellfish waters
Shellfish directive	Classification of production areas on the basis of bacteriological criteria
Environmental impact assessment directive	Part of the application and licensing procedures for development
Strategic environmental assessment directive	Identification and assessment of environmental consequences of aquaculture during preparation and before adoption
Species and habitat directive	Protection and conservation of natural habitats.
Wild birds directive	Protection and conservation of natural habitats
Water framework directive	Development of catchment management plans for implementation of integrated management. Operates with assimilation capacity.

Table 2.2 International conventions which apply to marine aquaculture operations

Conventions	Area of cover
OSPAR	Northeast Atlantic
Helsinki	Baltic Sea
Barcelona	Mediterranean Sea

fish produced or total release per farm per year, to encourage the producer to optimize the feed efficiency, as for example is the case in Denmark. Regulations on food standards (fish and shellfish products) may also apply such as maximum residue limits for pesticides and other contaminants in fish or shellfish flesh. Most countries require licenses for medicine and pesticide use, and in some countries, use of pesticides is not allowed at all (e.g., Denmark).

Monitoring strategies of aquaculture vary between countries, dependent on the regulatory control (Table 2.4). Self-monitoring applies in several countries, where the fish farmer collects and submits the results to the authorities or is supported by on-site Authority control at varying frequencies. Other countries have the monitoring done solely by the authorities. In most cases, water quality and benthic conditions are checked at regular intervals (2–12 times per year). Examples of European monitoring programmes are presented in detail for the North-Atlantic (Norway) and for the Mediterranean (Malta) below.

Most countries monitor the food quality of the products, in particular for shell fish, where accumulations of biotoxins and microbes are sampled up to twice a week during intensive production and periods of risks of contamination, e.g., during

Table 2.3 Regulations and legislation controlling marine aquaculture in Europe and Asia (updated from Fernandes et al. 2000)

		Regulatory control				
Country	Type of marine aquaculture	Carrying capacity	Environmental standards	Food standards	Medicines licenses	Pesticides licenses
Asia						
Cambodia	Finfish and shellfish	–	Sub-decree on water pollution control, Fisheries law	Proclamation on fisheries product hygiene	–	–
China	Finfish and shellfish	–	Water quality standards for fisheries	Food hygiene law, Standardization law	Administration of veterinary medicines (1988)	Environmental protection law, Law on the prevention and control of water pollution, Marine environment protection law, Regulations on the environmental management of first import of chemicals, Import and export of toxic chemicals (1994), Control over the safety of dangerous chemicals (2002)
Indonesia	Finfish and shellfish	–	Government regulation	Food Act, Fisheries Law, Decree of the Minister of Marine Affairs and Fisheries	Fisheries Law, Decree of the Minister of Agriculture	–
Korea	Finfish and shellfish	–	Water quality conservation act, Public water management Act, Fisheries Act	Fisheries act, products quality control Act	–	–

(continued)

Table 2.3 (continued)

		Regulatory control				
Country	Type of marine aquaculture	Carrying capacity	Environmental standards	Food standards	Medicines licenses	Pesticides licenses
Malaysia	Finfish and shellfish	–	Waters Act, Environmental Quality Act, Environmental Quality Regulations	Food Act, Food Regulations	Food Regulations establish limits of drug residues in fish products	–
Myanmar	Finfish and shellfish	–	Myanmar Investment Commission	Marine Fisheries Law covering hygiene, food additives, microbiological criteria, packaging, labeling, water standards, application of medical drugs, and HACCP system	Marine Fisheries Law and Drug Law	Pesticides Law
Philippines	Finfish and shellfish	–	Philippine clean water Act, Marine Pollution Decree	Fisheries Code, Consumer Act, FAO	Fish health section	Fertilizer and Pesticide Authority
Thailand	Finfish and shellfish	–	Enhancement and Conservation of National Environmental Quality Act, Fisheries Act	Food Act	Drug Act	Hazardous Substances Act
Australia	Finfish and shellfish	–	Environmental Protection Policy, Aquaculture Regulations	Australia-New Zealand Food Standards Code, Commonwealth Food Standards Australia-New Zealand Act, South Australia Food	Agricultural and Veterinary Chemicals Code Act, South Australian Agricultural	Agricultural and Veterinary Chemicals Act

				Act, South Australia Fisheries Act, Fisheries Regulations, South Australia Primary Produce Act	and Veterinary Products Act, Agricultural and Veterinary Products Regulations, South Australia Veterinary Practices Act, Fisheries (Exotic Fish, Fish Farming and Fish Diseases) Regulations	
Europe						
Denmark	Finfish (land-based and sea cage) (mainly rainbow trout); some shellfish (extensive)	Effluent released to marine waters (maximum of 560 tN and 54 tP)	Nutrient output (560 tN and 54 tP annually); feed type, feed convention ratio	None for finfish at producer stage. Bivalves checked for algal toxins	Prescribed by veterinarian; number permitted is limited	Not permitted
Finland	Finfish (mainly rainbow trout)	No figure	Discharge (8 g P/kg fish produced, 70 g N/kg fish produced)	In accordance with EU Directives	Granted by National Agency for Medicines	Not used due to favourable parasitic situation
France	Shellfish (mainly oyster and mussels); some finfish (mainly sea bream)	20 t $year^{-1}$ for finfish; No figures for shellfish	Shellfish growing water quality, EIS can dictate standards for finfish; threshold levels for chemical residues	Bacteria and algal toxins in shellfish flesh; none for fish	License given by veterinarian	License from Ministry of Health

(continued)

Table 2.3 (continued)

		Regulatory control				
Country	Type of marine aquaculture	Carrying capacity	Environmental standards	Food standards	Medicines licenses	Pesticides licenses
Germany	Shellfish (mainly mussels); some finfish (mainly rainbow trout)	No figure	Effluent water quality; closed season for mussel harvest; shellfish growing water quality	Shellfish flesh quality	Use must be recorded in "operation record book"	Use must be recorded in "operation record book"
Greece	Finfish (mainly sea bass and bream); shellfish (mainly mussels)	No figure	Cage location; stocking density; feed type, water and sediment quality; sanitary measures	Same as legislation for farmed domestic animals	Approved by National Drug Organizations	Approved by National Drug Organizations
Iceland	Finfish (mainly salmon) some char and rainbow trout	No figure	Cage location; water quality; nature conservation	Maximum residue limit for pesticides and other contaminants in flesh; standards for microbes and algal toxins	Permit from Veterinary Officer of fish diseases	Permit from Veterinary Officer of fish diseases
Ireland	Finfish (mainly salmon); shellfish (mainly mussels)	15 kg m^{-3} stocking density for finfish	Cage location, escape prevention, fallowing period; abundance of sea-lice; water and seabed quality	Algal toxins in shellfish flesh; disease and chemical residues in finfish and shellfish flesh	Issued by veterinarian	Issued by veterinarian Organotin antifoulants are not permitted
Italy	Finfish (mainly trout); some shellfish (mussels, clams)	No figure	Water quality	Heavy metals and organic compounds in bivalves	Issued by veterinarian	–

Malta	Finfish (mainly sea bream and blue fin tuna)	15 kg m^{-3} for sea bream and 3–4 kg m^{-3} for blue fin tuna	Water quality; sediment quality; benthic diversity; state of benthic habitats in the vicinity of farm	In accordance with EU Directives	Issued by Veterinarian at the Department of Fisheries and Aquaculture	None issued – pesticides not permitted
The Netherlands	Shellfish (cockles, mussels, oysters)	100,000 metric tons (fresh weight mussels 10,000 metric cockles	Safeguard of food reserves for wild birds; biotoxins and microbes in shellfish growing waters	Biotoxins and microbes in shellfish flesh	Not used in extensive mariculture	Not used in extensive mariculture
Norway	Finfish (mainly salmonids); shellfish (mainly mussels)	Site dependent	Receiving water body: nutrients, oxygen, sediment C and N, benthos, organic pollutants, metals. Site: sediment condition.	EU standards for fish and shellfish	Issued by veterinarian	Issued by veterinarian
Portugal	Shellfish (clams, oysters, cockles); some finfish (sea bass, sea bream, turbot)	No figure		In accordance with EU Regulations	Permission required from Direccao Geral da Veterinaria	Permission required from Direccao Geral da Veterinaria

(continued)

Table 2.3 (continued)

			Regulatory control			
Country	Type of marine aquaculture	Carrying capacity	Environmental standards	Food standards	Medicines licenses	Pesticides licenses
Scotland	Finfish (mainly salmon with some rainbow trout and turbot); shellfish (oysters, scallops, mussels)	Maximum weight of fish to achieve sustainability at a farm	Hydrological character, biological, sediment and water quality	Biotoxins in shellfish flesh, phytoplankton in shellfish growing water; maximum residue limit for chemical in finfish flesh	Permission required from Veterinary Medicines Directorate	Permission required from Veterinary Medicines Directorate
Spain	Shellfish (mainly mussels, cockles, clams); some (salmon, turbot)	No figure	Water quality for shellfish	EC Directives controlling production; maximum residue levels in flesh	Controlled by veterinarian	–
Sweden	Finfish some shellfish (mussels)	No figures for finfish, 10,000 t for shellfish	Water classification according to organic matter and metals in mussel tissue; farm sites, extraction and discharge of water, nutrients	Species (and strains) allowed	Use permitted	Use permitted

Table 2.4 Monitoring programmes used in marine aquaculture (updated from Fernandes et al. 2000)

		Monitoring			
		Environmental		Food	
Country	Type of marine aquaculture	Self monitoring	Authority monitoring	Self monitoring	Authority monitoring
Europe					
Denmark	Finfish (land-based and sea cages) (mainly rainbow trout); some shell fish (extensive)	Water quality monitored 12 times per year (finfish); sediment monitored twice per year (finfish)	Compliance with finfish standards (periodic)	Bivalves checked for algal toxins post harvest	Compliance with standards (periodic)
Finland	Finfish (mainly rainbow trout)	Water quality; farmers keep daily records of operation for inclusion in report	Water quality (chemistry, plankton and micro-algae); sediment quality (sediment, benthic macroinvertebrates)	–	Compliance with EU Directives
France	Shellfish (mainly oyster and mussels); some finfish (mainly sea bream)	Quality of effluent (finfish)	Benthic survey and nutrient analysis of the water column twice per year; chemical residues and microbiology of water twice per month; shellfish water quality (algal toxins and phytoplankton) monitored twice per month	Bacteria, phytoplankton toxins in finfish flesh (before marketing)	Bacteria, phytoplankton toxins in finfish flesh (without warning); bacteria, phytoplankton toxins in shellfish flesh (twice monthly)

(continued)

Table 2.4 (continued)

		Monitoring			
		Environmental		Food	
Country	Type of marine aquaculture	Self monitoring	Authority monitoring	Self monitoring	Authority monitoring
Germany	Shellfish (mainly mussels); some finfish (mainly rainbow trout)	–	Control of fishing vessels during mussel harvest and seed fishing. Water quality monitored five times in 3 years	–	–
Greece	Finfish (mainly sea bass and bream); shellfish (mainly mussels)	Regular water quality, sediment and benthic community monitoring	–	–	–
Iceland	Finfish (mainly salmon) some char and rainbow trout	All accessible knowledge collected to control environmental effects	Chemical residues and micro-organisms in fishing grounds	Internal quality control to ensure compliance	Chemical residues and micro-organisms in fish flesh; quality control
Ireland	Finfish (mainly salmon); shellfish (mainly mussels)	–	Sea lice monitored 14 times per year; water at salmon farms monitored monthly; benthic monitoring depends on tonnage (finfish)	–	Shellfish flesh monitored weekly for algal toxins, annual sampling for disease at salmon and oyster sites; analysis of farmed salmon for chemical residues
Italy	Finfish (mainly trout); some shellfish (mussels, clams)	–	macro-descriptor in water monitored seasonally (twice per week from June–September;	–	Bioaccumulation of heavy metals and organic compounds in bivalves twice per year

			ground parameters once per year; maximum trophic index of 5.5		
Malta	Finfish (mainly sea bream and blue fin tuna)	Water quality at some farms	Water quality; sediment quality; benthic diversity; state of benthic habitats in the vicinity of farm	–	Compliance with EU Directives
The Netherlands	Shellfish (cockles, mussels, oysters)	–	Aerial photography and ground counts of inter-tidal mussel stocks. Biotoxins and microbes in shellfish growing waters sampled 2 times per week	–	Biotoxins and microbes in shellfish flesh 2 times per week
Norway	Finfish (mainly salmonids); shellfish (mainly mussels)	–	Shellfish water quality once per month. Finfish farms monitored according to Norwegian Standard NS9410	–	10% of finfish licences sampled every year to check for chemicals in muscle and feed; all medicated fish are checked; shellfish waters monitored throughout the year for biotoxins, microbes and chemicals
Portugal	Shellfish (clams, oysters, cockles); some finfish (sea bass, sea bream, turbot)	–	Mollusc water quality once per month or once every 3 month depending on parameter	–	Compliance with food quality standards

(continued)

Table 2.4 (continued)

		Monitoring			
		Environmental		Food	
Country	Type of marine aquaculture	Self monitoring	Authority monitoring	Self monitoring	Authority monitoring
Scotland	Finfish (mainly salmon with some rainbow trout and turbot); shellfish (oysters, scallops, mussels)	Near-field sampling following an agreed programme (finfish)	Audit self-monitoring of finfish farms; consent compliance of medicines; environmental impact from finfish farms; shellfish growing waters quality (periodic)	–	Assays of biotoxins; microbial quality of water; trace compounds in shellfish flesh and fish tissue
Spain	Shellfish (mainly mussels, cockles, clams); some (salmon, turbot)	–	Periodic water quality, weekly red tide monitoring	–	Compliance with food standards monitored yearly (MRLs)
Sweden	Finfish some shellfish (mussels)	–	Water classification according to organic matter and metals in mussel tissue	Check for algal toxins in shellfish prior to harvest	Biotoxins in mussel meat normally sampled once per week; faecal coliforms, chemical residues in finfish flesh.

North-America					
Canada	Finfish (primarily Atlantic salmon and some Atlantic char, halibut and Atlantic cod). Shellfish (primarily blue mussels with minor production of scallops	Finfish: water temperature and dissolved oxygen measured daily throughout summer/fall periods of maximum growth. Shellfish: befouling (weekly to monthly), algal bloom status through phytoplankton monitoring	Finfish: Federal Government (DFO) annual monitoring of sediment geochemical variables to classify benthic enrichment status, no routine water mass column monitoring for any variable. Shellfish: No annual monitoring. Provincial governments track yield from all leases to record harvested biomass	Finfish and shellfish: Industry quality guidelines for fresh and frozen seafood products are followed	Canadian Depts. Of Food Inspection and Health Canada monitor products (random sampling on irregular basis) for substances such as bacteria and antibiotic residues. Compliance required for meeting exported seafood guidelines. Locally consumed fish (within Canada) is not routinely monitored

phytoplankton blooms. Finfish flesh is monitored for bacteria, chemical residues and phytoplankton toxins once every year, or in some countries, every year before marketing.

2.1.1 Research Support for Monitoring of Environmental Impacts

There have been several papers, reports and other documents dealing with the principles of monitoring and particularly so in the case of the monitoring of fish-farming impacts. The report by GESAMP (1996) addressed this issue by describing possible scenarios of fish farm locations and suitable monitoring programmes. Although the paper included a list of variables used for monitoring the ecological effects of coastal aquaculture, the authors realised that the information provided by some of these variables is of limited use in some situations. A comprehensive series of studies on monitoring and regulation was also undertaken in the framework of the MARAQUA project (The Monitoring and Regulation of Marine Aquaculture in Europe). These studies resulted in a series of papers on the scientific principles underlying the environmental monitoring of aquaculture (Fernandes et al. 2001), on the control of chemicals (Costello et al. 2001), on the genetic interactions between farmed and wild fish species (Youngson et al. 2001) and on the use of hydrodynamic and benthic models for the management of aquaculture impacts (Henderson et al. 2001). However, research on aquaculture-environment interactions has progressed remarkably during the last 5 years, particularly in the framework of EU-funded projects, which have provided useful information for the understanding of various ecosystem processes affected by the presence and operation of fish farms.

The effects of aquaculture on marine benthos, particularly on macrofauna, have been known for long (Gowen and Bradbury 1987), and in general, they seem to follow the pattern described by Pearson and Rosenberg (1978) regarding the succession of macrofaunal organisms along the benthic enrichment gradient. However, more than 40 articles in the scientific literature (review in Kalantzi and Karakassis 2006) have studied these affects using in total 120 biological and geochemical variables, most of which were highly intercorrelated. A meta-analysis of the most commonly used of those variables by Kalantzi and Karakassis (2006) showed that their values are determined by a combination of distance from the farm with bottom depth and/or latitude. Although the benthic effects are relatively easy to detect, there are some concerns regarding the cost of the associated faunal analysis, which becomes more and more difficult due to the rarity of experts in the taxonomy of benthic organisms (GESAMP 1996). A series of papers have addressed this issue by studying the potential use of surrogates and their effect on data quality. Karakassis et al. (2002) have used sediment profiling imagery (SPI) as a means for monitoring the effects of fish farms on silty bottoms and found that SPI can provide very reliable information on the state of the benthic environment. The use of different levels of taxonomic resolution (Karakassis and Hatziyanni 2000) and the use of various levels of taxonomic resolution, sieve mesh size and sample size (Lampadariou et al. 2005) can also be used as a basis for cost-effective monitoring protocols for the assessment of the state of the benthic environment in the vicinity of fish farms.

An alternative method for monitoring the effects on the benthic environment would be to focus on some geochemical variables that reflect the organic content of the sediment, such as total organic carbon (TOC) or organic matter, usually measured by means of the loss on ignition method (LOI) which provides straightforward results. Hyland et al. (2005) have shown that TOC can be used as an indicator of the quality of the benthic environment since it can predict quite reliably macrofaunal diversity. On the other hand, studies on the recovery process do not show good relationships between TOC and the benthic fauna community, but instead correlate with oxygen demand, indicating that it is the labile pool of TOC controlling faunal distribution (Pereira et al. 2004). It is also worth noting that TOC or LOI values in samples taken beneath fish farms could be misleading since their concentrations at the surface layer could remain fairly constant although the depth of the farm sediment measured through sediment profiles could change remarkably with season (Karakassis et al. 1998). Also, pools in the surface layer may show significant seasonal variation, as has been found in a Danish farm, where the TOC pools in the surface layer correlated with the seasonal changes in fish production (Holmer and Kristensen 1992). At larger spatial scales, the effects of fish farms on macrofauna are rather negligible and particularly so in the case of coarse sediment sites and, therefore, it could be expected that those effects are unlikely to disturb other (remotely located) uses of the coastal zone. In both Norway and Canada monitoring systems have been constructed with more detailed analysis of the sediments, but based on relatively simple measuring techniques, which allow the fish farmer himself to follow the benthic impacts at the farms (Hangen et al. 2001; Brooks and Mahnken 2003), but in Norway the authorities require that the monitoring is performed by an independent firm or institute. In Scotland, the authorities are implementing benthic monitoring along with modelling, which strengthens the field sampling (Cromey and Black 2005). If the sampling and models deviate, field conditions are up for a more detailed examination. Monitoring in both Norway and Scotland operates within different zones around the farms, where sites at increasing distance from the farms are allowed different degrees of benthic impact (Ervik et al. 1997; Cromey and Black 2005).

The effects on water quality are probably those causing more concern regarding the quality of the marine environment. It is well known that fish farms release large quantities of dissolved nutrients in the ambient water, particularly nitrogen and phosphorus (Holby and Hall 1991; Hall et al. 1992). Furthermore, these nutrients are mainly released during the summer period when light availability is high and therefore it could be expected that phytoplankton blooms are likely to occur in the vicinity of fish farms. However, numerous studies (Pitta et al. 1999, 2006; La Rosa et al. 2002; Soto and Norambuena 2004) have failed to detect significant changes in chlorophyll a or particulate organic carbon (POC) in the water column in the vicinity of fish farms. This paradox could be attributed to the dispersive nature of fish-farming sites, i.e., to the fact that phytoplankton cells do not stay long enough to capitalize on nutrients (Gowen et al. 1983), or to experience rapid grazing by zooplankton as suggested by Machias et al. (2005). Several studies have measured significant diel changes of nutrient concentrations in the vicinity of fish farms in oligotrophic waters (Karakassis et al. 2001; Pitta et al. 2006), indicating that dispersion is a very efficient mechanism at those sites. However, it has recently been shown (Dalsgaard and Krause-Jensen 2006) that *in situ* incubation of phytoplankton and of *Ulva* sp. can be used as a relatively low-cost monitoring strategy to document the distance from the farms where pelagic primary production is affected. This method has the advantage that it is not affected by episodic events such as those affecting concentrations of nutrients and particulate material in the water column, whereas the incubation period of the bioassays allows for estimates based on integration of the water quality conditions over several days. It is worth noting that even though these bioassays have been able to detect changes up to a distance of 200–300 m from the fish farms, the intensity of these effects decreases rapidly with distance. However, when several farms are aggregated in a fish-farming zone producing thousands of tonnes, it is reasonable to ask: what are the large scale effects of this aggregation which should be detectable despite the nutrient dispersion? A recent survey in the Mediterranean (Pitta et al. 2005) showed that most of the significant changes in nutrients as well as chlorophyll a or PON were found at the deepest layer of the water column below the thermocline, indicating that they are related to the remineralization of benthic organic material.

Wild fish communities are also affected by aquaculture. Partly, this effect is related to the attraction of some fish species to the floating structures (see Chapter 3 this volume), but fish communities can also be affected at large spatial scales (Machias et al. 2004, 2005, 2006; Giannoulaki et al. 2005) probably because of the changes in primary productivity in the area and the rapid transfer of nutrients up the food web. This effect has been documented in the Mediterranean where oligotrophic conditions and the structure of planktonic communities seem to favour this process. In this context it has been suggested (Machias et al. 2005) that fish communities are probably a good indicator of the increased material flux since they are long-lived organisms integrating processes over longer time periods, and their predators are unlikely to respond promptly to an increase in their biomass.

The effects of fish farms on seagrass meadows have been documented by many recent papers (Delgado et al. 1997; Holmer et al. 2003; Marbà et al. 2006; Diaz-Almela et al. submit). In the recently finished EU-funded project MedVeg (Effects of nutrient release from Mediterranean fish farms on benthic vegetation in coastal ecosystems) four sites were monitored along the Mediterranean for benthic fauna, sediment geochemistry, water quality and seagrass-related variables. The results showed that the distance of detectable effects varied greatly among the variables used. In particular, seagrass mortality seemed to be the indicator detected at greater distance than any of the others determined in this project (Marbà et al. 2006; Frederiksen et al. 2007; Diaz-Almela et al. submitted).

This is not surprising since it is well known that, in particular, *Posidonia oceanica* is a very sensitive endemic species in the Mediterranean that has been shown to suffer population reductions due to anthropogenic stress (Marbà et al. 2005).

All the above indicate that there are many different ecological processes and biotic communities affected by aquaculture. Some of these may be easily detected and monitored, such as the effects on macrofauna, although these are usually confined to a small area beneath and around fish farms. Others, such as water quality and plankton dynamics, need new protocols for assessing the degree of change imposed by aquaculture and further research to increase our understanding of the related processes. Monitoring of fish communities seems to be a promising tool for integrating the effects at larger spatial scales although there is need for defining exact protocols, while taking account of fisheries and habitat heterogeneity. In the Mediterranean and the tropics, the effects on seagrasses are probably the most important since they are related to key ecological species with prime importance for biodiversity. However, there is a need to study further these impacts and to gather long term monitoring data in order to have a conclusive picture of the processes and the risks involved. In any case, it should be emphasized that each one of these groups of variables indicates processes operating on different spatiotemporal scales and therefore monitoring focusing only on one group can hardly be a proxy for the entire health of the ecosystem.

2.2 Monitoring Environmental Impact from Norwegian Aquaculture

2.2.1 Introduction

During the last 30 years, Norway has developed an aquaculture industry based on production of marine fish, mainly Atlantic salmon (*Salmo salar* L). In 2005 580,000 metric tonnes of salmon and approximately 60,000 metric tonnes of other fish species and shellfish were produced. Norwegian aquaculture facilities are located along 2,000 km of coastline with numerous fjords and archipelagos and a temperature regime that is favourable for cultivation of cold-water species. More than 1,800 sites are located in the fjords and archipelagos where they are protected from the open sea but where water movement is sufficient to maintain production. Initially, fish farm facilities were placed in shallow areas but today many sites are located at a depth exceeding 100 m. Due to the natural conditions and a well-developed infrastructure, the coast is well suited for aquaculture. During the growth of the aquaculture industry and in concert with the increase in production, a number of environmental effects and problems have been encountered. Some of these have been minimized or resolved whereas others have increased in importance and new ones have emerged.

2.2.2 Environmental Objectives for Norwegian Aquaculture

In 1993 the Norwegian authorities decided on environmental objectives for Norwegian aquaculture, providing a national consensus (Anon. 1993). Defining the objectives was a joint project between the authorities concerned with aquaculture in Norway: the Directorate for Nature Management, the Directorate of Fisheries, The Norwegian Pollution Control Authority, The Norwegian Board of Health, The Norwegian Medicines Control Authority, and the Ministry of Agriculture Department of Veterinary Services. The report outlined the political objectives that the government and parliament had decided upon and which served as overriding objectives. The Stortings propositions no. 32 and no. 36 stated: "The development of the Norwegian aquaculture industry must be sustainable and based on respect for nature's thresholds of toleration." The report also presented the international conventions and treaties that Norway had agreed upon and which must be followed.

The environmental objectives for Norwegian aquaculture were divided into five major areas: escapees, diseases, medicines, chemicals and organic waste and nutrients. A description of each was provided and both short-term result goals and long-term environmental objectives for each type of impact were set. The report was followed by annual reports on the results achieved (e.g., Directorate for Nature Management 2000), and in 1997, the environmental objectives were reviewed (Directorate for Nature Management 1997).

The environmental objectives and the annual reports presented an important overview of the situation with regard to the environmental problem areas and provided a practical tool for following up on goals. These also made it possible to include changes in problem areas as well as to redirect focus to emerging issues, and have been used as guidelines for what should be monitored. However, they did not describe how to monitor the various effects, how often monitoring should take place, and which environmental quality standards (EQS) to use.

2.2.3 Environmental Impacts and Monitoring

A number of regulations, acts and laws administered by various ministries, directorates and other authorities regulate the Norwegian aquaculture industry with regard to licensing, production, food safety, disease control, the use of medicines and chemicals, and environmental impact (Maroni 2000).

The escape of salmon from farms is considered a serious problem since farmed fish may interact with wild salmon. To minimise the escape of fish from farms, a risk assessment must be carried out at each farm and all farms must comply with a standard for technical specifications (Anon. 2003). In the case of an escape event or suspicion of escape, the Directorate of Fisheries must be notified and recapture of escaped fish in a radius of 500 m from the farm initiated as stated in the Aquaculture Operation Regulations (Anon. 2004).

Diseases and ectoparasites have been a problem in the fish-farming industry since the beginning. Many of the major infectious diseases have however been combated by vaccination and improved hygiene, which has dramatically reduced the usage of antibacterial agents. However, sea lice infestations have proved difficult to overcome and the transfer of sea lice is still considered one of the major problems in Norwegian mariculture. At all fish farms, sea lice must be counted at least every second week when the water temperature exceeds 4°C and the results are reported to the Norwegian Food Safety Authority. If the number of lice per fish exceeds the threshold limits, the fish farmer is obliged to delouse at the farm (Anon. 2000).

All use of medicines is prescribed by a veterinarian and is registered by the Norwegian Medicines Control Authority and the Fisheries Directorate. Antibacterial agents, which were widely administered in the late 1980s, are presently only used in low amounts mainly on broodstock and early life stages (1,215 kg used in 2005, The Norwegian Institute of Public Health). Traditionally, sea lice medicines have been administered as bath treatments, first organophosphates and hydrogen peroxide and later pyrethroids, but in-feed medicines are becoming more widely used. Chitin synthesis inhibitors such as teflubenzuron and diflubenzuron were initially employed on a trial basis, but have not been used since 2002 due to their potential impact on non-target organisms. Instead, the use of avermectins has increased and 39 kg were sold in 2005 (The Norwegian Institute of Public Health). At the present there is no mandatory monitoring requirements for medicines and their residues in the marine environment. However, the environmental authorities may require monitoring with reference to The Pollution Control Act (Anon. 1981).

The most frequently used chemicals in Norwegian fish-farming are antifouling compounds for the net pens. The most common is copper, although this compound is meant to be phased out and of application should be significantly reduced before 2010 in accordance with the Declaration of The Hague of March 1990 (Anon. 1990). However, it has proven difficult to find a substitute, and there are still large amounts of copper in use. As is the case for medicines, there is no mandatory monitoring requirement for copper in the sediment but the environmental authorities may require monitoring with reference to The Pollution Control Act (Anon. 1981).

According to the Environmental Objectives of Norwegian aquaculture, organic wastes from fish farms must not result in unacceptable effects on the environment locally or regionally and permitted threshold levels of impact must be determined (Directorate for Nature Management 1997). Due to large variations in hydrographical conditions and depth at fish farm sites, the amount of organic waste that settles on the sediment will vary considerably. Furthermore, the size and the management of the fish farm will also influence the sedimentation. The impact, such as changes in sediment chemistry and in the benthic fauna community, will therefore also have a large variability between sites.

Overloading of sites and accumulation of organic material in the form of waste feed pellets and faeces can, besides the effects on the environment, be a cause of stress, poor growth and disease in the farmed fish, with the associated spread of infectious agents and need for medication. Organic material can therefore be influential for several types of environmental impact, even if the effect is greatest on the sediment under the cages.

2.2.3.1 Monitoring Benthic Impact

Parallel to the work of determining environmental objectives, a management system was developed which mainly focused on monitoring and modelling the impact of organic waste from fish farms. The system (MOM: Modelling – Ongrowing fish farms – Monitoring) combines modelling of potential impact with monitoring benthic impact and provides environmental quality standards (EQS, Ervik et al. 1997). The amount of monitoring carried out depends on the extent of the environmental impact and the EQS sets a limit for maximum allowable impact and makes it possible to distinguish between different impact levels. The monitoring programme of the MOM system (Hansen et al. 2001) has been used to make a Norwegian standard: "Environmental monitoring of marine fish farms NS-9410" (Norwegian Standards Association 2000). Mandatory environmental monitoring is performed according to NS-9410 as established in the Aquaculture Operation Regulations (Anon. 2004) and the responsible authorities are the Fisheries Directorate and the County Governor's Department of Environment. The standard describes methods for measuring bottom impacts from marine fish farms and gives detailed procedures on how environmental impacts from individual fish farm sites shall be monitored and includes EQS. All Norwegian standards are reviewed every 5 years and the Norwegian standard NS-9410 is currently under review with a new version scheduled in 2007.

NS-9410 focuses on methods for determination of sediment conditions at and in the vicinity of fish farms. Traditionally, monitoring of benthic impact at fish farm sites has been faunal community analysis. This type of monitoring is maintained in NS-9410, but mainly in the receiving water body, and at the site less time demanding and expensive surveys are used. The scientific benefit of the more advanced faunal community method was balanced against the advantage of a higher number of samples and more frequent surveys. Furthermore, due to smaller sampling gear, sediment samples can be retrieved from between net cages in compact net cage groups. Threshold values for environmental impact are set such that fish farm sites may be in use over a long period of time and aim to ensure favourable living conditions for the farmed fish as well as to prevent unacceptable impact on the surrounding area.

Presently NS-9410 describes monitoring of organic waste but sampling for medicines and chemicals in the sediment can conveniently be added.

2.2.3.2 NS-9410

The monitoring programme in NS-9410 includes three types of surveys (A, B and C investigation). The A- and B-investigations survey the potential and actual impacts on the sediment under and in the immediate vicinity of the fish farm. The C-investigation aims to obtain a picture of the impact on the receiving water body as a whole.

Table 2.5 The relationship between degree of exploitation and level of monitoring. The more severe the impact at the site, the higher the frequency of performing the A- and the B-investigations. Site condition 4 corresponds to overexploitation

	Level of monitoring (frequency of performing investigations)	
Degree of exploitation/site condition	A-investigation	B-investigation
1	every 3 months	every 2 years
2	every 2 months	Annually
3	monthly	every 6 months
4 (unacceptable)		eventual extended B-investigation

Two terms are employed to adjust monitoring depending on the impact at the site: the degree of exploitation and the level of monitoring. The degree of exploitation is an expression of the amount of impact from the fish farm compared with the holding capacity of the site. The site is overexploited if the holding capacity is exceeded and the division between acceptable and unacceptable sedimentary conditions is set as the highest level of accumulation within which burrowing bottom fauna can survive in the sediment. The higher the degree of exploitation at a site, the higher the level of monitoring that is required (Table 2.5).

The A-investigation consists of a simple measurement of sedimentation rate on the sea floor under a fish farm, and can give information on high point-source loading. The survey is easily done and is carried out by the fish-farmer himself. The survey gives information on potential bottom loading and is particularly useful in combination with the B-investigation. EQS are not used in the A-investigation.

The B-investigation comprises a simple trend monitoring of the bottom conditions under a fish farm. Because the survey is repeated regularly, at intervals determined by the extent of the environmental impact, the development of the environmental impact can be followed closely. At least ten grab samples are collected at the site and both the average condition at the site and the conditions under different parts of the fish farm are revealed. The B-investigation comprises three groups of sediment parameters: (1) presence or absence of animals larger than 1 mm in the sediment, (2) pH and redox potential, and (3) qualitative determination of outgassing, smell, consistency, colour of the sediment, grab volume and thickness of the layer of deposits. All parameters are assigned points according to the extent to which the sediment is affected by organic material. The points are added and the higher the sum the more affected the sediment. Since many parameters are used in concert, the survey is less sensitive to anomalies in individual parameters. EQS have been established which divide the sediment condition into four categories equivalent to the four degrees of exploitation (Table 2.5).

The C-investigation is a survey of the bottom conditions at the fish farm and outwards into the receiving water body. The main element is a survey of the bottom faunal communities, carried out according to another Norwegian Standard: "Water quality – Guidelines for quantitative investigations of sublittoral soft-bottom benthic fauna in the marine environment NS-9423", which describes guidelines for sampling and sample processing of macrofauna in soft sediments (Norwegian Standards Association 1998). In addition, information is obtained on other parameters that may be used to determine if organic material is of fish farm origin. The pollution control authorities have defined threshold values for environmental quality of fjords and coastal waters (Molvær et al. 1997), and these are applied to the C-investigation in the receiving water body. However, specific threshold values are provided in NS-9410 when the investigation is made close to the farm.

Both the B- and the C-investigations are carried out by private firms and research institutions.

2.2.4 Models and Coastal Zone Planning

The use of models is not compulsory in environmental regulation of Norwegian aquaculture, but models have been developed which may be helpful. In conjunction with the development of the MOM monitoring programme, a model was made to estimate the maximum production of fish that could be allowed at a site without exceeding the holding capacity at the site (Stigebrandt et al. 2004). The model comprises four sub-models (a fish model, a water quality model, a dispersion model and a benthic model) and is linked to a previously developed model on environmental quality in fjords (Aure and Stigebrandt 1990). The sub-models can be altered individually as new knowledge is acquired or as new management procedures or fish species are introduced. The scope of the model system may also be expanded to include other environmental effects of fish farming related to the use of chemicals and medicines. The model was developed so it can be utilised by both environmental administrators and fish farmers.

Additionally, a growth and advection model for pelagic sea lice copepods has been developed (Asplin et al. 2004). The dispersion of sea lice in coastal waters and fjords depends on the production of sea lice larvae, and thus is influenced by farmed fish at various locations, and by the hydrography of the waters and currents, which are in turn greatly influenced by the wind. The model is currently being tested and so far the results of the model have compared well with observations in a major fjord (Sognefjorden).

In the future, environmental impact is expected to gain increasing focus as the competition for space and resources in the coastal zone grows. Sustainability and integration with other coastal activities are therefore fundamental for the viability of the aquaculture industry. In Norway, a system is under development that covers both the planning and the operational phases of aquaculture, and which can ensure an efficient use of areas available for aquaculture and can adjust the environmental impact of the industry to the holding capacity of the area. Information on topography and hydrography, as well as an overview of allocation of different uses and environmental status, will be combined with simulation models to locate aquaculture activities and to adapt the environmental impact to local and regional conditions. Monitoring will be an important element, which will ensure that the holding capacity is not exceeded.

2.3 Monitoring the Environmental Impacts of Aquaculture in Malta

2.3.1 Introduction: Development of Aquaculture Activities in Malta

Aquaculture on an industrial scale started in Malta around 1991, following initial land- and sea-based experimental and pilot projects undertaken in the mid-1970s and early 1980s. During the period 1991 to 2000, the activity mainly involved culture of sea bass (*Dicentrarchus labrax*) and sea bream (*Sparus aurata*) in offshore cages located a few hundred metres or less from the shore. Production of these two species increased steadily from around 100 tons in 1991 to 2000 tons in 1998 (Axiak et al. 1999). The offshore sea bass and sea bream farms, owned by some six different operators (Schembri et al. 2002), were sited in eight localities, all of which were relatively sheltered and supported extensive seagrass (*Posidonia oceanica*) meadows. Water depth at the different fish farm sites ranged between 10 m and 22 m. On the other hand, land-based coastal aquaculture activities contributed only 2% (equivalent to an annual production of 50 tons of sea bream) to the total local aquaculture production. However, the land based operations also included two hatcheries for sea bream, one of which was located within the National Aquaculture Centre and which at peak production was contributing up to 2.5 million sea bream fry per year, most of which were exported to Europe (Axiak et al. 1999). In the late 1990s, strong competition from fish farms based on mainland Europe, and the high operational costs incurred by local farms, particularly freight and levy charges imposed on the exported product, led to a general decline in culture of sea bass and sea bream, and the attention of some local fish farmers turned to the relatively new and lucrative activity of tuna-farming (Schembri et al. 2002).

Tuna farming, also commonly referred to as "tuna penning", is a relatively recent but highly successful enterprise that was introduced to Europe in 1979 and adopted on a large commercial scale in the 1990s. Tuna farming is classified as capture-based aquaculture and differs from traditional aquaculture in that the farmed stock is derived from catches taken from wild populations, while the captive tuna are fed fresh fish (e.g., herring and mackerel) (Ottolenghi et al. 2004). In Europe, tuna penning has been (to date) restricted to the Mediterranean, where the main species farmed is *Thunnus thynnus*, the Atlantic Bluefin Tuna. The intensity of tuna farming has increased steadily in the Mediterranean over the last decade or so, reaching a current total annual production of around 16,000 tons. However, information on the influence of tuna farming on the marine environment, both outside and within European coastal waters, is somewhat lacking and there is a dearth of published data on the environmental impacts of the activity, while data on potential adverse effects resulting from indirect activities, for example, the impacts of the baitfish fishery that supplies the fresh feed for tuna, is unavailable. Moreover, in view of the large and increased catch effort of tuna fishers to meet the farms' demand, information on the potential adverse impact of tuna penning on wild stocks of *Thunnus thynnus* is unavailable and this has placed the activity at the centre of much controversy and debate, and harsh criticism has been levelled at it by national and international environmental NGOs (e.g., the World Wide Fund for Nature; see WWF 2004).

The advent of tuna farming in Malta in 2000 raised concerns at the Malta Environment and Planning Authority (MEPA – the local agency concerned with environmental protection in Malta) and the public, mostly because of the potential adverse impacts resulting from the large scale of operations of the projected tuna farms. As a result, MEPA stipulated that tuna cages must be sited at least 1 km offshore, in waters having sufficiently strong water currents, and distant from benthic habitats that have a high ecological value (e.g., seagrass meadows and maerl beds). Furthermore, prior to granting tuna farm operators a development permit, MEPA requested that appropriate surveys be carried out in offshore areas having the required characteristics in order to determine the specific location of the cages. The result was that all four tuna farms that started operations in the early 2000s were sited in waters having a depth of around 50 m, and had their cages located over a "bare sand" habitat. When initiated locally in 2000, tuna farming had an annual production of 300 tons that increased steadily to around 3000 tons in 2005, making the country one of the largest current producers of farmed tuna in Europe. The three farms that are currently operating are located off the northeastern coast of the island of Malta (Fig. 2.2) at a distance of around 1 km from the shore, on a seabed consisting mainly of bare soft sediment (muddy sand), and in waters characterised by strong currents and having a depth of between 46 m and 55 m. Tuna penning activities usually start around July and extend to December/February, after which there is a 4–6 month fallowing period.

Recently, the Fisheries Conservation and Control Division of the Ministry for Rural Affairs and the Environment (responsible also for local aquaculture) lodged an application with MEPA to designate an "Aquaculture Zone" located about 6 km off the eastern coast of the island of Malta (Fig. 2.2). The zone covers an area of some 9 km^2 and is located in waters having depths of between 65 m and 105 m. The aim is to locate future tuna penning installations in one offshore area that is distant from land in order to minimise impacts on the coast and the shallow waters off it. Following approval by MEPA, tuna penning operations within this zone started in July 2006 in a sea area of 3 km × 1.5 km.

2.3.2 Environmental Monitoring of Sea Bass and Sea Bream Farms

At around the same time that local aquaculture activities reached industrial production levels, the central government set up a Planning Authority as the main regulatory body for development, and this institution later merged with the then Environment Protection Department and changed its name to the Malta Environment and Planning Authority (MEPA) to become the local planning, development control and environmental

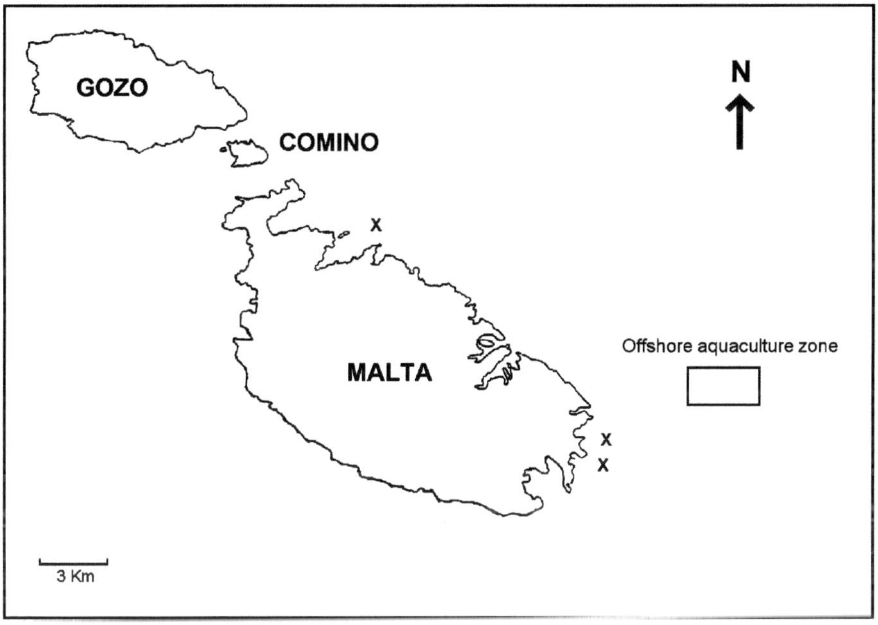

Fig. 2.2 Map of the Maltese islands showing the location of the three currently operating tuna farms (x) and the recently designated offshore aquaculture zone

protection agency. Since its establishment in 1992, MEPA, as its predecessors, was granted the overall responsibility of processing aquaculture development proposals and to oversee any required environmental monitoring of the activity. Consequently, in 1994, the then Planning Authority issued a set of Policy and Design Guidelines for Fish-farming (PDGF; see Planning Authority 1994). The "monitoring" chapter of the guidelines required an environmental monitoring programme for each fish farm to enable assessment of the impact of the activity on the environment. According to the PDGF, the environment monitoring programme should:

- measure changes, if any, in specific environmental attributes such as currents, temperature, dissolved oxygen, and levels of nutrients and bacteria;
- monitor the state of benthic assemblages and habitats, and accumulation of waste products in the vicinity of the farms;
- record material introduced in the environment by the fish farms (such as chemicals and physical items forming part of the cage structures);
- record other inputs and impacts on the environment in the general area of the fish farms, but which are not directly related to the fish-farming activities (e.g., discharges from outfalls and other major sources of pollution, and fishing activities).

The guidelines also stated the specific physico-chemical and biological attributes to be monitored, the frequency of data collection, and the number of sampling points, which were to vary depending on the size and location of the respective fish farm (Table 2.6).

Table 2.6 Details of specific attributes, sampling stations and frequency of the required environmental monitoring programme for aquaculture activities in Malta, as required by the Malta Environment and Planning Authority (source: Planning Authority, 1994)

Environmental attributes to be monitored	No of sampling stations	Monitoring frequency
Currents (speed and direction)	1–2 stations at various depths	Every 2 months
Water column: temperature; salinity; dissolved oxygen; turbidity; chlorophyll a; nitrates; phosphates; ammonia and total bacteria	Several stations at various depths	Every 2 months
Sediments: granulometric properties; and organic carbon and organic nitrogen content	Several stations	Every 6 months
Benthos: benthic habitats and communities	Mapping of all benthic communities within the area occupied by the cages and their moorings, together with collection of samples at stations as necessary to establish the species composition of benthic communities	Every 6 months

To enable assessment of changes in the monitored environmental attributes after initiation of the aquaculture activities, the guidelines required collection of baseline data before the start of the operations, so that these may be used as a reference against which to compare data collected during monitoring. Collection of baseline data would form part of the Environment Impact Assessment (EIA) as a requirement for the granting of a development permit. Furthermore, the guidelines specified that an environmental monitoring report, including all raw data collected, should be presented to MEPA and the Directorate of Veterinary Services. Submission of reports to the latter agency would ensure that appropriate practices in relation to health management of fish stock and product quality are in operation.

Environmental monitoring at sites supporting sea bass and sea bream farms was initiated around 1994, however, this was sporadic and certainly did not satisfy MEPA's guidelines to the full (Schembri et al. 2002). In some cases, the required baseline survey for a specific fish farm was made (as this could not be avoided since it formed part of the development application process), but no monitoring was undertaken following initiation of the fish-farming activities, while other farms claimed to have carried out monitoring of water quality at their own laboratories. Apparently, no data on currents has been collected at any of the fish farm sites. As a result of the irregular and incomplete environmental monitoring for sea bass and sea bream farms, the data available for these operations is scanty.

Since all sea bass and sea bream farms were located in the vicinity of *Posidonia oceanica* meadows, monitoring of benthic habitats and assemblages centred on assessing the spatial distribution, coverage and state of health of the seagrass within the area occupied by the cages and their moorings. This essentially consisted of mapping surveys of the seabed to assess potential changes in the spatial distribution and coverage of seagrass habitat resulting from the fish-farming activities.

In an attempt to fill gaps in knowledge of the environmental impacts of sea bass and sea bream farms, a number of studies were undertaken by the University of Malta, most of which formed part of undergraduate and postgraduate research projects (e.g., Cassar 1994; Dimech et al. 2002). Some of these studies included collecting data on environmental attributes that went beyond the minimum requirements set by MEPA's PDGF, for example, measurement of *P. oceanica* meadow and shoot attributes (shoot density, mean number of leaves and leaf length per shoot, shoot biomass and shoot epiphyte loading; see Cassar 1994; Dimech et al. 2002).

Overall, the results of environmental monitoring at sites used to farm sea bass and sea bream indicated that seagrass meadows located directly below the fish cages underwent severe regression or were completed decimated, and that the effects of the aquaculture activities on the monitored seagrass attributes (shoot density, mean leaf length, number of leaves per shoot, and epiphyte load) extended a considerable distance (in the case of one farm, around 200 m; Dimech et al. 2002) from the farm site. The results of a study aimed at assessing the impact of sea-based fish cages (located in waters having a depth of 10 m) on the decapod, mollusc and echinoderm fauna associated with *P. oceanica* beds indicated the presence of three distinct zones in the vicinity of the farm (Dimech et al. 2002):

(i) Zone 1, comprising the area occupied by the cages and an additional band of 30 m around the farm. The macrofaunal assemblages present within this zone were characterised by a low species richness and the dominant trophic groups were grazers and deposit feeders (decapods, polyplacophorans and gastropods), which exploit the abundant epiphytes and deposited organic matter present close to the cages.
(ii) Zone 2, comprising the area located at a distance of between 30 m and 90 m from the farm. This zone supported macrofaunal assemblages that had the highest species richness and abundance, while the fauna was dominated by the same trophic groups in Zone 1.
(iii) Zone 3, comprising the area located at distances exceeding 90 m from the farm. This zone supported macrofaunal assemblages having species richness and abundance values that were intermediate between those recorded from Zones 1 and 2, and in which the dominant trophic groups comprised grazers, deposit feeders, suspension-feeders (mostly bivalves) and predators.

This "zoning" pattern, consisting of differences in the species composition and structure of the benthic macrofaunal assemblages with increasing distance from the farm site, is very similar to that recorded in the vicinity of offshore salmon farms (e.g., Brown et al. 1987; Ye et al. 1991).

2.3.3 Environmental Monitoring of Tuna Penning Activities

As part of the permit conditions issued by MEPA for tuna penning activities, the farm operators were required to commission a comprehensive environmental monitoring programme to be carried out by independent consultants approved by MEPA. Since the environmental characteristics at the tuna penning sites were very different from those of the near-shore sites where sea bass and sea bream farms were located, while the type and scale of operations were also very different, it was immediately realised that MEPA's environmental monitoring guidelines contained in the 1994 PDGF could only be applied to tuna farms following modification. Given the circumstances, in 2001 MEPA amended the 1994 PDGF such that the revised guidelines stated that no aquaculture development would be considered in areas less than 1 nautical mile from the shore, or in sites having a water depth less than 50 m (give or take 5 m) (Planning Authority 2001).

In granting development permits for tuna-penning activities, MEPA requested that monitoring of tuna penning activities should include monitoring of: (1) sediment attributes; (2) benthic diversity; (3) the gross physical and biological characteristics of the seabed below the tuna cages through underwater videography; (4) the state of seagrass beds and of biological characteristics at important dive sites located in the vicinity of the farms using underwater mapping and videography (in some cases, even if these were present at distances exceeding several hundred metres from the tuna cages); and (5) water quality. The specific requirements for environmental monitoring of aquaculture operations in the PDGF of 2001 are given in Table 2.7. The environmental monitoring programmes for all tuna farms were initiated in 2000 and are still ongoing.

Samples to monitor sediments and benthic diversity have been collected annually (since 2000) at each tuna farm from a number of stations located: (1) adjacent to the tuna-pens, (2) at a distance of some 100 m from the tuna pens, and (3) at a number of reference sites; the sampling programme being mainly based on a Before-After-Control-Impacted (BACI) design (Borg and Schembri 2005). Using this design, an adverse impact is deemed to have occurred if a significant change (at the 0.05 level of significance) for one or more of the monitored attributes is recorded between the baseline condition and that following the tuna penning activities. In the case of benthic diversity, this would be a significant decrease in the total number of species and/or abundance of the selected indicator species.

Monitoring of the gross physical and biological characteristics of the seabed below the tuna cages is being undertaken through surveys carried out by SCUBA divers using direct observation and underwater videography. During initial surveys of the seabed below the tuna cages, it was immediately realised that the main impacts on the seabed resulted from the presence of large amounts of uneaten feed-fish that accumulated on the seabed below the cages. However, the amount of feed-fish below the cages varied greatly, even between cages within the same farm. It was therefore considered appropriate to develop an index to enable an objective semi-quantitative assessment of the amount of uneaten feed-fish present (Table 2.8).

Table 2.7 Details of specific attributes to be monitored, sampling stations, and frequency of the required environmental monitoring programme for aquaculture activities in Malta (source: Malta Environment & Planning Authority 2001)

Environmental attributes to be monitored	Number of sampling stations	Monitoring frequency
Water column: temperature; salinity; dissolved oxygen; turbidity; chlorophyll a; nitrates; phosphates; ammonia; faecal coliforms and total bacteria	A sampling site underneath each cage; Sampling at points along a perimeter around the cage site 25 m away from the cages; At least two sampling points 100 m away from the cage site (according to the direction of the prevailing currents) Sampling sites in areas that are of ecological, commercial, tourism, or recreational interest (this is to be decided on a site-by-site basis)	Monthly, for as long as the fish are kept in the cages
Sediments: granulometric properties; and organic carbon and organic nitrogen content	Several stations within the cage site	Annually, in the same month each year
Benthos: species diversity; photographic/video evidence regarding the state of the seabed; mapping of benthic communities; core samples for faunal, granulometric and sediment analysis as described for sediments above; seagrass morphological parameters (e.g., shoot and leaf density, shoot length, etc.) where applicable	Mapping of all benthic communities within the area occupied by the cages and their moorings, together with collection of samples at stations as necessary to establish the species composition of benthic communities	Annually

Table 2.8 The "uneaten food index" devised by Borg & Schembri (2001) for the purpose of quantifying and comparing the amount of dead uneaten feed-fish under the different tuna-pens

Index value	Description of amount of uneaten feed-fish present on the seabed
0	No uneaten feed-fish present
1	<1 uneaten feed-fish present per m^2 of seabed
2	>1 uneaten feed-fish present per m^2 of seabed, but the fish do not form a continuous layer covering the seabed
3	>1 uneaten feed-fish present per m^2 of seabed. Fish form a single, uninterrupted layer within at least a 1 m^2 area on the seabed.
4	>1 uneaten feed-fish present per m^2 of seabed. Fish form two or more uninterrupted layers on top of each other within at least a 1 m^2 area on the seabed.

The general state of seagrass beds and habitats, including those at popular dive sites, is being assessed through mapping surveys and underwater videography, carried out by SCUBA divers. During the surveys, the divers record the state of health and spatial extent of the main marine benthic habitats in the respective study area. Potential changes in the state of the benthic habitats and their spatial distribution are assessed by comparing maps of the situation recorded before initiation of the tuna-penning activities with that after each monitoring session.

Monitoring of water quality consists of surveys of the same physico-chemical and bacteriological attributes that have been monitored in the vicinity of sea bass and sea bream farms. Samples of water for these surveys are being collected at depths of 1 m and 5 m below the surface at several stations located in the immediate vicinity of the tuna farms and at reference stations located at a distance from the tuna cages.

Monitoring at the new offshore aquaculture zone (Fig. 2.2) commenced in June 2006 prior to the start of tuna penning activities there. The baseline survey for the sediments and benthic diversity monitoring components was based on the same design used at the other three tuna farms located closer to the coast, with samples being collected remotely using a standard 0.1 m^2 Van Venn grab. However, because of the deep waters that characterise the area, monitoring of the seabed using the same underwater videography and SCUBA diving techniques that have been used to date at the other tuna penning sites located in shallower waters, is not possible, and it is planned to use remotely operated video cameras instead.

Overall, the results of the various monitoring components undertaken since 2000 for the three tuna farms located 1 km offshore indicated that, where detected, the main adverse impacts resulted from accumulation of large amounts of feed-fish on the bottom under and in the vicinity of the cages. The results from the video surveys carried out near the tuna pens indicated that, towards the end of each penning season (in autumn), considerable amounts of dead uneaten feed-fish were present on the seabed directly below the tuna pens, and this resulted in alterations in the physical and biological characteristics of the seabed under the cages. The recorded changes in biological characteristics included the disappearance of certain megafaunal species (e.g., the irregular sea urchin *Spatangus purpureus* and the crinoid *Antedon mediterranea*) that prior to the start of the penning operations were characteristic of the soft sediment habitat where the tuna pens are located, and the appearance of high population densities of detritus-feeding and scavenging macroinvertebrates (e.g., the ophiuroid *Ophiura texturata* and the crab *Inachus* sp., and the fish *Gobius* sp.). Gross changes in physical characteristics of the seabed included the presence of large quantities of fish bones and a few anthropogenic items originating from the tuna farms. The video surveys also showed that the amount of feed-fish present varied considerably between different tuna farms and between cages within the same farm, with some cages only having a few fish beneath them and others having multiple layers. Overall, a consistent pattern was evident where a decrease in the amount of uneaten fish occurred only when tuna were no longer present in the pens during the fallowing period. The remaining uneaten fish decompose slowly and, where the uneaten fish are present in large numbers, form a continuous layer of decomposing organic material that continues to decay gradually. Sometimes, following storms and possibly due to strong bottom currents, this layer is admixed with the underlying mobile sediment. In places where the decomposition process is complete, the only remains are fish bones that eventually disperse in the sediment leaving little or no trace of the original uneaten fish on the surface. Once the source of the impact (periodic addition of new uneaten food) is removed, the slow recovery to the original state is signalled by the reappearance of some of the megafaunal species that formed part of the original benthic assemblage characterising the bare muddy sand bottom over which the tuna pens are located (Borg and Schembri, unpublished data).

The results of benthic diversity monitoring indicated that, at times, a significant decrease in species richness, and in the abundance of the indicator macrobenthic species, occurred in the vicinity of particular tuna farms, but this effect was mainly restricted to the area directly below the cages. Similarly, significantly higher levels of organic carbon and/or organic nitrogen and/or significant changes in mean sediment grain size were recorded in some of the monitoring sessions, but the observed changes were again mainly restricted to the seabed area directly below the cages.

The mapping and videographic surveys of important habitats and dive sites located in the vicinity of the tuna farms did not detect any changes in the physical and biological characteristics of the monitored sites. Likewise, the water quality studies did not show any consistent trend in the levels of the monitored variables that could be attributed to the tuna penning activities (Schembri et al. 2002). Lower levels of oxygen, reduced water transparency, and elevated nutrient levels were at times recorded at the tuna penning sites relative to the reference sites during the farming season (July – December), however, the observed changes in the monitored variables were sporadic and not statistically significant. Data collected in June 2006 from the new offshore aquaculture zone are still being analysed and consequently, results from the monitoring programme for tuna farms located within this area are not yet available.

2.3.4 Conclusions and Recommendations in Malta

Guidelines for environmental monitoring of aquaculture activities in Malta were issued by the responsible local agencies relatively early during the period of initiation and expansion of local fish-farming involving culture of sea bream and sea bass. However, most fish farms failed to adhere to the environmental monitoring requirements, at least on a regular basis, while it appears that enforcement was not effective (Schembri et al. 2002). As a result, few monitoring data on the impact of sea bream and sea bass aquaculture activities on the marine environment exist. Where data are available, the results of benthic environmental monitoring indicated an overall adverse impact on seagrass beds in the vicinity of sea bream and sea bass cages. However, site characteristics such as the current regime, water depth and exposure, together with the size of the fish-farming operation and the farm management programme at a specific locality, appear to be crucial in determining the magnitude of the adverse impact. The results of the water quality monitoring programmes for sea bream and sea bass farms did not indicate any large adverse changes in water quality attributes resulting from the fish-farming activities.

There is currently only a low level of production of sea bream and sea bass in Malta, as the attention of aquaculture operators is presently on tuna penning. However, some operators still retain a permit to culture these species in addition to penning tuna (Schembri et al. 2002), while it is likely that production of sea bream and sea bass, as well as of additional species that are being introduced into aquaculture in the Mediterranean, will increase in the future, depending on the vagaries of the market for tuna and other species, and as new operators enter the field and wild tuna stocks dwindle. The environmental impact of any new (non-tuna) farms is not likely to be as severe as that of the early sea bream and sea bass farms since it is unlikely that such farms will be allowed to locate inshore or close to sensitive habitats, particularly since the technology for siting farms in deep water now exists, and because the environmental impact monitoring requirements of aquaculture projects are nowadays much more rigidly enforced.

Overall, the results of environmental monitoring of tuna penning operations during the last 6 years (2000–2006) revealed a consistent pattern of a localised adverse impact that mainly resulted from the uneaten feed-fish which accumulate on the seabed during the tuna farming season (July to December). The amount of feed-fish present decreases only when all the tuna have been harvested, following which, any feed-fish remaining on the seabed continue to decompose slowly. These results are characteristic of a "pulse disturbance" where the physical and biological characteristics of the seabed are temporarily altered during the tuna penning season but return back to more or less the pre-disturbance condition before the start of the next tuna penning season. Nonetheless, repeated accumulation of feed-fish on the seabed in the vicinity of the tuna pens may prevent complete recovery of the benthic assemblages following each tuna penning season, potentially leading to a "press disturbance" where environmental conditions become permanently altered.

The observed differences in the amount of feed-fish present on the seabed below the cages indicate potential differences between different tuna farms and/or cages within the same farm in: (1) feed management, or (2) the rate of food intake by the tuna, or a combination of (1) and (2). It appears that the key to preventing this from happening is to implement a rigorous feed-management strategy that includes:

- careful monitoring of the feeding behaviour of the tuna and stopping the supply of food as soon as the tuna are satiated in order to avoid as much as possible uneaten food ending up on the bottom; and
- removal of dead uneaten feed-fish from the bottom should inordinate amounts accumulate below the cages either due to overfeeding or to accident.

On the other hand, the results of recent (2004–2005) monitoring surveys indicated an overall large improvement in feed-management at local tuna farms. For example, values of the index for uneaten feed-fish (Table 2.8) recorded in 2005 averaged 1, compared to values of between 3 and 4 recorded in 2001–2003. Furthermore, it should be emphasised that all tuna farmers have, in general, adhered to the environmental monitoring requirements, while one particular operator has even taken the initiative of including details of the monitoring and results obtained on their web site (e.g., http://www.ajdtuna.com/). The possibility of developing an alternative feed source for tuna should be explored, as this could potentially reduce adverse impacts on the seabed, while alleviating fishing pressure on wild stocks of feed fish.

While the accumulation of decomposing organic matter on the seabed is the key source of marine benthic impact of the Maltese tuna farming operations, it is not the only potential adverse factor. Mass deaths of tuna have occurred at least on two separate occasions, however, it seems that the farm operators have taken remedial action and recovered the carcasses from the seabed at the earliest opportunity; thus during the 6 years of monitoring, tuna carcasses were only encountered near the cages on two or three occasions, and then as single dead fish. The accidental introduction of anthropogenic items, most of which are related to the tuna-penning activities, is also of concern. This can be mitigated relatively easily by enforcing a strict policy of not throwing anything into the sea and by implementing periodic "clean-ups" of the seabed. Additional impacts result during feeding and harvesting of the tuna, when entrails and oily slicks transported by surface currents have been reported. These observations highlight the importance of guidelines for operational procedures and mitigation measures to reduce adverse environmental effects on the marine environment.

Acknowledgement We are grateful to Barry Hargrave and Patrick White for providing data on the regulations and monitoring of aquaculture operations, and to Marie B. Lauridsen for Fig. 2.1. We also thank Mr Michael Sant (Malta Environment and Planning Authority, Malta) and Prof Carmelo Agius (University of Malta, Malta) for supplying us with useful information.

References

Axiak V, Gauci V, Mallia A, Mallia E, Schembri PJ, Vella AJ (1999) State of the environment (summary) report for Malta 1998. Floriana, Malta: Environment Protection Department, Ministry for the Environment; 130 pp

Anonymous (1981) Act of 13 March 1981 No.6 concerning protection against pollution and concerning waste (the Pollution Control Act), most recently amended by Act of 20 June 2003 No.45. LOV-1981-03-13-6, Lovdata, Oslo, Norway

Anonymous (1990) Ministerial Declaration of the Third International Conference on the Protection of the North Sea, The Hague, 8 March 1990

Anonymous (2000) Forskrift om bekjempelse av lakselus (Act for combat of sea lice). FOR 2000-02-01 nr 70, Lovdata, Oslo, Norway (In Norwegian)

Anonymous (2003) Forskrift om krav til teknisk standard for installasjoner som nyttes til akvakultur (Regulations relating to technical standard for installations used in aquaculture). FOR-2003-12-11-1490, Oslo, Norway (In Norwegian)

Anonymous (2004) Regulations relating to operation of aquaculture establishments (Aquaculture Operation Regulations). FOR 2004-12-22 no. 1785, Lovdata, Oslo, Norway

Asplin L, Boxaspen K, Sandvik AD (2004) Modelled distribution of salmon lice in a Norwegian fjord. International Council for Exploration of the Sea ICES CM 2004/P: 11

Aure J, Stigebrandt A (1990) Quantitative estimates of eutrophication effects on fjords of fish farming. Aquaculture 90:135–156

Borg JA, Schembri PJ (2001) Report on a video survey of the sea bed in the vicinity of tuna pens, made on October 2001. Malta: Unpublished report, Independent Consultants, 7 pp.

Borg JA, Schembri PJ (2005) Impact of offshore tuna-penning on soft sediment macroinvertebrate assemblages in Malta (Central Mediterranean). Paper presented at the American Society for Limnology and Oceanography (ASLO) Summer 2005 meeting "A pilgrimage through global aquatic sciences" Santiago de Compostela, Spain 19–24 June 2005. [Abstracts book p. 20]

Brown J R, Gowen RJ, McLusky DS (1987) The effect of salmonid farming on the benthos of a Scottish sea loch. Journal of Experimental Marine Biology and Ecology 109:39–51

Brooks KM, Mahnken CVW (2003) Interaction of Atlantic salmon in the Pacific northwest environment II. Organic wastes. Fish Research 62:225–293

Cassar M (1994) An investigation of the environmental impact of a marine cage fish farm in Malta. Unpublished M.Sc. dissertation, Faculty of Science, University of Malta, Msida, Malta, 175 pp

Christensen PB, Rysgaard S, Sloth NP, Dalsgaard T, Schwaerter S (2000) Sediment mineralization, nutrient fluxes, denitrification and dissimilatory nitrate reduction to ammonium in an estuarine fjord with sea cage trout farms. Aquatic Microbial Ecology 21:73–84

Costello MJ, Grant A, Davies IM, Cecchini S, Papoutsoglou S, Quigley D, Saroglia M (2001) The control of chemicals used in aquaculture in Europe. Journal of Applied Ichthyology 17:173–180

Cromey CJ, Black KD (2005) Modelling impacts of finfish aquaculture. In Hargrave BT (ed.), The handbook of environmental chemistry, Vol. 5: Water Pollution Environmental effects of marine finfish aquaculture. Springer Berlin Heidelberg New York, p. 129–156.

Dalsgaard T, Krause-Jensen D (2006) Monitoring nutrient release from fish farms with macroalgal and phytoplankton bioassays. Aquaculture 256:302–310

Delgado O, Grau A, Pou S, Riera F, Massuti C, Zabala M, Ballesteros E (1997) Seagrass regression caused by fish cultures in Fornells Bay (Menorca, Western Mediterranean). Oceanologica Acta 20:557–563

Diaz-Almela E, Alvarez E, Santiago R, Marbà N, Holmer M, Grau T, Danovaro R, Argyrou M, Karakassis Y, Duarte CM (submitted) Benthic inputs as predictors of seagrass (*Posidonia oceanica*) fish farm-induced decline. *J. Exp. Mar. Biol. Ecol.*

Dimech M, Borg JA, Schembri PJ (2002) Changes in the structure of a *Posidonia oceanica* meadow and in the diversity of associated decapods, mollusc and echinoderm assemblages resulting from inputs of waste from a marine fish farm (Malta, Central Mediterranean). Bulletin of Marine Science 71(3):1309–1321

Directorate for Nature Management (1997) Miljømål for norsk oppdrettsnæring. Nye miljømål for perioden 1998–2000 (Environmental objectives for Norwegian aquaculture. New environmental objectives for the period 1998–2000). DN-note 1999–1, Trondheim, Norway, 34 pp. (In Norwegian, English abstract)

Directorate for Nature Management (2000) Environmental objectives for Norwegian aquaculture. Report on results achieved in 1999. DN-note 2000-3b, Trondheim, Norway, 44 pp

Ervik A, Hansen PK, Aure J, Stigebrandt A, Johannsen P, Jahnsen T (1997) Regulating the local environmental impact of intensive marine fish farming. I. The concept of the MOM system (Modelling–Ongrowing fish farms–Monitoring). Aquaculture 158:85–94

Fernandes TF, Eleftheriou A, Ackefors H, Eleftheriou M, Ervik A, Sanchez-Mata A, Scanlon T, White P, Cochrane S, Pearson TH, Read PA (2001) The scientific principles underlying the monitoring of the environmental impacts of aquaculture. Journal of Applied Ichthyology 17:181–193

Frederiksen MS, Holmer M, Diaz-Almela E, Marbà N, Duarte CM (in press). Sulfide invasion in the seagrass *Posidonia oceanica* at Mediterranean fish farms: assessment by stable sulfur isotopes. Marine Ecology Progress Series

GESAMP (IMO/FAO/Unesco-IOC/WMO/IEEA/UN/UNEP Joint Group of Experts on the Scientific Aspects of Marine Environmental Protection), 1996. Monitoring the ecological effects of coastal aquaculture wastes. GESAMP Reports and Studies 57: p 380.
Giannoulaki M, Machias A, Somarakis S, Karakassis I (2005) Wild fish spatial structure in response to presence of fish farms. Journal of the Marine Biological Association of the United Kingdom 85:1271–1277
Gowen RJ, Tett P, Jones KJ (1983) The hydrography and phytoplankton ecology of Loch Ardbhair: a small sea loch on the West Coast of Scotland. Journal of Experimental Marine Biology & Ecology 71:1–16
Gowen RJ, Bradbury NB (1987) The ecological impact of salmonid farming in coastal waters: a review. Oceanography and Marine Biology Annual Review 25:563–575
Hall POJ, Holby O, Kollberg S, Samuelsson MO (1992) Chemical fluxes and mass balances in a marine fish cage farm. IV. Nitrogen. Marine Ecology Progress Series 89:81–91
Hansen PK, Ervik A, Schaanning MT, Johannsen P, Aure J, Jahnsen T, Stigebrandt A (2001) Regulating the local environmental impact of intensive marine fish farming. II. The monitoring programme of the MOM system (Modelling – Ongrowing fish farms – Monitoring). Aquaculture 194:75–92
Hargrave BT (2005) The handbook of environmental chemistry, Vol. 5: Water pollution environmental effects of marine finfish aquaculture. Springer Berlin Heidelberg New York, 467 pp
Henderson A, Gamito S, Karakassis I, Pederson P, Smaal A (2001) Use of hydrodynamic and benthic models for managing environmental impacts. Journal of Applied Ichthyology 17:163–172
Holby O, Hall POJ (1991) Chemical fluxes and mass balances in a marine fish cage farm. II. Phosphorus. Marine Ecology Progress Series 70:263–272
Holmer M, Kristensen E (1992) Impact of marine fish cage farming on sediment metabolism and sulfate reduction of underlying sediments. Marine Ecology Progress Series 80:191–201
Holmer M, Perez M, Duarte CM (2003) Benthic primary producers – a neglected environmental problem in Mediterranean maricultures? Marine Pollution Bulletin 46:1372–1376
Holmer M, Wildish D, Hargrave B (2005) Organic enrichment from marine finfish aquaculture and effects on sediment processes. In: Hargrave BT (ed.), The handbook of environmental chemistry, Vol. 5: Water pollution environmental effects of marine finfish aquaculture. Springer, Berlin Heidelberg New York, pp.181–206
Hyland J, Balthis L, Karakassis I, Magni P, Petrov A, Shine J, Vestergaard O, Warwick R (2005) Organic carbon content of sediments as an indicator of stress in the marine benthos. Marine Ecology Progress Series 295:91–103
Kalantzi I, Karakassis I (2006) Benthic impacts of fish farming: meta-analysis of community and geochemical data. Marine Pollution Bulletin 52:484–493
Karakassis I, Hatziyanni E (2000) Benthic disturbance due to fish farming analyzed under different levels of taxonomic resolution. Marine Ecology Progress Series 203:247–253
Karakassis I, Tsapakis M, Hatziyanni E (1998) Seasonal variability in sediment profiles beneath fish farm cages in the Mediterranean. Marine Ecology Progress Series 162:243–252
Karakassis I, Tsapakis M, Smith CJ, Rumohr H (2002) Fish farming impacts in the Mediterranean studied through sediment profiling imagery. Marine Ecology Progress Series 227:125–133
Karakassis I, Tsapakis E, Hatziyanni E, Pitta P (2001) Diel variation of nutrients and chlorophyll in sea bream and sea bass cages in the Mediterranean. Fresenius Environmental Bulletin 10:278–283
La Rosa T, Mirto S, Favaloro E, Savona B, Sara G, Danovaro R, Mazzola A (2002) Impact on the water column biogeochemistry of a Mediterranean mussel and fish farm. Water Research 36:713–721
Lampadariou N, Karakassis I, Pearson TH (2005) Cost/benefit analysis of a benthic monitoring programme of organic benthic enrichment using different sampling and analysis methodology. Marine Pollution Bulletin 50:1606–1618
Machias A, Karakassis I, Labropoulou M, Somarakis S, Papadopoulou KN, Papaconstantinou C (2004) Changes in wild fish assemblages after the establishment of a fish farming zone in an oligotrophic marine environment. Estuarine Coastal and Shelf Science 60:771–779

Machias A, Karakassis I, Somarakis S, Giannoulaki M, Papadopoulou KN, Smith C (2005) The response of demersal fish communities to the presence of fish farms. Marine Ecology Progress Series 288:241–250

Machias A, Giannoulaki M, Somarakis S, Maravelias CD, Neofitou C, Koutsoubas D, Papadopoulou KN, Karakassis I (2006) Fish farming effects on local fisheries landings in oligotrophic seas. Aquaculture 261: 809–816

Malta Environment & Planning Authority (2001) Note on the situation regarding the proposed fish farm search areas in the Policy and Design Guidance for Fish Farming (1994). Malta: Malta Environment and Planning Authority, 3 pp

Marbà N, Duarte CM, Díaz-Almela E, Terrados J, Álvarez E, Martinez R, Santiago R, Gacia E, Grau AM (2005) Direct evidence of imbalanced seagrass (*Posidonia oceanica*) shoot population dynamics in the Spanish Mediterranean. Estuaries 28: 53–62

Marbà N, Santiago R, Díaz-Almela E, Álvarez E, Duarte CM (2006) Seagrass (*Posidonia oceanica*) vertical growth as an early indicator of fish farm-derived stress. Estuarine, Coastal and Shelf Science 62: 425–433

Maroni (2000) Monitoring and regulation of marine aquaculture in Norway. In: Rosenthal H, Schnack D, Hilge V (eds.), MARAQUA. The monitoring and regulation of marine aquaculture in Europe. Journal of Applied Ichthyology 16:192–195

Molvær J, Knutzen J, Magnusson J, Rygg B, Skei J, Sørensen J (1997) Klassifikasjon av miljøkvalitet i fjorder og kystfarvann (Classification of environmental quality in fjords and coastal waters). SFT Guidelines 97:03. TA-1467/1997, 36 pp. (In Norwegian)

Norwegian Pollution Control Authoritiy (1993) Environmental objectives for Norwegian aquaculture. Oslo, Norway, 17 pp

Norwegian Standards Association (1998) Vannundersøkelse – Retningslinjer for kvantitative undersøkelser av bløtbunnsfauna i marint miljø (Water quality – Guidelines for quantitative investigations of sublittoral soft-bottom benthic fauna in the marine environment) NS-9423. Available from firmapost@pronorm.no, 16 pp. (In Norwegian)

Norwegian Standards Association (2000) Environmental monitoring of marine fish farms NS-9410. Available from firmapost@pronorm.no, 22 pp

Ottolenghi F, Silvestri C, Giordano P, Lovatelli A, New MB, 2004. Capture-based aquaculture. The fattening of eels, groupers, tunas and yellow tails. FAO, Rome, 308 pp

Pearson TH, Rosenberg R (1978) Macrobenthic succession in relation to organic enrichment and pollution of the marine environment. Oceanography and Marine Biology Annual Review 16:229–311

Pereira PMF, Black KD, Mclusky DS, Nickell TD (2004) Recovery of sediments after cessation of marine fish farm production. Aquaculture 235, 315–330.

Pillay TVR (2004) Aquaculture and the environment. Blackwell, Oxford, UK, 208 pp

Pitta P, Apostolaki E, Tsagaraki T, Tsapakis M, Karakassis I (2006) Fish farming effects on chemical and microbial variables of the water column: a spatio-temporal study along the Mediterranean Sea. Hydrobiologia 563:99–108

Pitta P, Apostolaki ET, Giannoulaki M, Karakassis I (2005) Mesoscale changes in the water column in response to fish farming zones in three coastal areas in the Eastern Mediterranean Sea. Estuarine, Coastal and Shelf Science 65:501–512

Pitta P, Karakassis I, Tsapakis M, Zivanovic S (1999) Natural versus mariculture induced variability in nutrients and plankton in the Eastern Mediterranean. Hydrobiologia 391:181–194

Planning Authority (1994) Policy and design guidance – fishfarming. Malta: Planning Authority, 12 pp

Read PA, Fernandes TF, Miller KL (2001). The derivation of scientific guidelines for best environmental practice for the monitoring and regulation of marine aquaculture in Europe. Journal of Applied Ichthyology 17:146–152

Schembri PJ, Baldacchino AE, Mallia A, Schembri T, Sant MJ, Stevens DT, Vella SJ (2002) State of the environment report for Malta 2002: Living resources, fisheries and agriculture. In: State of the environment report for Malta 2002. Santa Venera, Malta: Ministry for Home Affairs and the Environment, pp. 162–346

Soto D, Norambuena F (2004) Evaluation of salmon farming effects on marine systems in the inner seas of southern chile: a large-scale mensurative experiment. Journal of Applied Ichthyology 20:493–501

Stigebrandt A, Aure J, Ervik A, Hansen PK (2004) Regulating the local environmental impact of intensive marine fish farming. III: A model for estimation of the holding capacity in the MOM system (Modelling – Ongrowing fish farm – Monitoring). Aquaculture 234:239–261

World Wid Fund for Nature (WWF) (2004) Tuna farming in the Mediterranean: the bluefin tuna stock at stake, Rome: WWF Mediterranean Program Office, 35 pp

Ye L, Ritz DA, Fenton GE, Lewis ME (1991) Tracing the influence on sediment of organic waste from a salmonid farm using stable isotope analysis. Journal of Experimental Marine Biology and Ecology 145:161–174

Youngson AF, Dosdat A, Saroglia M, Jordan WC (2001. Genetic interactions between marine finfish species in European aquaculture and wild conspecifics. Journal of Applied Ichthyology 17:152–162

Chapter 3
Aquaculture and Coastal Space Management in Europe: An Ecological Perspective

Tim Dempster[1,*] and Pablo Sanchez-Jerez[2]

Abstract Coastal aquaculture is widespread in Europe and there is a need for proper coastal space management among the different users of the coastal zone. Integration of aquaculture into coastal space entails both siting installations in physical space in relation to the existing network of coastal users, such as shipping, fishing, recreational activities and other industry, and ensuring that the extent of aquaculture does not lead to widespread changes to coastal ecosystems. Where the competition for space is particularly intense, political decisions, which simultaneously seek to minimize both environmental impacts and user conflict, may be the only mechanism to allocate space to new aquaculture installations. From an ecological perspective, better integration of aquaculture into European coastal space so that ecological carrying capacities are not exceeded requires knowledge-based management of the interaction of ecological impacts of aquaculture with those of other coastal users, particularly concerning nutrient loading, and modification to biodiversity and species that are important to fisheries. Geographical information systems (GIS) are proven tools for natural resource management and space planning and are suggested to be used for planning aquaculture's integration into European coastal areas.

Keywords Farm siting, fish aggregation, marine protected area, wild fish

3.1 Coastal Aquaculture in Europe

Coastal aquaculture farms are ubiquitous in many European countries. Sea-cages hold over 1 million tons of fish while hundreds of thousands of tons of mussels, oysters and clams are grown on suspended ropes, racks or trays (FAO 2006). The

[1]SINTEF Fisheries and Aquaculture, NO-7465 Trondheim, Norway.
E-mail: tim.dempster@sintef.no

[2]Department of Marine Sciences and Applied Biology, University of Alicante, P.O. Box 99, E-03080-Alicante, Spain. E-mail: psanchez@ua.es

*Corresponding author

culture of marine and anadromous fish in sea-cage farms is widespread in northern Europe (e.g. Norway, Scotland, Ireland, Denmark, Faeroe Islands). Norway is presently the leading producer of fish; in 2004 over 600,000 t of salmon and sea trout were produced in 870 concessions, while 285 concessions farmed other species such as cod, halibut and arctic char (Norwegian Fisheries Directorate 2005). The Mediterranean Sea supports over 500 sea-cage farms producing more than 160000 t year^{-1} of sea bream and sea bass in Greece, Spain, Italy, France, Turkey and numerous other countries. Tuna ranching operations exist in eight countries in the Mediterranean; approximately 225,000 t year^{-1} of small, wild caught pelagic fish are used to fatten an initial biomass of 15,000 t–20,000 t of wild-caught tuna (Borg and Schembri 2006). Extensive industries along the Atlantic coasts of Spain, Portugal and France (Goulletquer and Le Moine 2002) and in other parts of Europe exist for the culture of mussels, oysters and other shellfish. These occupy substantially greater coastal space than sea-cage fish farms, which typically occupy 1–5 ha per installation. For example, the 125,000 t of oysters and 20,000 t of mussels stocked in the Marennes–Oléron Bay region in France occupy 4,000 ha of coastal space and 3,000 ha of nearby wetlands (Goulletquer and Le Moine 2002).

The European Union plans to expand aquaculture further, to increase seafood supplies, create jobs and reduce the trade deficit of the EU in seafood products. As this expansion occurs, finding suitable locations for aquaculture installations in coastal areas, and managing the interaction of aquaculture with other users of the coastal zone will become increasingly important (Stead et al. 2002). Here, we discuss the concept of "competition for coastal space" as a combination of the competition for physical space and the competing activities of different users for coastal space.

3.2 Interactions of Aquaculture with Other Users of the Coastal Zone

There are two aspects to integrating the diverse array of aquaculture that exists throughout Europe within coastal space and managing its interactions with other users of the coastal zone: (1) planning of site allocation for aquaculture activities, and (2) management of the interactions of installations, the environment and other users once they have been set up. As the environmental effects of aquaculture will interact with those of other activities, integrated management, where the uses and environmental effects of all users of coastal activities are considered simultaneously, may reduce conflicts and minimise negative environmental effects.

Ecological requirements should dictate first and foremost the position and extent of aquaculture in the coastal zone (Costa Pierce 2002; Guneroglu et al. 2005). Siting criteria should be based on "ecological carrying capacities" or the ability of the ecosystem to absorb anthropogenic pressures with no major changes to ecosystem functions and processes. A suite of environmental impacts caused by coastal aquaculture must be incorporated into this process to determine how habitat and biodiversity modification caused by aquaculture can best be managed and mitigated. A substantial body of knowledge exists on the environmental impacts of aquaculture and this gives excellent insight for allocating aquaculture, while ensuring least adverse impact. Information on the genetic effects of sea-based aquaculture on wild populations through escapees (Naylor et al. 2005) and cross breeding of wild and cultured organisms (Wier and Grant 2005), nutrient loading (Karakassis et al. 2005), modification of benthic communities (Karakassis et al. 2000), heavy metal and persistent organic pollutant contamination (deBruyn et al. 2006; Mendiguchia et al. 2006; Sather et al. 2006), spreading of disease and parasites (Bjørn et al. 2001), impacts on seagrasses (Delgado et al. 1999; Ruiz et al. 2001; Marba et al. 2006), impacts on farm-associated wild fish (Dempster et al. 2002, 2006) and megafauna (Nash et al. 2000) and a range of other environmental impacts must be considered to assess the suitability of new sites. Capture of millions of tons of small pelagic fishes to make fish meal and fish oil for aquaculture feeds (Tacon and Forster 2003) can be considered an "oceanic" impact which requires fisheries management measures in addition to coastal zone management. Setting "carrying capacities" to determine the overall extent of aquaculture in particular coastal regions is a challenging task, as for many of these ecological effects, the level of information currently available is insufficient to determine the extent of the effect.

Once ecological criteria are established, aquaculture can then search for suitable space that minimizes conflict with the myriad of other users of coastal waters, such as shipping, fishing, recreational activities and industry (Fig. 3.1). As aquaculture is the "new kid on the block" in terms of its use of space, in many coastal areas it will struggle to obtain suitable sites that do not conflict with pre-existing users that

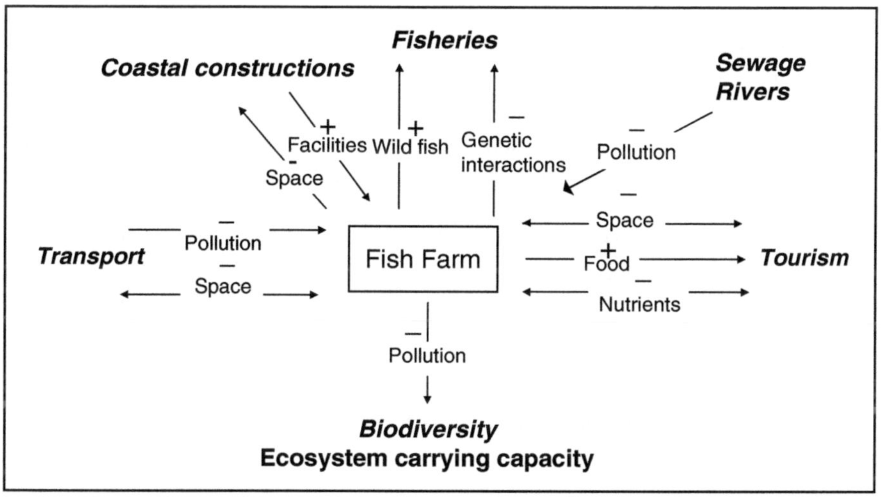

Fig. 3.1 Positive and negative interactions concerning space and environmental impacts between aquaculture and other coastal users

may be more important to the economics of the region (Staresinic and Popović 2004). In such instances, political decisions on the use of coastal space may be the only way that aquaculture may gain access to coastal waters.

Once an aquaculture installation is in place, regulatory measures are required to ensure that ongoing interactions between users of the coastal zone are ecologically sustainable. Certain environmental and spatial interactions among users are likely to vary over time or will not be evident until after a farm is established, such as changes in water quality characteristics or how commercial and recreational fisheries will interact with aquaculture activities (Dempster et al. 2005). Management measures may therefore need to be location-specific and adaptive.

3.3 Competition for Physical Space Between Aquaculture, Shipping, Tourism and Recreation

3.3.1 Shipping

Among the chief users of coastal waters is shipping, whether it is for commercial, recreational or defence related purposes. While the use of coastal space by ships in any particular area is relatively temporary, it nevertheless places considerable restrictions on the placement of aquaculture installations. Commercial shipping lanes and their immediate vicinity, together with military shipping areas, almost completely exclude aquaculture due to the risks posed by surface-based structures as navigational hazards. Numerous ships transport hazardous products, such as chemical and petrol-derived products, which affect the coastal environment adversely when accidental spillage occurs (Davis 1993) and pose environmental and health risks to coastal aquaculture. For example, the break-up of the oil tanker *Prestige* off the Galician coast of Spain in 2002 caused 9 million Euros of lost mussel production in the year following the accident (Garza-Gil et al. 2005). Recreational sailing and boating activities also challenge aquaculture for coastal space, particularly in areas where both operate from local ports. Space in the immediate sea areas surrounding ports is sought after by both aquaculture and recreational boating activities due to ease of access. Most sea-cage fish farms along the relatively featureless coastline of south-eastern Spain are sited less than 5 km from the coast and operate out of ports that are popular for recreational boating.

3.3.2 Tourism and Coastal Aesthetics

Coastal tourism is growing in popularity across Europe and as such, mariculture and tourist uses will compete strongly for coastal space. For instance, countries that border the Mediterranean are particularly popular tourist destinations. Of the world's 689 million international tourists in 2001, one third travelled to Mediterranean countries (World Tourism Organization 2002) and the largest concentrations of these tourists visited coastal destinations. Both the coastal tourism and mariculture industries require a marine environment of high quality to fulfil their business objectives (Staresinic and Popović 2004).

In most countries throughout Europe, coastal tourism is an economic force many times greater than mariculture. Staresinic and Popović (2004) compared the relative contributions of tourism and mariculture to the economy of Greece, which has the greatest production of maricultured fish of all Mediterranean countries and a substantial mussel industry. While mariculture was responsible for more than 6000 jobs in 2002, the 14.2 million foreign tourists who visited Greece in 2002 generated 293,000 jobs (WTTC 2003). Where such a discrepancy in the overall value of the industries to the economy exists, tourism, as the stronger competing force, may well dictate access to coastal space. Sea bream and sea bass farmers along the south-eastern Mediterranean coast of Spain state that interaction with tourism-oriented local authorities is the greatest barrier to development of aquaculture in this region and is their greatest concern for continuing existing operations in coastal areas.

Tourism developments and coastal aquaculture impact coastal areas through physical developments, such as construction of hotels, marinas and hatcheries, and deployment of sea-cages and mussels rafts. Further, both activities result in increased nutrient and pollution levels from disposal of sewerage or fish farm wastes into coastal waters (see Section 2.3 for discussion of this interaction). Competition between tourism and mariculture over long-term access to a high-quality marine environment has been documented in many European countries (Stephanou 1998; Conides and Papaconstantinou 2001; Staresinic and Popović 2004).

Sea-cage fish farms or mussel rafts typically have large surface structures that impact upon the aesthetics of seascapes viewed from the shore (Fig. 3.2). Land-based facilities to support coastal aquaculture may also create conflict by alteration of the coastal landscape, particularly where this occurs in ports or resorts, or near tourist beaches. Tourism, driven by the desire to have extensive, uninterrupted ocean views, may increasingly block development of aquaculture in particular areas which may otherwise be suitable on this premise alone, or even force aquaculture from particular regions of the coast. Staresinic and Popović (2004) outlined the problem in Croatia, and the same may be true of many areas of the Mediterranean that have a dense concentration of coastal tourism sites (e.g., the Balearic Islands; Valencia 2006). In the Canary Islands, extensive use of the coastal zone by tourism activities and the extremely narrow continental shelf that limit suitable depths in which to moor aquaculture installations combine to make the competition for coastal space particularly intense (Perez et al. 2005). Even in Norway, which has the longest coastline of any European country, and the largest and most economically important aquaculture industry, placement of aquaculture sites is sometimes considered in terms of the aesthetic impacts of installations. Movement from inner fiord locations to more exposed coastal locations is an industry-wide trend for the farming of salmonids and other species in sea cages (Sunde et al. 2003) and may in part be due to the aesthetic impacts of installations.

Fig. 3.2 Fish farms in the coastal seascape: a mixed sea bream (*Sparus aurata*) and sea bass (*Dicentrarchus labrax*) farm off the Mediterranean coast of Spain (left) and an Atlantic salmon (*Salmo salar*) farm in a Norwegian fiord (right)

3.4 Nutrient Loading in Coastal Areas – Interactions Between Aquaculture and Other Activities

The European Community has a total shoreline of 90,000 km. More than 20% of the population is economically dependent on the coastal zone, which is intensively used and settled by humans. Because of population pressure and economic development, water quality is declining throughout coastal areas due to an increase in nutrient loading, which can be attributed to several sources. Population increases over the last two centuries in coastal cities has lead to increased discharges from sewage treatment plants to the marine environment. Over the last 20 years, in addition to other anthropogenic pressures, marine aquaculture has expanded in many European coastal areas, increasing pressure on marine ecosystems.

In addition to the more traditional, extensive aquaculture of mussels and oysters, which use primary production from the marine ecosystem, intensive production of fish within sea-cages is increasingly occupying more coastal space. Sea-cage aquaculture in Europe produces mainly carnivorous species (salmonids, sea bass, sea bream) because of market demands. Cage aquaculture uses high protein pellets to feed these carnivorous species. The nutrients unassimilated by the caged fish introduce a large source of nutrients to coastal areas. For example, more than 800,000 t of feed was used to produces the 600,000 t of salmonids in sea-cages in Norway in 2004 (Norwegian Fisheries Directorate 2005). Occasionally, nutrient inputs from aquaculture can exceed the assimilative capacity of the local marine environment, leading to coastal eutrophication (Naylor et al. 2000). Fish production can generate considerable amounts of effluent, such as waste feed, faeces, medicinal substances, heavy metals and persistent organic pollutants, which can pollute the marine environment with a range of negative impacts varying in severity (Black 2001; Read and Fernandes 2003; Mendiguchia et al. 2006; Sather et al. 2006).

Organic waste products from aquaculture can be particulate or dissolved. Dissolved products include ammonia, phosphorus, dissolved organic carbon and lipids. The environmental impact of these dissolved products depends on the rate at which nutrients are diluted before being assimilated by the pelagic ecosystem (Fig. 3.3). Particulate discharges from farms derive mainly from lost food and faeces, which will sediment at different rates to the sea floor depending on local current regimes (Sara et al. 2003) and re-sedimenting processes (Cromey et al. 2002). Particulate waste products settle to the bottom around fish farm at a scale of tens to hundreds of metres (Karakassis et al. 2000) in areas with weak currents, but can disperse over 1000m (Sara et al. 2003) where current flows are greater. Differentiating between particulate and dissolved nutrients is important because their relative effects on benthic and pelagic systems differ.

Nutrient loading from feeding will depend on (i) feed wastage, (ii) solubility of nutrients from pellets and (iii) the rate of absorption by the cultured fish due to digestibility (Islam 2005). These in turn will depend on the farmed species, stocking density, and feeding regimen (Islam 2005). The amount of particulate matter entering the system will largely depend on the level of uneaten food, which can vary from 1 to 20% depending on the type of food, feeding strategy and stochastic factors (e.g., weather conditions). Solubility of nutrients from dry pellets is low because of technological advances in fish food production, and therefore leaching rates from food pellets is relatively low (Fernandez-Jover et al. 2007a).

Uneaten food is a major contributor of N and P to the environment. However, much of the uneaten food is removed by wild fish aggregated around fish farms before it sinks to the bottom, therefore, loading of nutrients to the system is reduced drastically (Vita et al. 2004). Lupatsch and Kissil (1998) studied the N and P budget of sea bream and determined that more that 70% of the total amount of N and P in feed was lost to the environment as waste. Islam (2005) calculated that between 68% –and 86% of the N consumed by fish was voided as dissolved N in the form of urea and ammonia, assuming that 4% of the N was lost in faecal pellets. In total, this amounted to approximately 32kg of ammonia for each ton of feed used. Additionally, during the sedimentation of faecal pellets, N leaches rapidly into the water column (Chen et al. 1999).

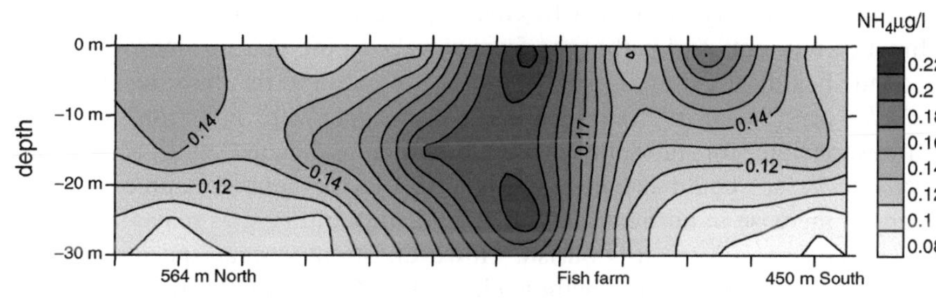

Fig. 3.3 Example of the spatial distribution of ammonium (NH_4 µg/l) throughout the water column around a sea bream and sea bass farm in the southwestern Mediterranean Sea several hours after feeding (Sanchez-Jerez et al., unpublished data)

While such nutrient inputs may have impacts at the local scale, the overall impact of increased nutrient loading from aquaculture may be unnoticeable at a macro-scale. Mediterranean aquaculture produces little detectable increase in nutrients in the entire Mediterranean, compared to the input of nutrients from other anthropogenic activities and from atmospheric and terrestrial sources (Karakassis et al. 2005). Similarly, the input of nutrients from fish-based aquaculture along the Norwegian coast represents only a small proportion of the overall nutrient budget of the coastal ecosystem (Ervik and Aure 2006). At the local scale, some studies have indicated that seasonal variation of nutrients can be more important that differences between impact and control locations, indicating that farm activities do not always produce a detectable increase of nutrients. Maldonado et al. (2005) found that neither surface nor bottom waters at the fish farms showed abnormal concentrations of nitrite and nitrate relative to controls. This result was unlikely to be the result of uptake by phytoplankton, as chlorophyll a values under the fish cages were low relative to control sites. Ruiz et al. (2001) found that the major differences in nitrate and nitrite concentrations along the SE coast of the Iberian Peninsula were due to seasonal changes in the environment rather than caused by fish-farming activities. Therefore, at a local scale, the probability of individual fish farms at their current sizes affecting themselves and their immediate environment is low (Pitta et al. 1998; Karakassis et al. 2005).

Urban development and human pressure in coastal areas can likewise affect aquaculture. Anthropogenic nutrients from waste water and agriculture run off are responsible for a large component of the nutrients that cause marine eutrophication (Costanzo et al. 2001). If sewage treatment for coastal cities is inadequate, the introduction to coastal areas of insufficiently treated waters can have large adverse effects, especially during summer when water temperatures are high. For example, eutrophication can affect fish production directly by reducing dissolved oxygen (Page et al. 2005). Alternately, decreases in primary production in coastal areas are also possible if freshwater runoff is dramatically reduced, which may affect coastal oyster and mussel production. The Marennes-Oléron Bay region in France is a major oyster and mussel production area. A decrease in the amount of freshwater entering the bay, due to a fourfold increase in the amount of irrigated land in the catchment area combined with greater water usage on a per hectare basis, led to reduced nutrient supply to coastal culture sites of oysters and mussels. This resulted in decreased survival rates of oyster and mussel spat which require reduced salinity (see review by Goulletquer and Le Moine 2002).

Input of nutrients and pathogens from sewage to the marine environment can reduce the health of cultured fish. Mortalities of cultured fish have resulted from *Vibrio harvey* (Saeed 1995) infections. *Streptococcus* sp. infections have also been responsible for mortality of wild mullet associated with aquaculture. *Streptococcus* sp. could originate from terrestrial or aquatic sources, and the associated increase in nutrient loading around aquaculture sites may allow indigenous *Streptococcus* sp. to flourish. However, streptococcal species typically introduced by sewage are not pathogenic to fish (Gilbert et al. 2002). Important human health concerns exist with regard to culture of shellfish in coastal waters, since sewage contains pathogens which can contaminate shellfish and may be passed on to the consumer; several diseases can be transmitted through human ingestion of contaminated shellfish (Hill 2005).

Recreational boating also impacts water quality (McGee and Loehr 2003). Disposal of untreated sewage has been defined as the most significant problem associated with recreational boating; significantly higher faecal coliform levels exist in coastal waters where recreational boating activity is high (Davies and Cahill 2000). Discharge of sewage from boats lowers water quality, which represents a problem in water bodies that undergo limited flushing or support shellfish beds located nearby. Special wastewater treatment procedures should be necessary for coastal tourist areas to limit the entry of effluent to the sea, while adequate planning for different activities, such as recreational boating, should take into account the water quality link to aquaculture.

From a regional point of view, the impacts associated with the development of tourism and aquaculture on water quality, and existing impacts such as agriculture, will be synergistic. The sum of the nutrients introduced from aquaculture, sewage, river discharges and agricultural run-off (e.g., fertilisers) to the marine environment has the potential to exceed the assimilative capacity of benthic and pelagic systems and cause eutrophication. Carrying capacities vary regionally, for example, Ervik and Aure (2006) state that modelling indicates that coastal sites in northern Norway have greater carrying capacities than sites in the south. Planning for regional economic development of coastal zones should therefore account for varying carrying capacities and set limits of "acceptable change" of environmental conditions for aquaculture. Potential changes of water quality will occur over different temporal and spatial scales. Therefore, it is necessary to investigate, using remote sensing and other geospatial and long-term data sources, coastal changes relating to coastal urban development, human demographic trends in coastal areas, increases in aquaculture facilities and other sources of pollution such as agriculture. Knowledge of these processes will enable tighter regulation of the allowable limit of N and P added to water bodies and marine sediments.

Technological advances in real time measurement can help monitoring and management of coastal water quality, particularly where many users simultaneously require high water quality yet reduce water quality through their activities. For example, the EU project I-MARQ (5th EFP, No IST-2001-34039; www.imarq.inf) developed a system for marine decision support, which aimed to simplify the common Environmental Decision Support Systems (EDSS) that help users assess the state of the marine ecosystem depending on different management regimes. This kind of technological advance can be used to evaluate microbiological and eutrophication risks related to aquaculture development.

3.5 Interactions of Wild Fish, Aquaculture and Fishing

3.5.1 Siting of Farms in Coastal Waters

Interactions of coastal aquaculture with fishing must be managed both before and after installations are in place. Before allocation of new aquaculture sites, consideration must be given to whether farms will displace fishers from existing fishing grounds through physically restricting access to fishing. Information on the importance of a specific area to fishing lies in its relative contribution to the overall catch of a particular fishery and finer detailed information on the scale of hundreds of metres can be obtained from the fishers themselves. Areas where catch rates are high, where catches are economically or socially important, or where particularly important habitats for juveniles of important fisheries species exist (e.g., seagrass meadows, macroalgal forests) should be deemed unsuitable as sites for aquaculture to avoid conflict between the fisheries and aquaculture sectors.

During farm placement, consideration must also be given to the proximity of the site to areas that may be of particularly high importance to wild fish stocks, such as known points of natural aggregation for feeding, spawning or migratory pathways of anadromous fish. For salmonid aquaculture, two substantial environmental effects are of concern: 1) escape of cultured fish and their subsequent mixing with wild stocks (see review by Weir and Grant 2005); and 2) that the large numbers of cultured fish held in coastal areas may increase parasite loads of their wild counterparts (Bjorn et al. 2001; Morton et al. 2004; Krkošek et al. 2005). Presently, much is known about the causes and environmental effects of escapes for salmonids (Naylor et al. 2005), while comparatively little is known for other species such as sea bream, sea bass and Atlantic cod (but see Moe et al. 2005). Inter-breeding and competitive interactions of escapees with wild salmon within rivers may have detrimental effects on wild populations. Likewise, high parasite loads on seaward-migrating salmon smolts have been implicated as a potential cause of high mortality at sea and reduced return of adults to rivers (Bjorn et al. 2001).

Assessment of the risk that escapees and other effects pose to wild populations when placing farms has been suggested (Naylor et al. 2005; WWF 2005). Declaration of the "national salmon fiords" throughout Norway in 2003 and the consequent restriction on placing new fish farms in these areas is an example of considering important wild fish stocks when locating farms (Sivertsen 2006). In response to concerns regarding escapees and parasite loads of seaward-migrating smolts, particular rivers flowing into coastal fiords in Norway were regarded as of such high importance to wild salmon populations that sea-cage salmonid farms were restricted or removed from these fiords.

An emerging issue regarding escapes is that certain fish species are being raised to sizes within sea-cages at which, if they become sexually mature, they are capable of spawning. This requires the concept of escape from mariculture to be redefined to include the escape of reproductive gametes into the environment. Jørstad and van der Meeren (2006) allowed 1000 gene tagged cod to spawn within a small fiord system in Norway. Upon sampling larvae in the waters surrounding the farm, 25% were traced back to caged parents. This indicates that if spawning occurs within commercial cod farms where numbers of animals are far greater, the contribution of "escaped" larvae to cod recruitment within fiords may be substantial. Spawning of sea bream within sea-cages has also been observed in Greece (Dimitriou et al. 2007). If breeding programmes shift the genetic diversity of aquacultured fish away from wild stocks, the extent of spawning within sea cages and whether larvae subsequently survive and recruit into natural populations in significant numbers will likely greatly affect siting of farms.

Coastal aquaculture sites have also been suggested as having the potential to disrupt the spawning of marine fish species if improperly placed, although little evidence of this presently exists. Bjørn et al. (2005) found that wild coastal Atlantic cod (*Gadus morhua*) avoided the smell of salmon farms in tank-based olfactory experiments, which suggests they may also avoid areas with farms. Atlantic cod are known to have high fidelity to specific spawning grounds (Wright et al. 2006). If farms deter fish from accessing spawning areas or impede migratory pathways to spawning areas, the success of spawning may diminish. Detailed information on fish movements in space and time is required to determine if some fish species avoid farm areas. If so, farms may best be placed away from known spawning areas.

3.5.2. Effects of Existing Fish Farms on Wild Fish

Once sea-cages have been deployed, they will attract wild fish to their immediate surrounds, which in turn are likely to attract fishers. Deciding on the appropriate level of interaction between aggregations of wild fish and commercial and recreational fisheries requires knowledge of the species and overall biomass of the wild fish aggregations through time, the extent to which they will be targeted by fishers, and importantly, the existing management regime of the fishery and the overall status of the wild fish stock.

Coastal aquaculture farms have considerable demographic effects on wild fish by aggregating large numbers in their immediate vicinity. Early studies by Carss (1990) in Scotland and Bjordal and Skar (1992) in southern Norway around marine salmon farms indicated that saithe (*Pollachius virens*) aggregated at farms in considerable numbers. Dempster et al. (2002, 2005) highlighted that Mediterranean sea-cage fish farms attracted wild fish assemblages that had up to 30 different species and estimated that the aggregation biomasses ranged between 10 and 40t at 5 of the 9 farms investigated (Dempster et al. 2004). Similarly large aggregations have since been noted in Greece (Smith et al. 2003; Thetmeyer et al. 2003) and the Canary Islands (Boyra et al. 2004; Tuya et al. 2005, Fig. 3.4). While mussel rafts in the Mediterranean Sea (Brehmer et al. 2003) are also known to aggregate wild fish, the majority of studies concerning demographic impacts of coastal aquaculture on wild fish have focussed on aggregations around sea-cage farms.

3.5.3 Composition and Variability of Wild Fish Aggregations

Although zoogeographic differences in the species of fish that aggregate around farms exists, pelagic planktivorous species dominate assemblages at most farms and these fish opportunistically feed upon food pellets lost from cages. In warm water areas, such as Mediterranean Spain and the Canary Islands, over 30 different

Fig. 3.4 Aggregations of wild fish near the bottom (left) and the surface (right) of a coastal sea-cage sea bream farm in the Canary Islands, Atlantic Ocean. (Photo courtesy of Arturo Boyra, www.oceanografica.com)

species of wild fish aggregate at farms, although only 1 to 3 taxa (principally Mugilidae, *Trachurus mediterraneus*, *Sardinella aurita* and *Boops boops*) dominate the assemblages (Dempster et al. 2002; Boyra et al. 2004; Tuya et al. 2005; Dempster et al. 2005). The sizes of aggregated planktivorous fish at farms in Mediterranean Spain are large and most are likely to be adult (85% adult: Dempster et al. 2002; 71% adult: Dempster et al. 2005). In cold-water areas, such as Scotland and Norway, fewer species have been noted to associate with farms (*Pollachius virens*; Carss 1990; Bjordal and Skar 1992), however, no extensive surveys of the wild fish that are attracted to high-latitude farms have been undertaken to date. Indirect evidence suggests that these fish are also predominantly adult, as a large proportion of wild saithe tagged after capture from beneath a Norwegian salmon farm migrated to offshore spawning grounds (Bjordal and Skar 1992).

Aggregations of demersal fish also occur beneath farms although aggregation size varies greatly between locations; few demersal fish occur beneath farms in Mediterranean Spain while large, multi-species aggregations occur under farms in the Canary Islands (Dempster et al. 2005). Large abundances of sparids such as *Pagellus* sp., large Chondrichthyid rays (8 species) and *Heteroconger longissimus* have been observed at farms in the Canary Islands (Boyra et al. 2004; Tuya et al. 2005; Dempster et al. 2005). Large, carnivorous fish, such as *Pomatomus saltator, Coryphaena hippurus,* and *Sphyraena* spp. aggregate around many farms (Dempster et al. 2002; Dempster et al. 2005). *P. saltator* commonly occurs in shoals of hundreds to thousands of individuals at farms along the south-east coast of Spain (Dempster et al. 2002) and feeds mainly on wild *Sardinella aurita* around the cages (Sanchez-Jerez et al. unpublished data).

Considerable spatial variability in wild fish abundance and biomass exists among farms located along the same stretch of coastline (Dempster et al. 2002). Aggregations are temporally stable over the scale of several weeks to months, both in relative size and species composition, indicating some degree of residency of wild fish at farms (Dempster et al. 2002). However, large seasonal differences in the species composition and biomass of wild fish assemblages have been noted around farms in the Spanish Mediterranean (Fernandez-Jover et al. 2007b; Valle et al. 2007), yet this pattern is not consistent for all locations since such strong seasonal differences have not been recorded from farms in the Canary Islands (Boyra et al. 2004).

3.5.4 Coastal Aquaculture Sites as Artificial Habitats

Coastal aquaculture sites may be considered as artificial ecosystems, where wild fish are subject to ecological processes which differ greatly from their natural habitats (Fig. 3.5). Both the size and persistence of aggregations of wild fish around farms suggests they may have a variety of ecological and physiological effects. These include modified diet, physiological condition, tissue fat content and fatty acid composition, reproductive condition, parasite load, exposure to predation and susceptibility to fishing pressure.

Diets of wild fish that are associated with farms are modified. Wild *Trachurus mediterraneus* associated with two farms on the coast of Spain (Fernandez-Jover et al. 2007b), and *Pollachius virens* associated with a farm in Norway (Skog et al. 2003),

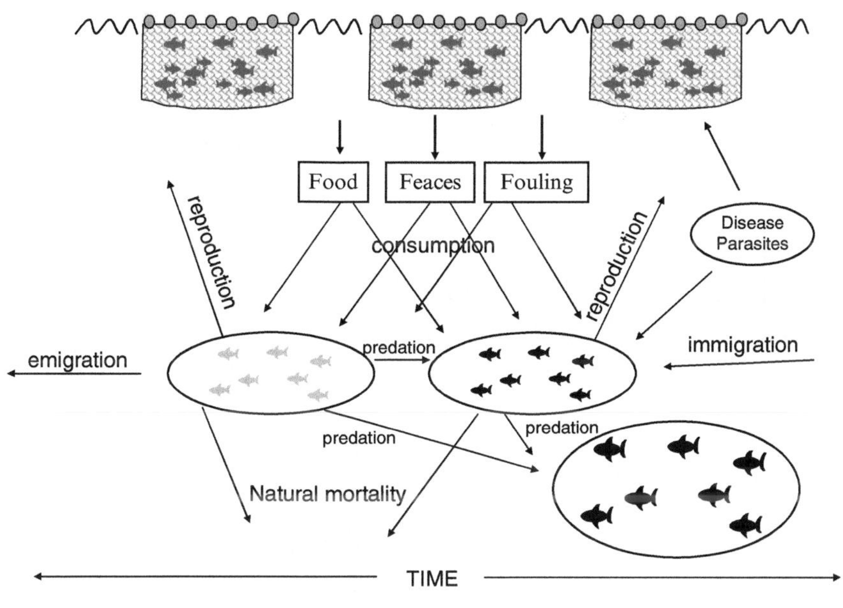

Fig. 3.5 Schematic diagram of the artificial wild fish habitat in the vicinity of a coastal sea-cage fish farm

have both been shown to feed predominantly on lost feed from farms when in their vicinity, compared to wild fish taken from control locations which fed mainly upon fish and invertebrates. Fernandez-Jover et al. (2007b) demonstrated that the modified diet in farm-associated *T. Mediterraneus* resulted in significantly higher body condition and significantly different fatty acid composition in their tissues compared to control fish that fed on natural diets. The higher body condition of farm-associated fish may translate to greater production of reproductive products and ultimately lead to improved spawning success, if egg quality is not adversely affected by the modified fatty acid composition of the fish (Fernandez-Jover et al. 2007b). Similar physiological effects of consumption of large amounts of fish feed appear to occur in saithe aggregated at farms. Their body form and liver size are markedly different to their wild counterparts caught distant from farms (Skog et al. 2003; Dempster et al. unpublished data, Fig. 3.6).

Perhaps the greatest likely impact of aggregation of wild fish at farms is the possible modification of natural population mortality rates through either greater exposure to predators that also aggregate around farms and/or increased susceptibility to fishing. Currently, little information exists on the level of fishing pressure on wild fish when they are aggregated around farms, although targeting of fish around farms by commercial and recreational fishers has been observed frequently in the Mediterranean Sea and appears to be increasing in intensity (Valle et al. 2007). Modified levels of parasites and disease in wild fish may be a further potential impact of such dense, temporally persistent aggregations present in close proximity to large biomasses of caged fish (Dempster et al. 2002).

Fig. 3.6 Marked difference in morphology between saithe (*Pollachius virens*) of similar length sampled from a control location (upper fish, 5 kg) and in the vicinity of a fiord-based salmon farm (bottom fish, 6.5 kg) in Norway

3.5.5 Wild Fish as "Natural Bio-filters" Beneath Sea-cage Fish Farms

Food originating from fish farms is available to wild fish in two forms: as large food pellets lost through the cage and as a "soup" of particulate organic matter (POM) of broken pellets and faeces from caged fish. Through consumption of the food available around farms, high abundances of wild fish may greatly influence the dynamics of nutrient flows. Two caging experiments that excluded fish from beneath farms showed that wild fish consumed a large proportion of the total sedimenting nutrients (Vita et al. 2004: 80%, Felsing et al. 2005: 40–60%). The extent to which waste food pellets and POM derived from a farm are consumed will depend largely on the biomass of wild fish around cages and the species composition of the assemblages (Dempster et al. 2005).

Wild fish may assimilate nutrients lost through the cages and disperse particles and nutrients that originate at farms. *Mugil cephalus* kept in experimental enclosures on the bottom reduced the impact of sea-cages by mixing, oxygenating and resuspending sediments and enhancing effluent dispersal (Katz et al. 2002). The abundant large Chondrichthyid rays beneath farms at the Canary Islands may play a similar role (Dempster et al. 2005). To harness the ability of wild fish to act as assimilators of wild feed and reduce the benthic impact of fish farms, Dempster et al. (2005) suggested that the large aggregations of planktivorous and demersal fish around farms could be protected from fishing.

3.5.6 Managing the Interactions of Aquaculture Sites and Fishing

Management of fisheries concerning species that interact with aquaculture sites will greatly affect the outcomes for wild fish populations. Allowing targeting of over-exploited fish stocks around fish farms in areas where regulation exists through gear or fishing time restrictions alone may add to overfishing. Such overfishing will not appear in estimates of catch per unit effort, since the attractive nature of farms to wild fish ensures that more fish will arrive to replace those caught. In this case, fish farms will acts as "ecological traps", as they have the ecological cues that fish recognise in preferred habitats, yet their association with the farms diminishes their rate of survival (Battin 2004). The Mediterranean Sea may be a good example of where this could occur, as many of the species associated with farms are currently assessed as fully exploited or over-exploited.

Where potential overfishing presents a problem, the best approach may be to protect aggregations around farms. Sea-cage fish farms are incompatible with MPAs designed to protect biodiversity, as assemblages shift away from those naturally observed. However, they may be compatible with the aims of MPAs designed to enhance fisheries (Dempster et al. 2006) as they concentrate large numbers of a variety of pelagic and demersal fish species which form a portion of the spawning stock (Dempster et al. 2002; Boyra et al. 2004; Dempster et al. 2005; Tuya et al. 2005; Tuya et al. 2006). Sea-cage fish farms should thus be designated as "no-fishing zones" and incorporated into the management of coastal areas along with MPA zones to protect biodiversity. Small fishing exclusion zones of hundreds of metres around aquaculture sites exist in some countries (e.g., south-eastern Spain, Norway), principally to avoid damage to fish farm equipment through boat strike or fishing gear entanglement. However, most European nations have no such restrictions at present. The effectiveness of such small exclusion zones (less than 5 hectares) in protecting wild fish has never been tested; detailed individual-based information on fish movements is required to assist effective management in this regard.

Allowing targeting of particular fish species that are not over-exploited in areas where fisheries management limits the overall amount of catch (such as through total allowable catch (TAC) systems), may mean that aggregations at aquaculture sites and fishing can interact without increasing the overall catch. In this scenario, the fishery may benefit economically from the interaction through spending less time searching for fish (thereby using less energy), less time fishing (thereby increasing efficiency) and possibly greater consistency in catch levels (thereby increasing profit). However, the level of by-catch of such fisheries and the quality of their catch must be examined further.

A separate important interaction of aquaculture with wild fisheries concerns the "ranching" of bluefin tuna (*Thunnus thynnus*) throughout the Mediterranean Sea (Fig. 3.7). Ranching remains dependent upon fattening fish caught from the wild. Tuna are fished from natural stocks by purse seine, transferred to sea-cages and then fed with whole fish (e.g., mackerel) for up to six months to optimize their fat content for the Japanese market (Gimenez and Sánchez-Jerez 2006). This practice affects tuna populations as adult spawning stocks are targeted for capture, which may add to overfishing of Atlantic bluefin stocks. Several authors have argued for

Fig. 3.7 A bluefin tuna (*Thunnus thynnus*) of approximately 250 kg harvested from a sea-cage ranch off the Spanish Mediterranean coast (left) and approximately 25 tons of whole wild fish on the deck of a fish farm boat on route to feed the caged tuna (right)

either abolition of tuna farming or stronger management measures to better control the practice (IUCN 2004; Volpe 2005; Borg and Schembri 2006).

3.6 Competition Between Aquaculture and Biodiversity Protection

The impacts of marine aquaculture on biodiversity are rarely positive (Beveridge et al. 1994), and as such, installations have been described as competing for space with future potential MPAs (www.wwf.org). Declaration of marine protected areas (MPAs) in European coastal waters to preserve biodiversity will likely preclude aquaculture activities that modify biodiversity. While the use of MPAs for coastal management in Europe has so far been limited and most are relatively small, the declaration of MPAs over extensive coastal areas elsewhere has restricted availability of sites for aquaculture. For example, the zoning of the Great Barrier Reef Marine Park in Australia in 2004 for biodiversity protection included approximately 40% of the area of the park within protection zones that exclude the establishment of aquaculture facilities (Fernandes et al. 2005).

For coastal areas where biodiversity is not or will not be protected through area restrictions on aquaculture and other activities through the presence of MPAs, which would concern the vast majority of European coastal space, constraints on the level of biodiversity modification that aquaculture may cause to natural systems remain important. EC Directives relevant to marine aquaculture implicate the integration of aquaculture management within the overall management strategy of the coastal zone, through Integrated Coastal Zone Management (ICZM) and requires Environmental Impact Assessments in licensing procedures for aquaculture developments (Fernandes and Read 2001). The EC Water Framework and the Species and Habitat Directives are the most important protocols for maintaining the integrity of the marine ecosystem structure in relation to marine aquaculture (Read and Fernandes 2003).

The Water Framework Directive places emphasis on ecological status, which is defined as the quality of the structure and functioning of aquatic ecosystems associated with surface waters. Ecosystem quality will be maintained through the control of water contamination through human activities such as aquaculture. In terms of "biological" contaminants, such as genetically modified or selected individuals from farmed stocks, the Water Framework Directive and the Habitat Directive aim to protect existing levels of genetic variability and diversity in natural populations.

The Species and Habitat Directive (92/43/EEC) promotes the protection of habitats and species with a holistic approach, that concerns the integrity of ecosystem characteristics and the protection of natural biodiversity. Hundreds of "Special Areas of Conservation" (SAC) have been defined to protect European marine biodiversity. A single management plan for each site is necessary and negative interactions of aquaculture within these areas should be avoided. An important concept common to both EU Directives is the consideration of assimilative capacities of water bodies. Carrying capacities of ecosystems should be modelled to estimate the acceptable limit for aquaculture development near protected sites (see Jiang and Gibbs 2005) for an example of such a model). Different international conventions (Convention for the Protection of the Marine Environment of the North East Atlantic, the Helsinki Convention for the Protection of the Marine Environment of the Baltic Sea Area and the Barcelona Convention for the Protection of the Mediterranean Sea against Pollution) propose good environmental management practice to limit pollution and protect biodiversity (Davies 2001; Read and Fernandes 2003).

A precautionary approach to aquaculture in coastal areas should be considered because of the risk of reducing biodiversity due to nutrient loading and elevated levels of organic matter in bottom sediments and in the water column. This involves standardising indicators of change and setting limits of acceptable modification to environmental parameters. For biodiversity conservation of marine ecosystems, a set of environmental quality standards should be set at the European level for the various directives, and then adapted for regional environmental conditions. These environmental quality standards will be a set of measurable parameters to detect environmental impact and biodiversity change. In Norway, environmental quality criteria for fiords and coastal waters were established in 1997 (Molvær et al. 1997; NSF 1998). These criteria are presented within a classification system for impacts of nutrients, organic matter, micropollutants, and fecal bacteria, and established water quality standards for various coastal uses. At present, the ECASA project (Ecosystem Approach for Sustainable Aquaculture, www.ecasa.org.uk), is attempting to identify, assess and develop indicators of the impact of aquaculture on a European-wide basis.

Monitoring programmes are necessary to ensure effective regulation and promote adaptive management of aquaculture in coastal areas (Carroll et al. 2001). At least seasonally, monitoring of water and sediment conditions should be routinely carried out by fish farms to ensure compliance with the Environmental Quality Standards. Monitoring programmes are often conducted by farmers. Where this is the case, auditing to determine the quality of self monitoring is required, particularly with regard to the use of regulated substances (antibiotics, disinfectants).

Effluents from fish farms can have undesirable impacts on local marine communities; these would vary depending on the quantity and composition of substances released, the temporal scale over which the release takes place, the assimilation capacity of the water mass and the sensitivity of the communities. The spatial distribution of fish farm installations can have substantially different effects on marine biodiversity according to habitat type. Less complex habitats, such as seabeds dominated by soft sediments, are well known to be affected by fish farming. Farming of salmonids, sea bass and sea bream produces anoxic conditions due to the increased load of organic matter, which produces hypoxia and facilitates the growth of specialized macrofauna which are tolerant to organic enrichment (Carroll et al. 2001; Wildish and Pohle 2005) and the proliferation of benthic microalgae on the seabed due to the benthic flux of nutrients (Karakassis et al. 1999). Changes to the macrofauna are marked, and include increases to the abundance of opportunistic species, such as *Capitella* cf *capitata*. Western Mediterranean fish farms, located in open areas, reduce the number of families of macrofauna and diversity compared with control areas (Maldonado 2005). However, the spatial extent of these impacts is limited. For Mediterranean fish farms, Karakassis et al. (2000) found a consistent spatial pattern where the benthic community approaches its normal characteristics at 25 m from the core of the fish farm. Wildish and Pohle (2005) reviewed a range of studies and found that most effects on the benthos were local or footprint-limited (0.05–0.5 km^2), even though fish farm wastes may spread over a greater range from the farm (e.g., up to 1 km: Sara et al. 2003, 2006).

Regions where seagrass meadows are present are more susceptible to significant changes in biodiversity than regions where sandy habitats prevail. Several studies around the Mediterranean Sea show that fish farms affect seagrass meadows, modifying habitat structure (content of organic matter on sediments) in the surrounding meadows at a scale of hundreds of meters (Ruiz et al. 2001; Marba et al. 2006). Changes to shoot morphology, shoot density, biomass, rhizome growth, nutrient and soluble sugar concentrations are possible impacts of fish farm activities near seagrass meadows (e.g., Dimech et al. 2002). Even after several years of cessation of the impact, the decline of seagrasses continues (Delgado et al. 1999). The results of a study on vertical growth of *Posidonia oceanica* suggest that these effects begin soon after the initiation of farming activities, hence suggesting a low resistance of seagrass meadows to fish farm impacts (Marba et al. 2006).

Fish farm activities may also impact other types of seagrass meadows, as has been recorded from *Cymodocea nodosa* meadows in the Canary Islands (Tuya et al. 2005), which can lead to a cascading effect on seagrass-associated fauna. *Cymodocea nodosa* seagrass meadows throughout the Canary Islands have been degraded by fish farming. Some fish species are strongly associated to this meadow, such as *Diplodus annularis*, *Spondyliosoma cantharus* or *Mullus surmuletus* (Tuya et al. 2005). In combination with the strong fishing pressure that exists in the Canary Islands, degradation of *C. nodosa* meadows may accelerate the reduction of these fish populations, increasing the problem of overfishing and stock depletion.

To avoid impacting seagrass meadows, aquaculture facilities are deployed in deeper waters and are recommended to be sited a minimum of 800 m from *Posidonia* beds in the Mediterranean Sea (EU project MEDVEG: Effects of nutrient release from Mediterranean fish farms on benthic vegetation in coastal ecosystems; www.medveg.dk). A problem generated by shifting the spatial arrangement of farms in the coastal zone is that other important biotic communities can be affected if management does not account for them. For example, *mäerl* beds occur worldwide and are formed by an accumulation of unattached calcareous red algae, growing in a superficial living layer on sediments within the photic zone (Fig. 3.8, Barberá et al. 2003). It is, as for *Posidonia* beds, a protected habitat under European legislation. Decreases in water quality affect the survival of *mäerl* beds; consequently, locating aquaculture facilities on *mäerl* grounds also entails negative consequences for marine biodiversity.

Fig. 3.8 A deepwater *mäerl* bed off Columbretes Island off the south-eastern coast of Spain showing the calcareous algae which dominate the benthic assemblage and a starfish *Equinaster sepositus* (left) and a view of the *mäerl* bed from 5 m above the bottom showing the banding pattern of sand and calcareous algae (right)

When fish farms are located relatively near the coast, the dissolved organic matter can affect intertidal communities. In the Canary Islands, *Caulerpa racemosa* and *Corallina elongata* were observed at fish-farm impacted locations at higher coverage rates than control locations (Boyra et al. 2004). *Caulerpa racemosa* is a weedy species that exhibits fast growth, with high dispersion and a broad tolerance to physiological conditions (Piazzi et al. 2005). Moderate nutrient increments might favour the development of *C. elongata* (Diez et al. 1999), as this calcareous red algae has been implicated as a pollution tolerant species, being associated with several types of environmental stress. The high level of organic matter input, caused by the waste products of fish farming activities, have been known to encourage the development of filter-feeding and detritivorous animals (Brown et al. 1990). The replacement of algae by filter-feeding animals can be considered as an indication of severe ecological disturbance (Diez et al. 1999). Boyra et al. (2004) found a significant increase of invertebrates such as *Anemonia sulcata* at farm-impacted areas. The presence of the filter-feeding *A. sulcata*, a sea-anemone that occurs frequently in areas with a high content of organic matter in the water, supports the suggestion that fish farming activities cause disturbances to intertidal areas around fish farms.

A variety of chemicals are also used in European marine aquaculture, including disinfectants, antifoulants and veterinary medicines (Costello et al. 2001; Read and Fernandes 2003). Some of the drug-impregnated food is ingested by scavengers, and may diffuse into the water column or become incorporated into sediments. The impacts of anti-microbial compounds can be summarised as effects on non-target organisms, effects on sediment chemistry and processes, and the development of resistance (Beveridge et al. 1997). The use of formaldehyde for the treatment of ectoparasites can have deleterious effects on biota. During summer, formaldehyde can be used frequently for fish bathing treatments and this may affect both pelagic and benthic communities.

3.7 Solutions for Development of Aquaculture in the Coastal Zone

3.7.1 *Effective Integration with Other Coastal Users*

Effective "ecological" integration of aquaculture with other users of the coastal area requires development and implementation of adequate regulatory systems, including regulation of nutrient loading and adverse impacts on mobile and non-mobile flora and fauna in the vicinity of farms. Stead et al. (2002) argue for more effective Integrated Coastal Zone Management (ICZM) and better use of new technologies, such as geographical information systems (GIS; see section 3.2), by coastal planners and managers to achieve this. Integration into the social and economic aspects of use of the coastal zone may require a different approach; one that must be industry-led. Staresinic and Popović (2004) argue that aquaculture and tourism have solid collaborative potential in certain areas, in particular regarding the consumption of seafood by tourists. Strong leadership from within the industries that use the coastal zone is required to drive open collaboration on the use of space and environmental issues between the two sectors.

Political decisions to create space for aquaculture in areas with suitable environmental characteristics may be the only route to allocate space in areas where user conflicts are too extensive to allow new sites for aquaculture. As an example, we can present an initiative undertaken in the Murcia region of Spain, where the local authority decided to produce a regional plan for aquaculture development. After consulting with all involved sectors (tourism, environment, agriculture, navy, transport) and asking them to produce spatial information of their coastal uses and to indicate where aquaculture could subsequently develop, no unused portion of coastal space was identified. Thereafter, the planning process stalled at this stage for 2 years. Finally, new aquaculture farms were concentrated in an offshore area and only after considerable political support for a national aquaculture development plan (IUCN 2004).

3.7.2 *Move Offshore or Submerge*

Offshore aquaculture has been touted as a solution to both increasing the production of seafood and reducing the need for positioning aquaculture in coastal waters (e.g., Marra 2005). While competition for offshore space is far reduced, the costs of offshore aquaculture, both human and economic, are likely to be higher than inshore operations and thus only very large farms may be feasible and competitive (Ryan 2004). Until several technological and operational advances are made and offshore aquaculture can compete economically with current inshore operations, in the medium term (next 10–20 years), it is unlikely that a major proportion of the aquaculture industry will move to truly offshore locations.

Fig. 3.9 Schematic diagram of the open ocean aquaculture demonstration site off New Hampshire, USA (left) and a submersible cage being put in position at the same demonstration site (right) (Reproduced by permission of the University of New Hampshire Open Ocean Aquaculture Project (www.ooa.unh.edu))

The use of submersible sea-cages and shellfish installations may reduce conflicts with other uses of coastal space. Removal of surface structures eliminates the aesthetic impact of aquaculture on coastal seascapes and the source of conflict with coastal populations, while also reducing the extent to which they act as navigational hazards. Submersible structures avoid the strong physical forcing at the ocean surface caused by storms as most surface wave energy (95%) dissipates within the first 10 m in the open sea. Thus, they may allow use of a range of offshore sites distant from the coast and could also reduce the number of escapes of cultured fish, which are principally due to storm damage (Naylor et al. 2005). At present, surface cage technologies are cheap and dominate the marketplace. Submersible or semi-submersible cages are currently used for the culture of sea bream (*Sparus auratus*) in Italy (Refa-med leg tension cages), Pacific threadfin (*Polydactylus sexfilis*) in Hawaii (SeaSpar cages) and cod (*Gadus morhua*) off New Hampshire (in Sea Station cages, Fig. 3.9) (Ryan 2004). Widespread adoption of submerged cage technology by industry, however, will require solutions to several technological and operational obstacles. Further, it must be rigorously demonstrated that their use does not diminish growth rates, food conversion ratios or the welfare of the cultured fish in comparison to standard surface systems.

3.8 GIS and Methodology for Integrated Coastal Zone Management

While development and implementation of ICZM policies is now an established concept, the tools and methodologies for achieving such goals are still under development (Henocque 2003). It is clear, however, that for coastal management to be effective, policies must be based on informed decisions. This, in turn, requires ready access to appropriate, reliable and up to date data and information in a suitable form. Since much of this data is likely to have a spatial component,

Wright and Bartlett (2000) suggested that geographical information systems (GIS) are relevant to this task, and have the potential to contribute to coastal management in a number of ways. In the case of aquaculture, the use of GIS not only provides a visual inventory of the physical, biological and economical characteristics of the environment, it also allows rational management without complex and time-consuming manipulations. Despite this, the use of GIS to integrate aquaculture into coastal space has been modest.

The Economics and Social Committee of the EU (2001/C155/05) recommend that development of ICZM should integrate long-term changes, be an interactive and dynamic process, and incorporate all factors to facilitate development planning. Monitoring should be done concomitantly with information transfer, and facilitated by technologies such as remote sensing and GIS. Aquaculture should be incorporated into ICZM at a European level though the Common Fisheries Policy, where different activities such as fishing and aquaculture can be integrated for sustainable development.

Geographical Information Systems are excellent tools for both monitoring and management applications. GIS allows organisation of the existing users and interactions in the coastal zone and can help integrate the development of future activities in relation to the existing users, thereby reducing competition for space and potentially limiting environmental impacts. GIS can be used to relate the spatial variability of oceanographic and ecological processes to recognise spatial patterns along a determined area. To model the particulate waste distribution around aquaculture facilities, GIS can be used at a single location or a regional scale. For example, Hassen and Prou (2001) used GIS procedures to assess nutrient loading related to aquaculture activities along the Atlantic coast of France. Modelling of input and distribution of wastes and discharges is a cost-effective tool that can assist in predicting impacts and thereby aid decision-makers. Particulate waste distribution models can be developed to predict the total particulate organic carbon lost from a fish farm as uneaten food and faecal material by mass balance and can also estimate the distribution of particles (Gowen et al. 1989; Perez et al. 2002). Prediction of the distribution of carbon on sediments using GIS reflected real sediment characteristics for farmed Atlantic salmon using GIS combined with a spreadsheet (Pérez et al. 2002). Such models can be applied to Environmental Impact Assessments (EIA), designing monitoring programmes, site selection, continuing farm management and development of future scenarios (GESAMP 1996; Pérez et al. 2002).

GIS systems can also organize and present spatial data in a way that allows effective environmental management planning. For example, regulatory agencies should decide the location of aquaculture facilities in coastal space, with detailed knowledge of biophysical and socio-economic characteristics, to best integrate aquaculture among other users. ICZM should be based on GIS to deal with the complexity of interactions and the enormous quantity of data involved. Sources of necessary data are extremely diverse, and include remote sensing data, field measurements, meteorological data, and socio-economic parameters. Examples of siting aquaculture based on decisions made using GIS exist for areas throughout Europe and elsewhere, including the Marennes-Oléron Bay region in France (Goulletquer and Le Moine 2002), the Balk Sea (Guneroglu et al. 2005), the Canary Islands (Pérez et al. 2005), the Moroccan coast (Arid et al. 2005), and Scotland (Ross et al. 1993; Nath et al. 2000).

Nath et al. (2000) reviewed existing case studies of the application of GIS for spatial decision support in aquaculture. Basic steps for a GIS study comprise: (1) identifying the project requirements, (2) formulation specifications, (3) developing the analytical framework, (4) locating data sources, (5) organizing and manipulating data for input, (6) analysing data, and (7) verifying outcomes and evaluating outputs. Once an activity has been modelled and quantified, it will invariably have some potential to conflict with other users of the space or resource. This calls for trade-off decisions to be made so that different activities can coexist. These decisions typically require consideration of economic, environmental and social ramifications of alternative space use practices. The "layers" of information taken into account for selection of sites for potential aquaculture developments include environmental data (water currents, habitats distribution, bathymetry, coastline, primary production), restricted areas (marine protected areas, important areas for the protection of species, sewage outfalls, navigation, ports) and potential user competition (fisheries grounds, leisure zones, other fish and shellfish farms). Some GIS packages have included decision support tools, for example, multi-objective land allocation (MOLA) and multi-dimensional decision space (MDCHOICE) tools in IDRISI software (Nath et al. 2000).

3.9 Conclusions

Integration of aquaculture into coastal space entails both siting installations in physical space in relation to the existing network of coastal users, such as shipping, fishing, recreational activities and other industry, and ensuring that the extent of aquaculture does not lead to widespread changes to coastal ecosystems. As aquaculture is a recent entrant into the competition for coastal space in many European countries, successful integration into the social and economic aspects of coastal regions will require management strategies that enable coexistence of users. Where the competition for space is particularly intense, political decisions, which simultaneously seek to minimize both environmental impacts and user conflict, may be the only mechanism to allocate space to new aquaculture installations. From an ecological perspective, better integration of aquaculture into European coastal space so that ecological carrying capacities are not exceeded requires knowledge-based management of the interaction of ecological impacts of aquaculture with those of other coastal users, particularly concerning nutrient loading, and modification to biodiversity and species that are important to fisheries. Geographical information systems (GIS) are proven tools for natural resource management and space planning and should be used extensively for planning aquaculture's integration into European coastal areas.

References

Arid H, Moudni H, Orbi A, Talbaoui M, Idrissi JL, Massik Z, Masski H, Littaye A, Paoli R (2005) Remote sensing and GIS for integrated management of aquaculture potentiality. European Space Agency (Special Publication), ESA SP (584): 121–129

Barbera C, Bordehore C, Borg JA, Glemarec M, Grall J, Hall-Spencer JM, De La Huz CH, Lanfranco E, Lastra M, Moore PG, Mora, J, Pita ME, Ramos-Espla AA, Rizzo M (2003) Conservation and management of northeast Atlantic and Mediterranean maerl beds. Aquatic Conservation: Marine and Freshwater Ecosystems 13(1): 65–76

Battin J (2004) When good animals love bad habitats: ecological traps and the conservation of animal populations. Conservation Biology 18(6): 1482–1491

Beveridge MCM, Ross LG, Kelly LA (1994) Aquaculture and biodiversity. Ambio 23(8): 497–502

Beveridge MCM, Ross LG, Stewart JA (1997) The development of mariculture and its implications for biodiversity. In: Omond RFG, Cage J, Angel M (eds.), Marine biodiversity: patterns and processes. Cambridge University Press, Cambridge, pp. 372–393

Bjordal Å, Skar AB (1992) Tagging of saithe (*Pollachius virens*) at a Norwegian fish farm: preliminary results on migration. ICES Council Meeting Paper 1992/G:35

Bjørn PA, Finstad B, Kristoffersen R (2001) Salmon lice infection of wild sea trout and Arctic char in marine and freshwaters: the effects of salmon farms. Aquaculture Research 32(12): 947–962

Bjørn PA, Sæther BS, Dale T, Michaelsen K, Svåsand T (2005) Behavioural responses in wild coastal cod exposed to salmon farms: possible effects of salmon holding water. A field and experimental study. Final report to the Fisheries and Aquaculture Research Fund. Fiskeriforskning, Tromsø, 26 pp (In Norwegian)

Black KD (2001) Environmental Impacts of Aquaculture. Sheffield Academic Press, Sheffield, UK, 214 pp

Borg JA, Schembri PJ (2006) Environmental aspects of tuna farming in Europe. Environmental Status of Marine Aquaculture in Europe Workshop, Menorca, Spain, 2006, Extended abstract, 4 pp

Boyra A, Sanchez-Jerez P, Tuya F, Espino F, Haroun R (2004) Attraction of wild coastal fishes to Atlantic subtropical cage fish farms, Gran Canaria, Canary Islands. Environmental Biology of Fishes 70(4): 393–401

Brehmer P, Gerlotto F, Guillard J, Sanguinède F, Guénnegan Y, Buestel D (2003) New applications of hydroacoustic methods for monitoring shallow water aquatic ecosystems: the case of mussel culture grounds. Aquatic Living Resources 16: 333–338

Brown VB, Davies SA, Synnot RN (1990) Long-term monitoring of effects of treated sewage effluent on the intertidal macroalgal community near Cape Schanck, Victoria, Australia. Botanica Marina 33: 85–98

Carroll ML, Cochrane S, Fieler R, Velvin R, White P (2001) Organic enrichment of sediments from salmon farming in Norway: Environmental factors, management practices, and monitoring techniques. Aquaculture 226(1–4): 165–180

Carss DN (1990) Concentrations of wild and escaped fishes immediately adjacent to fish farm cages. Aquaculture 90: 29–40

Chen YS, Beveridge MCM, Telfer TC (1999) Settling rate characteristics and nutrient content of the faeces of Atlantic salmon, *Salmo salar* L., and the implications for modeling of solid waste dispersion. Aquaculture Research 30(5): 395–398

Conides A, Papaconstantinou C (2001) The basis of coastal fishery management for small fishing communities: the case of the shrimp (*Penaeus kerathurus*) fishery in Western Greece. Naše More 48(5–6): 231–236

Costa-Pierce BA (2002) Ecological Aquaculture. Blackwell, Oxford

Costanzo SD, O'Donohue MJ, Dennison WC, Loneragan NR, Thomas M (2001) A new approach for detecting and mapping sewage impacts. Marine Pollution Bulletin 42: 149–156

Costello MJ, Grant A, Davies IM, Cecchini S, Papoutsoglou S, Quigley D, Saroglia M (2001) The control of chemicals used in aquaculture in Europe. Journal of Applied Ichthyology 17(4): 173–180

Cromey CJ, Nickell TD, Black KD, Provost PG, Griffiths CR (2002) Validation of a fish farm waste resuspension model by use of a particulate tracer discharged from a point source in a coastal environment. Estuaries 25: 916–929

Davies IM (2001) The influence of international conventions for the prevention of marine pollution on mariculture in European countries. In: Read PA, Fernandez, TF, Miller, KL, Eleftheriou. A., Davis, IM, Rodger, GK (eds.), The implications of Directives, Conventions and Codes of Practice on the Monitoring and Regulation of Marine Aquaculture in Europe. Proceedings of the Second MARAQUA Workshop, 20–22 March 2000, Institute of Marine Biology, Crete. Scottish Executive, Aberdeen, UK

Davies T, Cahill S (2000) Environmental Implications of the Tourism Industry. Resources for the Future Ed. Washington, 14 pp

Davis WJ (1993) Contamination of coastal versus open ocean surface waters. A brief meta-analysis. Marine Pollution Bulletin 26(3): 128–134

deBruyn AMH, Trudel M, Eyding N, Harding J, McNally H, Mountain R, Orr C, Verenitch S, Mazunder A (2006) Ecosystemic effects of salmon farming increase mercury contamination in wild fish. Environmental Science & Technology 40: 3489–3493

Delgado O, Ruiz J, Pérez M, Romero J, Ballesteros E (1999) Effects of fish farming on seagrass (*Posidonia oceanica*) in a Mediterranean bay: Seagrass decline after organic loading cessation. Oceanologica Acta 22(1): 109–117

Dempster T, Sanchez-Jerez P, Bayle-Sempere JT, Giménez-Casalduero F, Valle C (2002) Attraction of wild fish to sea-cage fish farms in the south-western Mediterranean Sea: spatial and short-term temporal variability. Marine Ecology Progress Series 242: 237–252

Dempster T, Sanchez-Jerez P, Bayle-Sempere JT, Kingsford M (2004) Extensive aggregations of wild fish at coastal sea-cage fish farms. Hydrobiologia 525(1–3): 245–248

Dempster T, Fernandez-Jover D, Sanchez-Jerez P, Tuya F, Bayle-Sempere J, Boyra A, Haroun RJ (2005) Vertical variability of wild fish assemblages around sea-cage fish farms: implications for management. Marine Ecology Progress Series 304: 15–29

Dempster T, Sanchez-Jerez P, Tuya F, Fernandez-Jover D, Bayle-Sempere J, Boyra A, Haroun RJ (2006) Coastal aquaculture and conservation can work together. Marine Ecology Progress Series 314: 309–310

Diez I, Secilla A, Santolaria A, Gorostiaga JM (1999) Phytobenthic intertidal community structure along an environmental pollution gradient. Marine Pollution Bulletin 38(6): 463–472

Dimech M, Borg JA, Schembri PJ (2002) Changes in the structure of a *Posidonia oceanica* meadow and in the diversity of associated decapods, molluscs and echinoderm assemblages resulting from inputs of waste from a marine fish farm (Malta, Central Mediterranean). Bulletin of Marine Science 71(3): 1309–1321

Dimitriou E, Katselis G, Moutopoulos DK, Akovitiotis C, Koutsikopoulos C (2004) Possible influence of reared gifthead sea bream (Sparus aurata, L.) on wild stocks in the area of the Messolonghi lagoon (lorian Sea, Greece). Aquaculture Research 38:398–408

Ervik A, Aure J (2006) How much fish aquaculture do we want in Norway? Fisken og havet. 2006(2): 97–101 (In Norwegian)

FAO (2006) Fishstat Plus. Aquaculture production: quantities 1950–2004. FAO, Rome

Felsing M, Glencross B, Telfer T (2005) Preliminary study on the effects of exclusion of wild fauna from aquaculture cages in a shallow marine environment. Aquaculture 243: 159–174

Fernandes L, Day J, Lewis A, 25 others (2005) Establishing representative no-take areas in the Great Barrier Reef: large-scale implementation of theory on marine protected areas. Conservation Biology 19(6): 1733–1744

Fernandes TF, Read PA (2001) Aquaculture and management of coastal zones. In: Read PA, Fernandez TF, Miller KL, Eleftheriou A, Davis IM, Rodger GK (eds.), The implications of Directives, Conventions and Codes of Practice on the Monitoring and Regulation of Marine Aquaculture in Europe. Proceedings of the Second MARAQUA Workshop, 20–22 March 2000, Institute of Marine Biology, Crete. Scottish Executive, Aberdeen, UK

Fernandez-Jover D, Sanchez-Jerez P, Bayle-Sempere J, Carratala A, Leon VM (2007a) Addition of dissolved nitrogen and dissolved organic carbon from wild fish faeces and food around Mediterranean fish farms: implications for waste-dispersal models. Journal of Experimental Marine Biology and Ecology 340(2): 160–168

Fernandez-Jover D, Lopez-Jiminez JA, Sanchez-Jerez P, Bayle-Sempere J, Giminez-Casalduero F, Martinez-Lopez FJ, Dempster T (2007b) Changes in body condition and fatty acid composition of wild Mediterranean horse mackerel (*Trachurus mediterraneus*, Steindachner, 1868) associated with sea cage fish farms. Marine Environmental Research 63(1): 1–18

Garza-Gil MD, Prada-Blanco A, Vazquez-Rodriguez MX (2005) Estimating the short-term economic damages from the *Prestige* oil spill in the Galician fisheries and tourism. Ecological Economics 58: 842–849

GESAMP (1996) The contribution of science to integrated coastal management. *Reports and Studies GESAMP*. 61, 66 pp

Gimenez-Casalduero F, Sanchez-Jerez P (2006) Fattening rate of bluefin tuna *Thunnus thynnus* in two Mediterranean fish farms. Cybium 30(1): 51–56

Gilbert P, Landsberg, JH, Evans JJ, Al-Sarawi MA, Faraj M, Al-Jarallah MA, Haywood A, Ibrahem S, Klesius P, Powell C, Shoemaker C (2002) A fish kill of massive proportion in Kuwait Bay, Arabian Gulf, 2001: the roles of bacterial disease, harmful algae, and eutrophication. Harmful Algae 1: 215–231

Goulletquer P, Le Moine O (2002) Shellfish farming and coastal zone management (CZM) in the Marennes-Oléron Bay and Chatentais Sounds (Charente Maritime, France): a review of recent developments. Aquaculture International 10: 507–525

Gowen RJ, Bradbury NB, Brown JR (1989) The use of simple models in assessing two of the interactions between fish farming and the marine environment. Aquaculture – a biotechnology in progress, Pp. 1071–1080

Guneroglu A, Kose E, Eruz C, Basar E, Erkebay S, Karsli F (2005) Use of geographic information system (GIS) to select fish cage farming sites in Surmene Bay, Black Sea. The Israeli Journal of Aquaculture. Bamidgeh 57(2): 81–89

Hassen MB, Prou J (2001) A GIS-based assessment of potential aquacultural nonpoint source loading in an Atlantic bay (France). Ecological Applications 11(3): 800–814

Henocque Y (2003) Development of process indicators for coastal zone management assessment in France. Ocean & Coastal Management 46 (3–4): 363–379

Hill BJ (2005) The need for effective disease control in international aquaculture. Developments in Biologicals 121: 3–12

Islam MS (2005) Nitrogen and phosphorus budget in coastal and marine cage aquaculture and impacts of effluent loading on ecosystem: Review and analysis towards model development. Marine Pollution Bulletin 50(1): 48–61

IUCN (2004) Mediterranean Aquaculture and the environment. Identification of issues. www.iucn.org. 34 pp

Jiang W, Gibbs MT (2005) Predicting the carrying capacity of bivalve shellfish culture using a steady, linear food web model. Aquaculture 244: 177–185

Jørstad KE, van der Meeren T (2006) Cultured cod 'escape' by spawning in cages. http://www.imr.no/aktuelt/nyhetsarkiv/2006/mai/oppdrettstorsk_rommer_ved_a_gyte

Katz T, Herut B, Genin A, Angel DL (2002) Grey mullets ameliorate organically enriched sediments below a fish farm in the oligotrophic Gulf of Aqaba (Red Sea). Marine Ecology Progress Series 234: 205–214

Krkošek M, Lewis MA, Volpe JP (2005) Transmission dynamics of parasitic sea lice from farm to wild salmon. Proceedings of the Royal Society B: Biological Sciences 272: 689–696

Karakassis I, Hatziyanni E, Tsapakis M, Plaiti W (1999) Benthic recovery following cessation of fish farming: a series of successes and catastrophes. Marine Ecology Progress Series 184: 205–218

Karakassis I, Tsapakis M, Hatziyanni E, Papadopoulou KN, Plaiti W (2000) Impact of cage farming of fish on the seabed in three Mediterranean coastal areas. ICES Journal of Marine Science 57(5): 1462–1471

Karakassis I, Pitta P, Krom M (2005) Contribution of fish farming to the nutrient loading of the Mediterranean. Scientia Marina 69(2): 313–321

Lupatsch I, Kissil GWM (1998) Predicting aquaculture waste from gilthead seabream (*Sparus aurata*) culture using a nutritional approach. Aquatic Living Resources 11(4): 265–268

Maldonado M, Carmona MC, Echeverría Y, Riesgo A (2005) The environmental impact of Mediterranean cage fish farms at semi-exposed locations: Does it need a reassessment? Helgoland Marine Research 59(2): 121–135

Marba N, Santiago R, Diaz-Almela E, Alvarez E, Duarte C (2006) Seagrass (*Posidonia oceanica*) vertical growth as an early indicator of fish farm-derived stress. Estuarine, Coastal and Shelf Science 67: 475–483

Marra J (2005) When will we tame the oceans? Nature 436: 175–176

McGee CD, Loehr LC (2003) An assessment of fecal coliform bacteria in cruise ship wastewater discharge. Oceans Conference Record (IEEE) 2: 733–736

Mendiguchia C, Moreno C, Manuel-Vez MP, Garcia-Vargas M (2006) Preliminary investigation on the enrichment of heavy metals in marine sediments originating from intensive aquaculture effluents. Aquaculture 254: 317–325

Moe H, Gaarder R, Sunde LM, Borthen J, Olafsen K (2005) Escape-free nets for cod. SINTEF Fisheries and Aquaculture Report SFH A 054041, Trondheim, Norway (In Norwegian).

Molvær J, Knutzen J, Magnusson J, Rygg B, Skei J, Sørensen J (1997) Classification of environmental quality in fiords and coastal waters: A guide. Norwegian State Pollution Control Authority (SFT) Publication 97:03, Oslo, Norway, 36 pp

Morton A, Routledge R, Peet C, Ladwig A (2004) Sea lice (*Lepeophtheirus salmonis*) infection rates on juvenile pink (*Oncorhynchus gorbuscha*) and chum (*Oncorhynchus keta*) salmon in the nearshore marine environment of British Columbia. Canadian Journal of Fisheries and Aquatic Sciences 61: 147–157

Nash CE, Iwamoto RN, Mahnken CVW (2000) Aquaculture risk management and marine mammal interactions in the Pacific Northwest. Aquaculture 183: 307–323

Nath SS, Bolte JP, Ross LG (2000) Applications of geographical information systems (GIS) for spatial decision support in aquaculture. Aquacultural Engineering 23: 233–278

Naylor RL, Goldburg RJ, Primavera JH, Kautsky N, Beveridge MCM, Clay J, Folke C, Lubchenco J, Mooney H, Troell M (2000) Effect of aquaculture on world fish supplies. Nature 405: 1017–1024

Naylor R, Hindar K, Fleming IA, Goldburg R, Williams S, Volpe J, Whoriskey F, Eagle J, Kelso D, Mangel M (2005) Fugitive salmon: assessing the risks of escaped fish from net-pen aquaculture. BioScience 55(5): 427–437

Norwegian Fisheries Directorate (2005) Statistics for Aquaculture 2004. http://www.fiskeridir.no/fiskeridir/kystsone_og_havbruk/statistikk

NSF (1998) Water quality guidelines for quantitative investigations of sublittoral soft-bottom benthic fauna in the marine environment. NSF, Oslo, Norway, 16 pp

Page FH, Losier R, McCurdy P, Greenberg D, Chaffey J, Chang B (2005) Dissolved oxygen and salmon cages culture in the southwestern New Brunswick Portion of the Bay of Fundy. In: Hargrave B (ed.), Environmental effects of marine finfish aquaculture,. Springer, pp. 1–28

Pérez OM, Telfer TC, Beveridge MCM, Ross LG (2002) Geographical information system (GIS) as a simple tool to aid modelling of particulate waste distribution at marine fish cages sites. Estuarine, Coastal and Shelf Science 54: 761–768

Pérez OM, Telfer TC, Ross LG (2005) Geographical information systems-based models for offshore floating marine fish cage aquaculture site selection in Tenerife, Canary Islands. Aquaculture Research 36(10): 946–961

Piazzi L, Balata D, Ceccherelli G, Cinelli F (2005) Interactive effect of sedimentation and *Caulerpa racemosa* var. *cylindracea* invasion on macroalgal assemblages in the Mediterranean Sea. Estuarine, Coastal and Shelf Science 64: 467–474

Pitta P, Karakassis I, Tsapakis M, Zivanovic S (1998) Natural vs. mariculture induced variability in nutrients and plankton in the eastern Mediterranean Hydrobiologia 391: 181–194

Read P, Fernandes T (2003) Management of environmental impacts of marine aquaculture in Europe. Aquaculture 226: 139–163

Ross LG, Mendoza QM, Beveridge MCM (1993) The application of geographical information systems to site selection for coastal aquaculture: an example based on salmonid cage culture. Aquaculture 112(2–3): 165–178

Ruiz JM, Pérez M, Romero, J (2001) Effects of fish farm loadings on seagrass (*Posidonia oceanica*) distribution, growth and photosynthesis. Marine Pollution Bulletin 42(9): 749–760

Ryan J (2004) Farming the Deep Blue. Report to the Irish Sea Fisheries Board and the Irish Marine Institute, 67 pp

Saeed MO (1995) Association of *Vibrio harveyi* with mortalities in cultured marine fish in Kuwait. Aquaculture 136(1–2): 21–29

Sara G, Scilipoti D, Mazzola A, Modica A (2003) Effects of fish farming waste to sedimentary and particulate organic matter in a southern Mediterranean area (Gulf of Castellammare, Sicily): a multiple stable isotope study ($\delta^{13}C$ and $\delta^{15}N$). Aquaculture 234: 199–213

Sara G, Scilipoti D, Milazzo M, Modica A (2006) Use of stable isotopes to investigate dispersal of waste from fish farms as a function of hydrodynamics. Marine Ecology Progress Series 313: 261–270

Sather PJ, Ikonomou MG, Haya K (2006) Occurrence of persistent organic pollutants in sediments collected near fish farm sites. Aquaculture 254: 234–247.

Sivertsen A (2006) The national salmon fiords. Fisken og havet 2006(2): 55–57 (In Norwegian)

Skog TE, Hylland K, Torstensen BE, Berntssen MHG (2003) Salmon farming affects the fatty acid composition and taste of wild saithe *Pollachius virens* L. Aquaculture Research 34(12): 999–1007

Smith C, Machias A, Giannoulaki M, Somarakis S, Papadopoulou KN, Karakassis I (2003) Diversity study of wild fish fauna aggregating around fish farm cages by means of remotely operated vehicle (ROV). Abstract, 7th Hel Symposium on Oceanography & Fish, 227 pp

Staresinic N, Popović A (2004) The links between tourism and mariculture: the case of Croatia. Proceedings of the IX World Congress of Rural Sociology, Trondheim, Norway, July 2004

Stead SM, Burnell G, Goulletquer P (2002) Aquaculture and its role in integrated coastal zone management. Aquaculture International 10: 447–468

Stephanou D (1998) Marine aquaculture development and tourism: The case of Cyprus. In: Aquaculture planning in Mediterranean countries. Cahiers Options Méditerranéenes 43: 35–40

Sunde LM, Heide MA, Hagen N, Fredheim A, Forås E, Prestvik Ø (2003) Review on technology in the Norwegian aquaculture industry. SINTEF Fisheries and Aquaculture Report STF 80 A034002, 32 pp

Tacon AJG, Forster IP (2003) Aquafeeds and the environment: policy implications. Aquaculture 226: 181–189

Thetmeyer H, Pavlidis A, Cromey C (2003) Development of monitoring guidelines and modelling tools for environmental effects from Mediterranean aquaculture. Newsletter 3: Interactions between wild and farmed fish, 7 pp. <www.meramed.com>

Tuya F, Boyra A, Sanchez-Jerez P, Haroun R (2005) Non-metric multivariate analysis of the demersal ichthyofauna along soft bottoms of the Eastern Atlantic: comparison between unvegetated substrates, seagrass meadows and sandy bottoms under the influence of sea-cage fish farms. Marine Biology 147: 1229–1237

Tuya F, Sanchez-Jerez P, Dempster T, Boyra A, Haroun RJ (2006) Changes in demersal wild fish aggregations beneath a sea-cage fish farm after the cessation of farming. Journal of Fish Biology 69: 682–697

Valencia JM (2006) Location and management of coastal areas for marine aquaculture in the Balearic Islands. Environmental Status of Marine Aquaculture in Europe Workshop, Menorca, Spain, 2006, Extended abstract, 5 pp

Valle C, Bayle-Sempere JT, Sanchez-Jerez P, Giminez-Casualdero F, Dempster T (2007) Temporal variability of wild fish assemblages associated with a sea-cage fish farm in the south-western Mediterranean Sea. Environmental Biology of Fishes 70(4): 393–401

Vita R, Marín A, Madrid JA, Jiménez-Brinquis B, Cesar A, Marín-Guirao L (2004) Effects of wild fishes on waste exportation from a Mediterranean fish farm. Marine Ecology Progress Series 277: 253–261

Volpe JP (2005) Dollars without sense: the bait for big-money tuna ranching around the world. BioScience 55(4): 301–302

Weir LK, Grant JWA (2005) Effects of aquaculture on wild fish populations: a synthesis of data. Environmental Reviews 13: 145–168

Wildish DJ, Pohle GW (2005) Benthic macrofauna changes resulting form finfish mariculture. In: Hargrave BT (ed.), Handbook of environmental chemistry, Vol. 5, Part M (Environmental effects of marine finfish aquaculture) Springer Berlin Heidelberg, pp. 275–304

Wright D, Bartlett D (2000) Marine and coastal geographical information systems. Taylor & Francis, London, UK, 319 pp

Wright PJ, Galley E, Gibb IM, Neat FC (2006) Fidelity of adult cod to spawning grounds in Scottish waters. Fisheries Research 77: 148–158

World Tourism Organization (2002) Mediterranean countries unite efforts for recovery. http://www.world-tourism.org/market_research/recovery/home

WTTC (2003) Greece, Spain, & Turkey: Travel and tourism, a world of opportunity. Travel and tourism economic research

WWF (2005) On the run – escaped farmed salmon in Norwegian waters. WWF Norway Report 2/2005. 44p http://www.wwf.no/core/pdf/wwf_escaped_farmed_fish_2005.pdf

Chapter 4
Detrimental Genetic Effects of Interactions Between Reared Strains and Wild Populations of Marine and Anadromous Fish and Invertebrate Species

T.F. Cross[1,*], G. Burnell[1], J. Coughlan[1], S. Culloty[1], E. Dillane[1], P. McGinnity[2], and E. Rogan[1]

Abstract Cultured strains of marine and anadromous species reared for aquaculture can be either inadvertently (as in farm escapes) or deliberately (as in stocking/ranching) introduced into the wild, where they may interact with wild conspecifics or other species. This chapter concentrates on the potentially detrimental genetic aspects of these interactions, largely in the context of species cultured in Europe but considering general principles, which have worldwide applicability. Most previous experimental work in the area has involved Atlantic salmon, which has the highest production of any finfish produced in Europe. These investigations have shown generally detrimental results for wild salmon populations, when interactions occur with reared strains. The various European species which might be affected by cultured introgressions (the major aquaculture species) are then considered under several headings: genetic composition of cultured strains compared with wild populations; modes of introduction into the wild; direct and indirect genetic interactions with wild populations/species; consequences of such interactions; establishing the severity of effects of wild/reared interactions with different species, utilising opportunist situations and field experiments; and, methods such as induction of sterility in reared strains to reduce detrimental effects. Relative risks for wild populations of the major aquaculture species are then considered, and general and specific genetic recommendations are presented.

Keywords Conservation of biodiversity, genetic interactions, molecular methods, reared strains, wild populations

---

[1] Department of Zoology, Ecology and Plant Science/Aquaculture and Fisheries Development Centre, University College Cork, Ireland. Tel: + 353–21–4904652; Fax: + 353–21–4904664; E-mail: t.cross@ucc.ie

[2] Marine Institute, Aquaculture and Catchment Management Services, Newport, County Mayo, Ireland

* Correspondence: t.cross@ucc.ie

4.1 Introduction

As noted in earlier chapters, worldwide aquaculture production is increasing rapidly and currently exceeds 45 million tonnes of marine and freshwater animals and aquatic plants (FAO 2007). Since capture fisheries seem to have plateaued at around 95 million tonnes and, as there is an ever-increasing demand for aquatic food, the expansion of aquaculture is likely to continue. The western European aquaculture industry produces a total of 1.6 million tonnes of fishery products a year (FAO 2007). The European Union strongly endorses aquaculture, aiming to reduce reliance on imported fish and shellfish products. Also, aquaculture plays a significant role in creating employment opportunities in rural and coastal communities where job alternatives are often scarce.

New animal species are continuously being assessed in terms of their biological and economic feasibility for culture, but there is a universal tendency to concentrate development on relatively few optimal species (e.g., certain penaeid prawn, oyster, cyprinid and salmonid fish species). The current chapter concentrates on marine and anadromous animal species. The majority of production in this area is from partial or whole life-cycle captive rearing for food, ornamental or medicinal purposes (defined here as "farming"). Although, currently, a relatively smaller proportion of total production comes from stocking/ranching programmes, this source is predicted to increase rapidly in the future. Both of these types of aquaculture can have detrimental genetic effects on wild populations, although, due to the potential for increased interactions with wild animals, stocking or ranching programmes may pose greater risks.

In finfish ranching, the juveniles are usually produced in hatcheries and then released, whereas for most shellfish the seed (eggs or larvae) are initially collected from the wild. The latter may then either undergo a period of intensive cultivation (e.g., scallop and lobster) before being released or are transferred directly to the ongrowing area (e.g., mussels). If there is ownership of the transplanted stock, then according to FAO definitions the activity is categorised as aquaculture (e.g., mussel farming in The Netherlands). However, if there is a socio-economic reason for the activity then it would probably be classed as fisheries management/stock enhancement (e.g., lobster stocking in United Kingdom and Norway). The source of the broodstock and the final destination of the juveniles are obviously important when considering potential genetic effects on conspecific indigenous populations.

Survival of ranched shellfish is relatively low and varies from 2% to 8% in lobsters up to 30% in scallop. For this reason the cost of juveniles becomes a key factor in deciding whether ranching or intensive culture is more economic. At the moment, for example, mussel seed can be taken from the wild. This situation is unlikely to continue and seed scarcity will force the industry to turn to hatcheries. When this happens it will be more profitable to ongrow on longlines, where high survival is guaranteed. However, such projections are somewhat academic regarding the risk of contaminating wild populations with cultured genes. Bivalve molluscs are broadcast spawners and, providing the correct environmental cues of temperature and light are present, they will reproduce whether in cages or on the seabed.

Increased production of cultured strains greatly increases the potential for huge escapes or deliberate introductions into the wild. Added to this, is the trend in marine finfish culture to farm large piscivorous fish (e.g., cod, tunas), to use bigger cages (both for performance and cost reasons) and to site these cages further offshore (so as to rapidly dilute waste and unused food). The accidental break up of one of these cages will result in the release of a very large number of reared fish.

A major, yet rarely addressed, question is: "How detrimental is the present level of culture to natural populations, where ecological, epidemiological or genetic interactions occur in the wild, and additionally, will these problems increase as production expands?"

The generalised genetic issues for marine and anadromous animals are addressed in the current chapter. The majority of previous work on interactions between wild populations and conspecific reared strains has been on Atlantic salmon, *Salmo salar*, an anadromous species where reared production, chiefly from sea cage farming, now approaches 1.5 million tonnes. The findings of these investigations may be summarised as follows:

- Cultured salmon are introduced inadvertently (as in farm escapes) or deliberately (as in stocking/ranching) to the wild, where they can interact genetically with natural populations (of the same or closely related species), either directly (by interbreeding) or indirectly (by ecological competition or disease introduction).
- Cultured strains of salmon usually have lower Darwinian fitness in the wild, compared with natural native populations.
- With direct effects, hybrid progeny of interbreeding may have reduced fitness (resulting in reduced survival and overall productivity), whereas indirect effects may drastically reduce the size of natural populations, exponentially increasing genetic drift, and possibly leading to inbreeding depression and to loss of local adaptation, where the latter occurs.
- Although hybridisation of Atlantic salmon with the close congener *Salmo trutta* occurs at low levels in wild populations, the incidence of hybridisation can also increase greatly following reared fish intrusions, leading to inter-specific hybrid progeny of very low reproductive fitness.
- Indirect genetic effects, having an ecological basis, result from the fact that reared salmon are usually better competitors in the short term (faster growing and being more aggressive) than their wild relatives, but survive substantially less well, leading to an overall loss of production per unit area of suitable habitat.
- There are several examples of indirect effects involving diseases:

1. Furunculosis, the bacterial disease caused by *Aeromonas salmonicida*, was accidentally introduced to Norway with farmed smolts and spread to wild salmon, which were naive and highly susceptible to the disease, resulting in high mortalities.

2. The monogenetic trematode, *Gyrodactylus salaris*, was introduced to Norway on infested salmon parr used in a stocking exercise. These parr came from the Baltic area (a different genetic grouping-see below), where salmon are relatively resistant, and this has led to massive mortality and drastic population reductions in many Norwegian rivers.

Here, the likely severity of these effects in other major marine and anadromous fish and invertebrate species, used in both contemporary and emerging aquaculture ventures, is discussed with consideration being given to how differences in genetic composition and life cycle in various species may influence these effects. Consideration is then given to experiments that have investigated the extent of these problems with different species and methods to reduce detrimental genetic effects are discussed. Most examples are of eastern North Atlantic and Mediterranean native species or of species introduced into these areas (e.g., Pacific oyster *Crassostrea gigas*), but since ecological analogues occur in other areas it is felt that the principles discussed will have a wider applicability. Recommendations are then presented which, it is hoped, will be of particular interest to Governmental and regional policy makers and to environmental managers.

4.2 Genetic Composition of Cultured Strains Compared with their Wild Progenitors

4.2.1 Within Population Intra-Specific Comparisons

Reared (cultured) strains* often differ genetically from their wild progenitor populations*, both in levels of genetic variability (usually reduced) and in genetic composition (usually different from progenitors and often temporarily unstable between reared cohorts). These effects have been demonstrated in numerous species over the last three decades, using an array of molecular techniques (see Box 4.1).

Genetic variability is usually expressed as heterozygosity (proportion of heterozygotes at a polymorphic gene locus) or as allelic richness (an estimate of the number of alleles at a specific gene locus). The latter can be a more sensitive indicator of loss of genetic variability and, is thus, more commonly invoked. However, loss of genetic variability in terms of heterozygosity may result in poor growth and performance. The most utilised measure of genetic composition is allele frequency (the proportion of each allele at a specific polymorphic locus in a wild or reared sample). One of the main causes of reduced genetic variability in reared strains is the use of much smaller numbers of parents as broodstock than are common in wild

---

*In this chapter the term "wild population" is used for genetically-distinct statistically-defined sub-specific groupings, usually reproductively isolated from one another. Reared groupings derived from wild populations are referred to as "reared strains".

Box 4.1 Molecular genetic methods

The development of molecular techniques has been likened to the stocking of a toolbox, with ever more powerful tools. The molecular methods used in interaction studies began to be developed in the 1960s, when the allozyme technique was first used. This method focuses on enzymes and other specific proteins, which are the products of functional genes, and as such, is an indirect genetic technique. It involves protein electrophoresis, usually on starch gels, where products are separated on the basis of charge, and has generally been superseded by techniques such as microsatellite analysis (see below), which concentrate directly on genomic DNA. The latter techniques show much more genetic variability (alleles per locus) both because of intrinsic differences such as high mutation rate, but also because proteins are affected by code redundancy and similarity in charge between products of different genetic composition. In certain cases, however, the allozyme technique is still useful, as with brown trout *Salmo trutta* in Spain where most reared strains used in stocking are effectively fixed for a *LDH-C** allele (or the now more commonly used underlying nucleotide sequence (McMeel et al. 2001)) which is almost absent in wild populations, since the former originate from a northern European population grouping. Other allozyme loci such as *MEP-2** in Atlantic salmon *Salmo salar* are influenced by natural selection and thus may have an important role in local adaptation, therefore further investigation of either these enzymes or the genes that code for them, should prove fruitful.

In the 1970s mitochondrial DNA, studied using restriction (specific-cutting) enzymes, was added to the suite of methods. Restriction enzymes could not in general be used with the nuclear genome, because its much greater size resulted in so many fragments that the results were usually ambiguous or uninterpretable. MtDNA, because of its haploid nature and greater propensity to accumulate mutations, offered certain advantages, but it was not exploited in the context of interactions until the polymerase chain reaction (PCR) was developed in the 1980s. The PCR allows the localisation and amplification of a specific segment of DNA, using primers (short segments of DNA which define either end of the target fragment) and a DNA polymerase, to produce millions of new duplicate copies *in vitro*. The PCR can be applied to nuclear or mitochondrial DNA and has enabled sequencing of specific DNA fragments and investigation of microsatellite and SNP variability (see below). Microsatellites have become the "marker of choice" for interaction studies because of their high variability and relatively high frequency throughout the genome (one locus every 10,000 bp). Box 4.1 Figure 1 shows a typical microsatellite locus from Atlantic salmon. Microsatellites, the loci currently used in human forensics, consist of tandemly repeated arrays of two, three or four bases (di-, tri- and tetranucleotides) of largely unknown function (most do not code for

(continued)

Box 4.1 (continued)

proteins), which can be localised and amplified by PCR from tiny tissue samples. At variable (polymorphic) loci, the different alleles vary in repeat number, with mutations increasing or decreasing the number of repeats. Most loci appear to be "neutral" (not affected by natural selection) and so are ideal as population markers. However, a minority of microsatellite loci are tightly linked to functional genes and can be used as markers of these genes in adaptational studies (see Box 4.2 on MHC). Using several polymorphic microsatellites and appropriate statistics, an individual can be assigned to its population or strain of origin. Progeny can also be assigned to parents and thus microsatellites have high utility when dealing with interactions.

A more recently developed marker is the so-called single nucleotide polymorphism (SNP), which usually consists of a point mutation at a given site, commonly with just two alternative bases and thus alleles. SNPs are much more common in the genome than microsatellites (at least one per 1,000 bp) and have the great advantage of transferability between laboratories, so complex and expensive intercalibration is not required. In any species where there has been considerable genome work using several individuals, very large numbers of SNPs (and their exact genome location) will be known. A number of different techniques can be used to identify alleles (nucleotide) and genotypes. The only current problem with SNPs is that there is no single/cheap technique for their detection. Once this is resolved they may supplant microsatellites as the "marker of choice" though it is recognised that it will be necessary to screen larger numbers of loci since most SNP loci are bialleic (whereas microsatellites typically have over 10 alleles). With PCR, it is possible to isolate large quantities of specific DNA fragments for sequencing. While initially very expensive, sequencing has now become a very rapid and cheap process, in the wake of the human and other genome projects. Thus, it may shortly be economical to identify large suites of SNPs for novel species, rapidly.

DNA sequencing (using automated techniques and the di-dedoxy method (Sanger et al. 1977), is also being increasingly used to investigate functional genes (those coding for proteins) usually be isolating mRNA and producing cDNA, using reverse transcriptase. Such functional genes, if polymorphic, will be of great importance in future interactions studies, since captive breeding will often change allele frequencies at these loci, potentially reducing fitness in the wild.

Functional genomics are also starting to be applied to studies of reared strains and wild populations. Using microarray technology to study multiple gene expression, Roberge et al. (2006) have shown that many of the same genes are up- or down regulated in entirely separate reared Atlantic salmon strains compared with native wild populations, in both Norway and Canada.

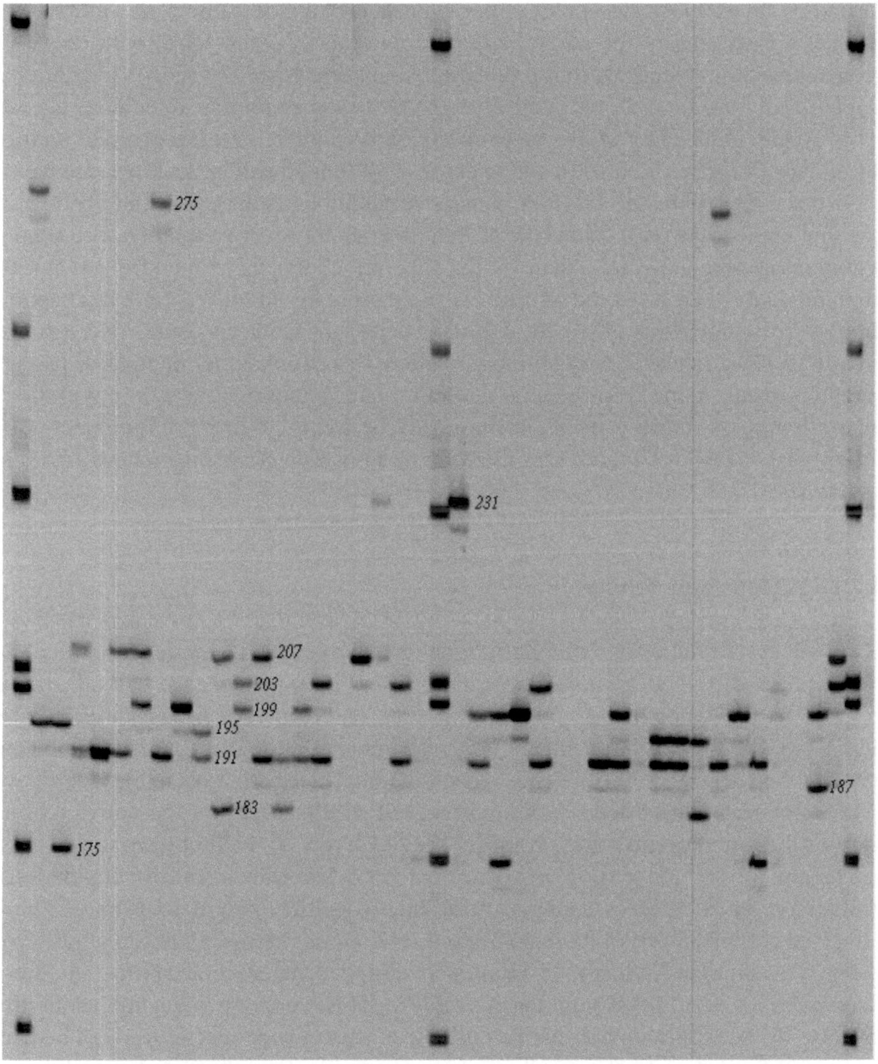

Fig. B.1 Automated sequencer polyacrylamide gel electrophoresis image of typical microsatellite locus from Atlantic salmon. Each individual fish is either homozygous or heterozygous for alleles ranging in size from 175 to 275 base pairs. Three identical size markers are present at either end and in the middle of the gel

spawning populations. (However, it is recognised, that particular mating strategies may substantially decrease effective number of parents.) Most aquatic animals (especially those of commercial fisheries and aquaculture interest) have very high fecundity (from thousands of eggs per female in Atlantic salmon to millions per female in cod) and, in the natural situation, mortality is very high (with only two progeny needing to survive to reproduction, assuming equal sex ratio, to maintain a stable population size). In the controlled and protected conditions of culture, mortality is usually reduced by one or more orders of magnitude, so a few parents can produce many progeny, often sufficient to entirely fill limited culture facilities. However, such practises ignore the fact that genetic variability can be exponentially lost as parental number decreases. In order to minimise such loss of variability or to keep it within reasonable limits (less than 1% per generation), it is recommended that at least 50 individuals of each sex are utilised (assuming that the animals are not hermaphroditic) (Cross and King 1983). In addition, departure from an equal sex ratio can greatly increase rate of loss of variability and in this case a larger number of parents should be used. Aside from genetic depletion and, in contrast to wild populations where temporal stability in allele frequency between cohorts predominates, different year-classes of reared strains often demonstrate large differences in allele frequencies.

4.2.1.1 Molecular Studies

Early allozyme studies on Atlantic salmon revealed reduced genetic variability and inter-cohort differences in allelic composition, in ranched and farmed strains compared to wild populations, in both European and North American studies (Cross and King 1983; McElligott et al. 1987; Verspoor 1988; Cross and niChallanain 1991; Cross et al. 1993). More recent studies using microsatellite loci have demonstrated similar effects (e.g., Norris et al. 1999). Many of the native marine species that are currently farmed in Europe (such as cod, halibut, turbot, sea bass, sea bream, lobster and scallop) also show similar differences from wild populations. Stefansson et al. (2001), using several microsatellite loci, demonstrated these effects in certain strains of halibut *Hippoglossus hippoglossus* but not in others, and there were similar findings, in relation to reduced genetic variability, in turbot *Scophthalmus maximus* (Coughlan et al. 1998). This reduction in genetic variability can also be the case with non-native introduced aquaculture species, though here an additional factor can be the small number of individuals in the initial introduction. An example is the abalone species *Haliotis discus hannei* that is farmed in Ireland. This species originates in Japan, and a comparison of wild Japanese individuals with Irish broodstock shows substantial reduction in the number of alleles in the latter group, at three microsatellite loci (Coughlan, Burnell and Cross-unpublished). Since the mussel *Mytilus edulis/galloprovincialis* and flat oyster *Ostrea edulis* farming industries in Europe have almost completely relied on wild-collected "seed" (juveniles), up to the present, rather than hatchery intervention, there is unlikely to be a similar problem with these species. However, in the case of mussels, since a complex and extensive hybrid zone occurs along the western European coastline from Portugal to the Faeroe Islands (Gosling 2003), there may be fitness differences between individuals reared in different locations from their natal site. In Bantry Bay, Ireland there is a natural hybrid population of mussels (*Mytilus edulis/galloprovincialis*) and the local aquaculture industry collects seed both on ropes and from intertidal rock surfaces. A molecular ecology study, using allozyme markers, has shown that the seed collected on suspended rope collectors was mainly the *Mytilus edulis* type and those scraped off rocks were predominantly *Mytilus galloprovincialis* (Fischer 1995). However, when ropes originating from both sources were tested just before harvest, the *Mytilus galloprovincialis* type predominated. The reason would appear to be due to the latter morph's stronger byssal attachment, allowing it to survive better in longline culture. Approximately 2,000–3,000 t of mussel are farmed within this bay yearly, and it is reasonable to assume that the culture operations are exerting a continual selection pressure in favour of *M. galloprovincialis*.

Another example occurs in the Netherlands, where wild mussel seed (*Mytilus edulis*) are transferred from one area of the Dutch Waddensee to more productive sites in the same region. A comparable industry occurs in Ireland where the seed is dredged from offshore sites and relocated to coastal areas within the Irish Sea. Until recently it was assumed that indigenous mussel populations in both countries were not exposed to new genetic material, however, shortages of seed in the Dutch Waddensee (due to conservation measures) have resulted in Dutch importations of Irish mussel seed. Although previous studies have shown that Irish Sea mussels (the new source of the Dutch imports) are *Mytilus edulis* and not affected by hybridisation with *Mytilus galloprovincialis*, the effects of introducing conspecifics from another geographic region is not known. There is some circumstantial evidence that even within the Irish Sea there may be adaptations of mussel populations to local ecology. A recent attempt to restore a N.E. Ireland estuarine mussel fishery with sub-tidal mussel seed from the southern Irish Sea was only partially successful, where despite importing a total of 3,700 t of mussels over 4 years, the restored stock only reached about one fifth its original level (Burnell, unpublished data). It is conceivable that the original estuarine population, which was removed to deepen a shipping channel, was physiologically adapted to the fluctuating salinity of the estuary. Hydrographic modelling has demonstrated the retention of larvae close to the mouth of the estuary, which supports the hypothesis of a self-recruiting population.

4.2.1.2 Breeding Programmes for Species Other than Salmonids

For cultured strains of European marine or anadromous species other than Atlantic salmon, breeding programmes (where they exist), are at a much earlier stage. This might be taken to indicate that there will be less genetic difference between wild populations and reared strains (since they have been less generations in captivity and therefore less accidental or deliberate anthropogenic selection has been applied). Certain cultured species such as cod *Gadus morhua* and Atlantic halibut *Hippoglossus hippoglossus* show substantially better survival of progeny following mass spawning situations, rather than when stripping and single-pair mating is used. However, it is becoming clear that in mass-spawning situations, only a small proportion of the total number of adults contribute to the resulting progeny, and of the parents that do contribute, fewer still dominate in terms of overall reproductive success. Such results are either due to failure to spawn by certain individuals, for physiological or behavioural reasons, or because of differential survival of progeny of different parents, or from some combination of these factors. It is obviously vital to quantify such effects. This can be achieved by screening both putative parents and progeny for an appropriate number of microsatellite DNA loci (Table 4.1), and then utilising a parental identification programme (Jones and Ardren 2003). Whatever the exact reasons for the failure of all potential parents to contribute progeny in mass spawning situations, very low numbers of families are often being used to found strains when mass spawning is utilised. This may be detrimental to the industry in the longer term (because of inbreeding effects minimising performance), as well as meaning that any reared animals that escape to, or are introduced into, the wild, will be substantially less genetically variable than wild individuals. Low variability may have detrimental fitness implications. There is evidence from many studies in a range of species that shows a positive relationship existing between genetic variability and performance, in terms of desired traits such as fast growth.

It should be noted that the vast majority of these studies have focussed on so-called neutral loci (non-adaptive loci) rather than adaptive loci (with fitness implications), which are generally less well understood. However, the assumption is often made that reduction in genetic variability at neutral loci is indicative of genome wide reduction in variability, which will therefore also affect variation at adaptive genes (but see Beebee and Rowe 2004).

Local adaptation appears to occur in salmonids and foreign populations perform less well than natives under natural conditions, when introduced into the stream occupied by the latter in the juvenile freshwater stage. Such reduction in

Table 4.1 Numbers of surviving cod offspring (n = 57) in each of 49 possible families from a mass spawning event in a mesocosm involving seven females (F1–7) and seven males (M1–7). Progeny were sampled at 3 months-old and typed for four microsatellite loci, for which the brood-stock had been previously screened. Parental assignment was carried out using the PAPA programme (Duchesne et al. 2002). Also included in the table are the total numbers ascribed to each dam (right hand column) and numbers per sire (bottom row). It can be seen that there is very uneven contribution from each dam and sire both to individual families and in total. (Data from Armitage, 2006)

	M1	M2	M3	M4	M5	M6	M7	Total/dam
F1		2	13		1		24	40
F2								–
F3		2	3				1	6
F4								–
F5		3	1				7	11
F6								–
F7								–
Total/sire		7	17		1		32	**57**

performance seems to occur even with salmon introduced as eggs from a nearby river (McGinnity et al. 2004). It is presently unclear whether there are similar local adaptational effects in other species, although this is likely to be the case, and if there are, non-native translocated or reared strains may have lower Darwinian fitness in the wild, compared with natural native populations (de Eyto et al. 2007).

Different broodstock strategies apply to producing animals for farming or for stocking. For farming, a closed cycle will usually be used and all life-stages will be in captivity. Such activities may also have associated breeding programmes, where the major goal will be faster growth, with other aspects such as delayed sexual maturity, carcass quality and disease resistance also being included as objectives (Gjedrem 1999). Thus, farm strains will diverge genetically from their wild progenitors and this deliberate divergence will increase as generations in the breeding programme progress. Recent microarray results from Roberge et al. (2006) have demonstrated changes in gene expression over generations caused by breeding programmes. In examinations of entirely separate Atlantic salmon breeding programmes in Norway and eastern Canada, it was noted that many of the same genes were up- or down regulated (genetic expression increased or decreased) when these fish were compared with wild individuals from either area. Such studies target the functional aspects of expressed genes, rather than focussing on neutral loci as in many previous molecular studies.

It should be noted that incorporation of breeding programmes into production of animals for sea farming is regarded an economic imperative by the industry, so it is unlikely that any Government-initiated protection measure for wild populations will suggest abandoning such activities because of potential detrimental effects for wild animals. As alternatives, better containment of farmed fish or invertebrates will be required and conservation or farm-free areas instigated to protect particularly vulnerable wild populations.

Genetically Modified (GM) individuals, where the definition of GM used here is intra- and/or interspecific transgenics (Devlin et al. 1994), will differ from wild ancestors from the time the transgene is successfully incorporated. European Union regulations currently prohibit the release of such organisms into the wild, and high levels of containment are used when they are being developed. However, transgenesis can greatly increase the growth potential of aquatic animal species (Devlin et al. 1994), so there is likely to be aquaculture-industry pressure to use such individuals in the future. If so, there will be the potential for escapes to the wild. In western Canada, where the deliberate release of transgenic salmon to the wild is prohibited, desk studies have been carried out to estimate the environmental impact of escapes of GM Pacific salmon (Devlin et al. 2004).

4.2.1.3 Breeding for Stocking and Ranching

In the preparation of animals for stocking, a totally different breeding strategy is utilised. Here, the imperative is to keep the strain as near wild as possible, to avoid any deliberate selection and to minimise the length of the period in culture (Cross et al. 2007). However, domestication selection, an incompletely understood phenomenon (see Bekkevold et al. 2006), still takes place even in very short periods of culture and Reisenbichler et al. (2004) argue that it is impossible to totally avoid genetic modification of animals being reared for stocking.

Breeding of strains for ranching (the preparation of juveniles for deliberate release into the sea, for growth to harvest size in the wild then subsequent recapture, usually in a commercial exercise) is a contentious area. Since recapture efforts are rarely totally successful, some individuals will be free to interact with animals from wild populations, and thus it might be wise to avoid or minimise anthropogenic selection. However, economic forces may dictate otherwise. It has been shown with Atlantic salmon ranching in Iceland that breeding for favourable traits (such as high return-rate) is feasible, at least in the commercial context (Jonasson et al. 1997).

4.2.2 Between Population Intra-Specific Comparisons

Wild Atlantic salmon exhibit a highly defined population structure (Verspoor et al. 2007), probably resulting from disjunct geographical distribution of freshwater spawning habitats, propensity for accurate natal homing and typically small population size relative to many marine species (the latter meaning that genetic drift has a much more profound effect in promoting structure). There are three major population groupings (rivers in eastern North America, western Europe and around the Baltic Sea respectively), but also a high degree of population structure at regional, and between- and within-river catchment levels among these groupings (Verspoor et al. 2007). Since there are also strong indications of local adaptation even between nearby rivers (McGinnity et al. 2004), then the provenance and particularly, the domestication history, of reared strains will be of major importance in considering the potential implications of interactions.

Other native marine species that are currently farmed in Europe (cod, halibut, turbot, sea bass, sea bream, lobster, scallop, mussels and flat oysters) differ from Atlantic salmon in that genetic population (stock) structure is much less defined and the extent of local adaptation has not been fully established (Waples 1998; deWoody and Avise 2000; Conover et al. 2006). It is not clear whether the results concerning population structure result from lower philopatry in these species or from exponentially lower genetic drift, or from both-see Bekkevold et al. (2005). Even though there are much smaller, though often statistically significant, genetic differences between most groupings than in salmon, there is often evidence of genetically different population groupings in large geographical areas and of different spawning populations in more local areas, e.g., Atlantic cod (see review by Imsland and Jonsdottir 2002) appear to differentiate into at least three major groupings corresponding to the western North Atlantic, eastern North Atlantic and Baltic respectively (O'Leary et al. 2007). In addition, there is also evidence of differences between Arctic and coastal cod off Norway (Fevolden and Pogson

1997) and of cod within the North Sea and English Channel in the eastern Atlantic (Hutchinson et al. 2001) and off eastern Canada (Ruzzante et al. 1999). Although presently unclear, these population groupings or local populations may be locally adapted.

As mentioned earlier, there is evidence from Atlantic salmon of adaptational differences between major population groupings. Bakke et al. (1990) demonstrated that salmon from the Baltic are resistant to the monogenetic trematode *Gyrodactylus salaris,* whereas western European salmon are highly susceptible to infestation, which usually results in mortality. Thus, it is recommended that for *Salmo salar,* there be no movement between major population groupings, at least for stocking (movement of salmon for farming has taken place from Europe to North America, but this is now strongly discouraged by relevant Governments). As a precautionary approach, the same stricture should be applied to other cultured species (e.g., when sea bass are moved from the Atlantic to the Mediterranean as broodstock, see recommendations). Less effort has been applied to investigating population structure in other species cultured in Europe than with salmon, cod and lobsters, and further studies are urgently required.

4.2.3 Marine Translocated Species

The native population structure of species such as Pacific oyster, *Crassostrea gigas* and Manila clam, *Ruditapes philippinarum,* translocated from other regions is not relevant in the present context, except that it is important to know the provenance and pre-translocation history of these reared strains (whether they were taken directly from the wild, in what numbers and at what life stage; what pathogens occur in a particular area; whether a hatchery generation was used prior to translocation, etc.). It is also important to know whether individuals originated from one or more natural populations. If the latter (or if a new translocation is planned from a different wild population) there may be problems with outbreeding depression, resulting in decreased fitness of progeny of crosses between genetically different populations, which may have performance implications in the area of introduction.

The Pacific oyster, *Crassostrea gigas*, is now farmed throughout the world with an annual production of about 3.6 million tonnes. This ubiquitous species was imported from Japan into British Columbia, Canada in the 1950s where it was naturalised and from there it was moved to France in the 1970s where, once again it established breeding populations (Gosling 2003). In 1964 Dutch oyster farmers imported it from British Columbia to augment native *Ostrea edulis* stocks. This introduction was carried out on the premise that these oysters would not reproduce at the latitude of Dutch coastal waters. However by 1980s *C. gigas* was able to extend naturalised populations from the Oosterschelde estuary to the Wadden Sea area near Texel where it replaced native *O. edulis* and *M. edulis* on the intertidal beds (Rajagopal et al. 2005). It is now posing both a severe ecological and economic threat to native species and traditional fisheries, respectively, in this region.

A similar situation is apparent with aquaculture of the Manila clam, *Ruditapes philippinarum*. Approximately 2.75 million tonnes are cultured worldwide each year, with 90% of the production in China. In several southern European sites (where it has been introduced) it appears to be flourishing at the expense of local clam species. This has been documented in the Italian lagoons (Mantovani et al. 2006), where it out competes the native *Tapes decussate*, and similarly, in Portugal where it has been blamed for the demise of this species (Gosling 2003). The problem with the spread of this exotic clam is that it may initially go unnoticed due to its benthic habitat and its superficial similarity to other clam species, and presently its range is expected to extend northwards.

4.3 Introduction to the Wild

Cultured strains can be introduced inadvertently to the wild, as in farm escapes, or deliberately, from stocking or ranching exercises. While the mode of introduction is different, the genetic consequences for wild conspecifics or other species with which the reared animals interact are similar.

4.3.1 Cultured Strains Inadvertently Introduced to the Wild (Farm Escapes–see Ferguson et al. 2007)

Farmed marine or anadromous species can escape from onshore tanks or sea cages (pens) due to equipment failure or human error. Escapes from cages are more likely (though direct escapes from tanks may also occur) and usually occur due to storm damage, commonly in winter or from predator attack (e.g., seals). It is important to note that escapes can occur at any life stage, and the age and season at which animals escape can greatly influence their subsequent behaviour and survival (e.g., for salmon, Hansen (2006) and references therein). There is usually increased mortality (over natural levels for that stage in the life-cycle) directly after escape, as cultured animals must adapt to capturing wild food and avoiding predators. Also, the ultimate impact of an escape incident will be strongly influenced by the life stage involved. It is clear from consideration of the regime of natural mortality, that a million immature juveniles will have a much lesser effect than the same number of animals nearing maturity (assuming similar levels of increased post-escape mortality at the different life stages).

Recapture of individuals subsequent to escape using various fishing methods is difficult because little is known about post-escape behaviour. In the case of Atlantic salmon, seine netting following a simulated escape (deliberate release) from Norwegian sea cages was largely unsuccessful, because it appeared that escaped fish aggregated deeper than the level fished (as shown by the use of sonar tags)

(Oystein Skaala, Institute of Marine Research, Bergen, Norway-personal communication). Another important factor for recovery measures is that there is rapid reporting of the escape incident, before the escapes disperse (Jorstad et al. 2006). It was previously assumed that a direct escape of cultured animals was the only inadvertent way of influencing wild populations but recent results have highlighted other ways. Cod (Jorstad et al. 2006) and sea bass (Youngson et al. 2001) can spawn in cages and their fertilised eggs subsequently drift into the wild (as shown with genetically marked individuals (Jorstad et al. (2006) for cod). Another possibility is that disease organisms can be transferred from cultured to wild animals without physical contact between the host animals (e.g., furunculosis in salmon in Norway) with subsequent detrimental effects to wild populations either ecologically or at a molecular level (see below). In several diseases of shellfish, transmission of pathogens has occurred when movements of shellfish between culture sites have taken place, e.g., *Bonamia ostreae* in the flat oyster *Ostrea edulis* and *Haplosporidium nelsoni* and *Perkinsus marinus* in the eastern oyster *Crassostrea virginica.* With *Perkinsus marinus* and *Bonamia ostreae,* transmission of the parasite can occur from oyster to oyster, via the water column (Andrews 1988; Culloty et al. 1999). Additionally, when *Bonamia ostreae* has been introduced into an area, eradication of the disease has failed, as it appears that *Bonamia ostreae* can be maintained in other benthic invertebrate species and later infect relayed oysters, even after the area has been left fallow for a number of years (Van Banning 1987).

4.3.2 Cultured Strains Deliberately Introduced to the Wild (As in Stocking/Ranching)

Cultured strains are deliberately introduced to the wild in stocking/enhancement or ranching exercises (referred to in Japan as "culture enhanced fisheries"). The minimum requirement for reducing the risk of introducing foreign genetic material to wild populations is to ensure that broodstock are from the same area as the proposed release site. In Norway, the Institute of Marine Research (IMR) has been releasing tagged juvenile lobsters (*Homarus gammarus*) around the islands of Kvitsoy since 1990 (Agnalt et al. 1999, 2004), with returns of up to 8% at market size (5–6 years after release). The juveniles in this case were obtained from wild captured Kvitsoy "berried"-females, where the attached eggs were already fertilised. The resulting larvae were hatchery reared through several moults before being released into areas that had previously been identified as suitable nursery sites. Some of these restocked lobsters are now sexually mature and IMR conducted an experiment to investigate the performance of reared F_1 offspring against wild lobster larvae from the same area (Jorstad et al. 2005). One of the main concerns about interactions between wild populations and reared strains is that interbreeding is likely to cause reduction in population fitness under natural conditions (as in salmon, McGinnity et al. 2003). Alarmingly, the survival of the F_1 cultured lobster larvae was 40% lower than the natives, when grown under identical conditions in the hatchery. It would appear that during the hatchery period of the reared parents there had been artificial (though-inadvertent) selection pressure, which resulted in reduced fitness of the reared progeny.

In Hokkaido, Japan, scallop seed (*Patinopecten yessoensis*) are settled on mid-water collectors, half grown in suspended culture and then released locally onto the seabed for final ongrowing. Stock enhancement has been carried out in this region for over 30 years and now yields about 300,000t per year (Uki 2006). Because the pelagic larvae are retained by local gyres and the juveniles usually remain within 0.5km of the release site, the area could be considered as an extensive marine farm since reseeding areas are rotated annually and predators are removed by dredging. As a result starfish (*Asterias amurensis* and *Asterias pectinifera*) and the sea urchin *Glyptocidaris crenularis* have been almost eliminated from the ongrowing areas. In many areas there is almost a monoculture of scallop on the seabed.

The situation in Europe is very different. Despite over 30 years of research and technology transfer from Japan, scallop culture is still in its infancy with total production of less than 1,000t. A small proportion of this is obtained from seabed ranching. The most successful projects are in France (150–200t per annum), Ireland (50–100t per annum) and Norway (50–100t per annum) (Shumway 2006). In each case, the industry has been careful to use local broodstock for their hatchery programme but, as has been demonstrated with lobster restocking, the hatchery part of the process will inevitably induce some genetic selection with possible loss of fitness in the F_1 and subsequent generations.

As mentioned above, a very different rearing strategy is generally employed for enhancement exercises, than when producing animals for farming. While it is generally assumed that such animals will have greater survival in the wild than farm strains, this may in fact be more detrimental to wild populations, as will be discussed below. Apart from genetic considerations, the success of stocking exercises is crucially dependent on the strategy employed, viz. the number and life stage of the animals used for stocking, the location/s and timing of where the introductions take place, and whether the exercise is undertaken once or repeated on a regular (annual) basis. As noted in Cross et al. (2007), there has been very little detailed follow-up monitoring and it is generally presumed that the aim of stocking is a larger self-sustaining "wild" population. However, this may not be possible because of limits to environmental carrying capacity or because of environmental constraints, such as the presence of dams on salmon rivers, which inundate natural spawning areas (Cross et al. 2007). In these cases, the stocking will have to be repeated on a regular basis and the exercise becomes, in effect, a type of ranching. Unfortunately, not all returning adults from such an exercise will be caught in terminal fisheries, so the potential for large numbers of ranched individuals to be introduced into the wild is high and this situation will usually be maintained by continued regular introductions, even if the reproductive fitness of reared animals or their hybrid progeny is somewhat lower than "pure" wild individuals. While with commercial ranching, the aim is to recover or capture all animals of marketable size, this is probably never achieved so "ranched wanderers" or strays must be considered as a potential threat to wild populations.

4.4 Genetic Interactions with Wild Populations/Species

Direct and indirect genetic interactions are possible between reared and wild individuals in nature. Direct interactions occur with the same or closely-related species by interbreeding (producing intra-specific or inter-specific hybrids). Indirect genetic interactions may occur where either ecological competition or disease introduction by the reared strain (which may lead to modification of immune-response genes) causes reduction in size of wild populations, leading to much reduced population size (and effective population number, N_e) and thus increased genetic drift, which can both decrease genetic variability and alter genetic composition.

In shellfish populations where disease has substantially reduced population size, breeding programmes to increase resistance have been undertaken with a view to supplementing or replacing the wild stocks, e.g., for increased resistance to *Bonamia ostreae* in *Ostrea edulis* (Naciri-Graven et al. 1998, 1999; Culloty et al. 2004), *Haplosporidium nelsoni* in *Crassostrea virginica* (Ford and Haskin 1987) and *Marteilia sydneyi* and *Bonamia roughleyi* in the Sydney Rock oyster *Saccostrea glomerata* (Nell and Perkins 2006). In a study of hatchery propagated populations of the flat oyster with increased resistance to *Bonamia ostreae,* although heterozygosity was still high in the resistant and control populations, the number of alleles in the selected population was significantly reduced compared with the control population, which appeared to be mostly due to a loss of rare alleles (Launey et al. 2001). As a result of this loss in variability a decrease in performance, for both growth and survival, was predicted from the second generation onwards.

4.4.1 Direct Genetic Interactions

Reared animals may breed among themselves in the wild (assortative mating) but it is likely they will also breed with wild conspecifics where these occur. In the case of Atlantic salmon, it is clear that interbreeding takes place between reared and wild fish, and F_1 hybrid offspring are produced. Subsequent generations have been produced in the hatchery and then reintroduced into experimental situations in a field experiment (F2 hybrids and parental × F1 backcrosses) (McGinnity et al. 2003). Early Norwegian experiments on genetic tagging in cod using an allozyme locus, indicate that some interbreeding occurs between wild and ranched fish, when the latter were released into fjords (Jorstad et al. 1994). These animals, which are evidence of further introgression, have been rarely identified in unmanipulated situations. The extent of interbreeding of reared and wild conspecifics (or assortative mating of reared animals when introduced into the wild), needs to be quantified for all aquaculture species with substantial production from farming, stocking or ranching, although Youngson et al. (2001) suggest that there will be far less introgression in sea bass and sea bream than in Atlantic salmon. Fleming et al. (2000) have shown, with Atlantic salmon, that spawning success of farm fish is considerably lower than wild fish, and that the success of farmed males is particularly low. These results have been obtained from detailed observational work of sexually-mature individually-tagged wild and farm salmon in large freshwater arenas with suitable spawning substrates. Such experiments are required with other species and although it is recognised that they may be more difficult to set up, these are being attempted with cod in Iceland (Gudrun Martinsdottir, Marine Research Institute, Iceland- personal communication).

The presence of reared animals in nature also increases the propensity of interspecific hybridisation with congenerics, certainly for reared Atlantic salmon and wild brown trout, *Salmo trutta* (Youngson et al. 1993; Hindar and Balstad 1994). Whether this is the case with other European aquaculture species has not been established but since inter-specific hybrids are rarely fully fertile, this could result in another potential problem for wild populations when reared aquaculture species enter the wild.

4.4.2 Indirect Genetic Interactions

Certain bivalve diseases have had major ecological consequences for native populations, for example, the overall European aquaculture production of flat oysters *Ostrea edulis* fell from 29,595 t in 1961 to 5,921 t in 2000 due to epizootics caused by *Bonamia ostreae* and a second protistan *Marteilia refringens*. Also, in the early 1970s, the Portuguese oyster (*Crassostrea angulata*) was dramatically depleted within Europe by an iridovirus (Marteil 1976). It has been speculated that the uncontrolled transfer of *Crassostrea gigas* introduced this iridovirus to *Crassostrea angulata*, which was highly susceptible (Boudry et al. 1998).

Despite this evidence of ecological effects, there are relatively few examples of indirect genetic interactions between cultured strains and wild populations, primarily because this aspect has not been investigated in detail in species other than Atlantic salmon. However, conditions undoubtedly exist where such interactions are possible. For Atlantic salmon "common-garden" experiments in Ireland (McGinnity et al. 2003), the farmed strain involved grew significantly faster in freshwater than the wild population (presumably since the farmed strain had been subject to several generations of selection for fast growth). As substantially more of the wild population migrated downstream out of the experimental stretch, competitive displacement of wild fish by farmed was considered likely. Several other authors have cited examples of ecological interactions in salmonids, e.g.,

- farmed salmon feeding on natural prey (Hislop and Webb 1992)
- potential for feeding competition at sea (Jonsson and Jonsson 2006)
- competition for mates (Fleming et al. 2000)
- competitive displacement of juveniles (McGinnity et al. 1997, 2003; in press)
- predator avoidance (Einum and Fleming 1997; Fleming and Einum 1997)

GM individuals for growth hormone, providing they survive, could act as super competitors/predators, though it has been suggested that these individuals will have lower reproductive fitness. For other aquaculture species, there are undoubtedly examples of major ecological effects of translocated species (see above in relation to *Crassostrea gigas* and *Ruditapes phillipinarum*). However, whether this is also a problem with native finfish and invertebrate aquaculture species other than salmon, remains to be investigated. Effective population size (N_e) of wild populations of species such as cod is likely to be exponentially larger than for salmon (assuming healthy populations that have not been severely reduced by overexploitation- but see Hutchinson et al. (2003) concerning cod), but this has not been established for all the major aquaculture species. Furthermore, current aquaculture production is much lower than for salmon and it seems unlikely that N_e of wild populations of these species will be reduced sufficiently to make genetic drift a major factor in reducing variability or altering genetic composition. However, a very rapid rise in production is anticipated and it should be noted that where wild conspecifics occur and interbreeding is going on, it may be difficult to identify or distinguish indirect from direct genetic effects.

Introduced diseases are a major concern in marine biology in general and can have profound ecological effects in the present context (as mentioned earlier, there are examples in salmon of the major effects of furunculosis and *Gyrodactylus* in Norway). Translocated species or sub-species may be more likely to cause damaging effects in this respect, since the diseases they carry are likely also to be genetically different or exotic to local taxa. Exotic diseases may be carried by introduced species, which have a relatively minor effect on their normal host, but can have a serious impact on naïve and often highly sensitive native species. In *Bonamia ostreae* infection of the oyster *Ostrea edulis,* it appears that prior to initial exposure, all naïve oysters are susceptible to infection resulting in heavy mortalities (Culloty et al. 2004). This has recently been demonstrated, with an extension of the range of this parasite from North America into Canada (Marty et al. 2006), within Europe to Scotland (http://www.scotland.gov.uk/News/Releases/2006/07/27154609), and from Europe into Morocco (http://www.oie.int/eng/info/hebdo/AIS_43.HTM). For Dermo disease in *Crassostrea virginica* (caused by *Perkinsus marinus*), the pathogen is now ubiquitous along the Atlantic and Gulf coasts of the USA, having recently extended its range to the Maryland portion of Chesapeake Bay and northward along the Atlantic coast from New Jersey to Maine. In addition, it has been found that different regions can possess unique assemblages of genetic strains of the parasite (Reece et al. 2001). Furthermore, comparisons of clonal and parental culture genotypes indicates that cultures initiated from a single oyster can be polyclonal, showing that an individual can be infected with multiple strains, thus making any control measures more difficult. Recently, in the study of bivalve diseases, methods such as suppression subtractive hybridisation have been used to look at gene expression in susceptible and resistant bivalves to such pathogens as *Perkinsus marinus* in *Crassotrea virginica* and *Crassostrea gigas* (Tanguy et al. 2004) and to bacteria-challenged *Crassotrea gigas* (Gueguen et al. 2003), to determine the role of particular immune components in response to infection. One limitation to investigating gene expression in invertebrates is that for a number of species, the full complement of immune components has still to be determined.

A more recently demonstrated indirect genetic effect appears to be mediated by disease organisms, which can be transferred from reared to wild fish species. This effect is observed in the Major Histocompatibility Complex (MHC) genes of wild *Salmo salar* and *Salmo trutta* putatively challenged by diseases carried by reared salmon (Box 4.2). Because of high-density rearing conditions, reared fish and invertebrates often have much higher disease challenges and/or loads than their wild conspecifics or congenerics, but these diseases can be controlled in captivity using anti-bacterial compounds or vaccination. However, when such individuals escape or are introduced into the wild, they may act as highly virulent carriers and cause outbreaks of the disease in wild fish or invertebrates, with subsequent demonstrateable effects at MHC genes (at least in salmonid fishes). The example described in Box 4.2 constitutes another potentially damaging effect of aquaculture, which should be investigated in other cultured species.

Box 4.2 MHC genes in interaction studies between reared strains and wild populations of teleost fish

In teleost fish, MHC Class I and II loci are not physically linked (Sato et al. 2000) and thus can evolve independently (hence termed MH). Members of the genus *Salmo* are found to possess single classical Class I (*UBA*) and II (*DAA/DAB*) loci (Shum et al. 2001; Stet et al. 2002; Aoyagi et al. 2002; Grimholt et al. 2002; Miller et al. 2006) greatly simplifying salmonid MHC studies.

Pathogen-driven balancing selection, with overdominance or heterozygote advantage, is believed to underpin high levels of polymorphism observed in MHC loci (Wegner et al. 2003). Challenge experiments in which domesticated salmonid stocks were exposed to a number of pathogens (Langefors et al. 1998; Lohm et al. 2002; Arkush et al. 2002; Grimholt et al. 2003) uncovered differential survival rates mediated primarily by MH heterozygosity and/or overdominant selection.

Grimholt et al. (2003) deliberately infected two distinct groups of post-smolt *S. salar* with furunculosis bacteria and ISAV virus, respectively. MH genotyping of mortalities and survivors demonstrated genotypic and allele effects, at class I for ISAV challenge and at class II for furunculosis. Mass screening was facilitated by the use of polymorphic VNTRs located in the 3'UTRs of the *Sasa-UBA* and *Sasa-DAA* genes (Grimholt et al. 2002; Stet et al. 2002), which exhibited simple linkage.

These findings led to an EU project (Salimpact) on MH genes in wild *S. salar* and brown trout, *Salmo trutta*, populations, where diseases carried by co-habiting, reared salmon were considered as challenge agents. Simultaneous screening of several neutral loci, (unaffected by disease challenge), with the MH-linked marker loci provided the opportunity to examine for selective effects on the MH marker loci in a number of interaction situations in Ireland

and Norway. In the Burrishoole river system in the west of Ireland, long-standing ranching and nearby sea cage farming occur alongside native salmonid populations.

Using archival scales, statistically significant changes in gene diversity occurred at the MH class I marker over time in wild *S. trutta,* but no changes were observed at neutral microsatellite loci (Coughlan et al. 2006). In an experimental natural stream in the same system, eyed eggs from native salmon and salmon from a neighbouring river (derived from wild broodstock and only retained in the hatchery until eyed egg stage) were introduced. Significant selective effects were evident at MH class II after eight months in freshwater, in the non-native population, but not in natives. No significant results were observed in either group at the MH class I locus or at eight neutral microsatellite loci (deEyto et al. 2007). Thus, it appears that variation at MH loci may be a feature of local adaptation, as well as influencing survival in native trout when challenged by diseases carried by reared salmon. Therefore, reared salmon may negatively impact on wild salmonid populations, putatively via disease transmission, in addition to having direct and other indirect genetic effects (McGinnity et al. 2003).

These data, demonstrating another way that aquaculture practices can detrimentally effect wild salmonid populations, suggest that further studies of such interactions in the context of fish immunity (e.g., MHC genes) and disease prevalence are required. Similar, MHC-based studies in other teleost species subjected to aquaculture are also advisable.

4.5 Consequences of Interactions

4.5.1 Direct Effects

Following direct interactions it is likely that hybrid progeny of interbreeding between farm and wild strains, as well as "pure" parental types, will result. The proportions of each of these types among first generation progeny will depend on the relative proportion of wild and reared parents, the relative spawning success of each sex of each type and, ultimately, the relative fitness of parental types and hybrids (with either type as dam and sire). The various field experiments with Atlantic salmon in Ireland and Norway have shown a trend for graduated spawning success and reproductive fitness; highest in pure wild, intermediate values in hybrids and lowest in reared (Einum & Fleming 1997; McGinnity et al. 1997, 2003; Fleming et al. 2000). Assuming that this is a general trend in interactions between wild populations and reared strains of most species, then following a single reared incursion, reared influence in the wild will gradually decrease over generations albeit at the expense of reduced fitness in this period. However it is recognised that incursions will usually be continuous, i.e., there will be regular injection of reared animals. Hindar et al. (2006) have modelled this sort of situation, varying the parameters referred to above and shown a worrying persistence of hybrids. Under this scenario, less introgressed animals than wild individuals can occur per unit area at carrying capacity, so that productivity of the system will decease. In addition, if hybrids come to predominate there may be a loss of local adaptation. Since there are usually only a limited number of reared strains of each species (for practical reasons, such as cost of rearing facilities incorporating breeding programmes), the genetic variation which is endemic to wild populations and which may indicate adaptive differences, will be lost.

The situation becomes more complicated in the F2 and subsequent generations. In the F2 generation, for example, assuming random mating, several types of F2 hybrids (depending on the sex of either type of parent and grandparent) and also of parental backcrosses (F1 X wild or reared parent) are possible. In a series of field experiments with Atlantic salmon, McGinnity et al. (2003) have demonstrated, that relative survival acts as a quantitative trait, with lower survival correlated with higher proportion of reared genes in the genome of hybrids and backcrosses (Fig. 4.1). Assuming this pattern of reproductive fitness and also the variation in spawning success demonstrated in Norway (Fleming et al. 2000), the most persistent reared types in the wild in subsequent generations in other species, will be wild X F1 hybrid backcrosses.

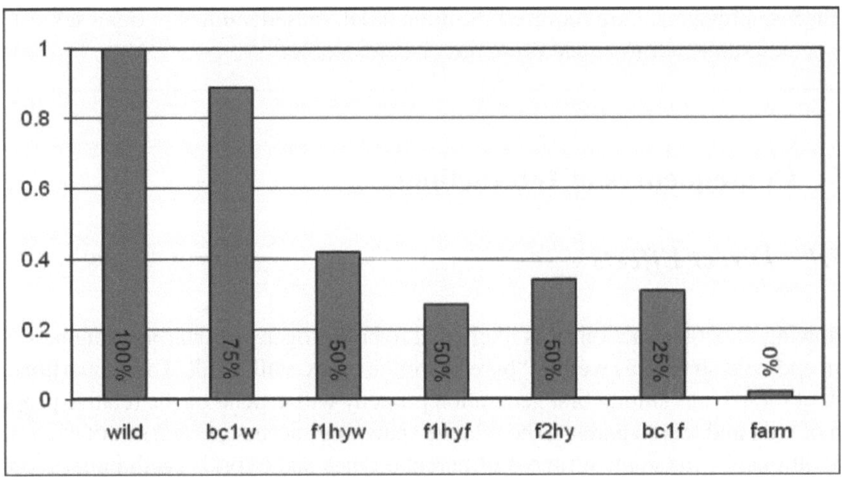

Fig. 4.1 The relative lifetime fitness (egg to adult over two generations) compared to wild of the progeny of farm and wild salmon and their hybrids with a measure of the proportion of "wild-parent" genes in the genome of each group (indicated as percentages). The relative estimated lifetime success ranged from 2% (farm) to 89% (BC1 wild) of that of wild salmon, the various hybrids having intermediate levels of fitness, indicating additive genetic variation for survival. It is assumed here that displaced parr captured in the Srahrevagh river trap have the same survival as parr of the same group remaining in the experiment river, i.e., that the river is not at its parr carrying capacity and spare habitat is available for displaced parr. Notes: bc1w = F_1 hybrid X wild; fihyw = F_1 wild X farm hybrid; fihyf = F_1 farm X wild hybrid; f2hy = F_2 hybrid; bc1f = F_1 hybrid X farm

4.5.2 Indirect Effects

Presuming there is a drastic reduction the size of natural populations (and also of N_e), exponential increase in genetic drift may result in a loss of potentially adaptative population structure, due to loss of alleles and alteration of allele frequencies. In other cases it may be difficult to distinguish between direct and indirect effects of interactions and it is conceivable that indirect effects may be masked by the consequences of interbreeding. Because of this problem, together with a lack of knowledge about functional genomics (e.g., immune response genes) in many species and a poor understanding of long-term ecological implications, many of the consequences of indirect interactions for wild populations are unknown.

4.6 Establishing the Severity of the Problems Caused by Wild/Reared Interactions with Different Species

Two types of scenarios are recognised where the consequences of interaction between wild and cultured individuals can be assessed, opportunist and experimental situations.

4.6.1 Opportunist Situations

These are defined as situations where escapes or deliberate introductions have already occurred, and the aim is to quantify the extent of the subsequent direct or indirect interactions. In this case, a range of molecular markers (Box 4.1) are investigated in the wild population/s and reared strain/s involved, searching for marker loci either with completely different alleles or haplotypes in the wild and reared groups (referred to as an absolute or qualitative marker), or at least loci which show substantially different allele frequency differences (termed quantitative markers). While a proportion of absolute markers can usually be found between congeneric or more distantly related species, they are rare within species, unless reared and wild individuals come from different major population groupings (but see Clifford et al. 1998). Thus, in the conspecific case it is usually necessary to rely on allele frequency differences at quantitative markers as defined above. With this type of marker, the discriminatory power increases with the number of individual loci or haplotypes included. Using a number of quantitative markers, individuals can be assigned to one or other group although certain markers will be intrinsically better when seeking high levels of discrimination. Microsatellite loci because of their high mutation rate and high allele number are particularly useful in this respect, as is the 5′ end of the d-loop region of the mitochondrial genome. With mtDNA, in addition to rapid mutation rate, there is a four times lower N_e (assuming equal sex ratio and fitness) than nuclear DNA because of its haploid nature, making the mitochondrial genome more likely to be affected by genetic drift when population size is reduced.

The greater the genetic difference between wild and reared morphs, the easier it will be to establish good levels of discrimination. With Atlantic salmon, because of the high level of population structure and also because of relatively low N_e, it is usually easy to find discriminatory markers, even when the native population is used to provide the progenitors of the reared strain (due to the use of limited broodstock numbers which results in a reduction of genetic variation and alteration in genetic composition). With other species cultured in Europe, e.g., cod (Hutchinson et al. 2001), lobsters (Triantaphyllidis et al. 2005), the relatively low levels of population structure observed suggests that it may be much more difficult to find suitable discriminatory markers. Genes coding for functional proteins (e.g., MHC-Box 4.2) may be more useful as markers of short term effects than neutral genes, since frequencies may be rapidly changed by the different selection regime experienced in culture. Identification of hybrids in opportunist situations in species where the level of discrimination between wild and reared morphs is very low, is not likely to be easy or even feasible and some type of experimentation may be necessary (see below). Modelling studies are urgently needed to determine the most appropriate approach.

4.6.2 Experimental Situations

One of the simplest experimental approaches to investigating the fate of reared individuals accidentally or deliberately introduced to the wild is by tagging. Physical tagging has previously been used (e.g., Hansen (2006) investigating the behaviour of farmed salmon in simulated escapes in Norway) but this method cannot track offspring, which requires genetic tagging. In the latter approach, a rare allele at a specific locus is chosen, and two heterozygotes (likely the only genotype available containing the rare allele, assuming Mendelian autosomal inheritance) are crossed to produce rare homozygotes (~25%) for release. "Rare" homozygotes recovered from the wild will most likely be reared individuals, and heterozygotes (at a frequency above "background") will be F_1 offspring of wild X reared matings. However, there are certain limitations with this method;

1. Unless many crosses of heterozygotes are undertaken *ab initio,* the marked individuals may show extremely limited genetic variation, and detecting many heterozygotes to use as parents will be difficult if the allele is rare.
2. There may be functional differences in the fitness of genetically marked and unmarked individuals. To establish whether there is equivalent fitness in marked and other reared individuals, tank experiments are often undertaken, where growth and survival are compared. However, differences that might become apparent in the much harsher conditions in the wild are unlikely to be observed

under these circumstances (McGinnity et al. 1997) so that the growth and survival of the marked fish may not accurately reflect the performance of the reared group, in general.

However, the method has proved useful in tracking cod released for ranching in Norwegian fjords (Jorstad et al. 1994) and more recently has been used to establish that spawning of farmed cod in sea cages leads to the progeny of reared fish being present adjacent to the cages which can disperse from these locations (Jorstad et al. 2006). Some combination of genetic tagging with robust statistical techniques for individual assignment should provide increasingly powerful methods for tracking reared individuals in the wild.

A more complicated and expensive, but ultimately far more informative, experimental approach to investigating interactions is to set up "common-garden" experiments. These have proved very useful in studying interactions in anadromous Atlantic salmon in freshwater in Ireland (McGinnity et al. 2003), Norway (Fleming et al. 2000), Scotland (Eric Verspoor, Freshwater Fisheries Services, Scottish Office, Pitlochry, Scotland-personal communication) and Spain (Carlos deLeaniz, University of Wales Swansea, Wales- personal communication), and in the marine phase (Box 4.3). Most of these experiments were carried out in a single site, where the performance of natives and an imported strain or population (and sometimes the hybrids between them) were compared. While it is recognised that the most comprehensive results would be obtained using a reciprocal design, this has not proved economically or practically possible in most situations. These experiments combined field ecology with molecular genetics. In Norway and Scotland different groups were batch marked using different allozyme genotypes or mitochondrial haplotypes, while in other experiments, VNTR screening of broodstock and parental assignment were used to identify progeny to family, which were then accumulated into the different groups (McGinnity et al. 1997). The difficulties associated with conducting such experiments in "open" wild situations with European marine species are exponentially greater, in that a reasonably restricted area, where the test groups of wild and reared individuals can be compared, is required. None-the-less, it may be possible to design such experiments using isolated or semi-wild situations, with some degree of genetic or non-genetic (e.g., oxytetracycline bath) batch marking of reared larvae and also parental assignment. The high information content of the results from such experiments (e.g., in relation to adaptive differences and individual variation), certainly justify the extra time and expense involved in their design and execution.

Evaluation of susceptibility of bivalve molluscs to various pathogens has been carried out in a number of field trials and the relative susceptibility of Irish and European population of *Ostrea edulis* to *Bonamia ostreae* has been evaluated by relaying the oysters in areas where the parasite is endemic (Culloty et al. 2004). Resistance of *Crassostrea virginca* of different heritage to *Perkinsus marinus* has been evaluated by comparing, for example, North Carolina and Chesapeake Bay oysters, using standard tray culture conditions, at several sites in both regions (Brown et al. 2005a). Furthermore, nine groups of oysters consisting of five regional strains and four hybrid strains were evaluated at three sites within

Chesapeake Bay (Brown et al. 2005b). Results of both trials indicated that performance was related to level of resistance, salinity of water and virulence of *Perkinsus marinus*.

Box 4.3 Burrishoole experiments

A two generation experiment was undertaken in the Srahrevagh river, a natural spawning tributary of the Burrishoole system, in western Ireland to measure the relative lifetime reproductive success of the progeny of wild salmon, escaped farmed salmon and their hybrids in the wild (McGinnity et al. 1997, 2003). The experiment was conducted as a "common garden" experiment in a natural river. Wild and farm salmon, and first and second generation hybrids and backcrosses between them, were planted as eyed eggs, thus removing the influence that hatchery rearing might have on performance if the fish were introduced at a later life stage. The study was also designed to eliminate behavioural differences between spawning adults and to examine the effect of solely genetic differences on survival and performance. Offspring of farm and "hybrids" (i.e., all F_1, F_2 and BC1 groups) showed reduced survival compared with wild salmon. The relative estimated lifetime success ranged from 2% (farm) to 89% (BC1 wild) of that of wild salmon, the various hybrids having intermediate levels of fitness, indicating additive genetic variation for survival (see Fig 4.1). There was also clear evidence of out-breeding depression in the F_2 hybrids. The progeny of farm salmon grew faster as juveniles and displaced wild parr, which as a group were significantly smaller. The offspring of farmed salmon showed a reduced incidence of male parr maturity compared with native fish. The latter also showed a greater tendency to migrate as autumn pre-smolts. Growth of hybrids were generally either intermediate or not significantly different from the wild fish. Wild salmon primarily returned to fresh water after one sea winter (1 SW), but farm and "hybrids" produced proportionally more 2 SW salmon. However, due to an overall reduced survival, this would result in reduced recruitment despite increased 2SW fecundity. The experiment showed that the interaction of farm with wild salmon results in lowered fitness, with repeated escapes causing cumulative fitness depression and potentially an extinction vortex in vulnerable populations.

An additional experiment has since been carried out in the Srahrevagh river using the "common garden" approach. This study was a comparison of the relative lifetime success and performance characteristics of communally reared offspring of wild native Burrishoole, ranched native and non-native salmon from the Owenmore River; a river that is in the same geographic area as the Burrishoole (McGinnity et al. 2004).

In this experiment, 0+parr from the Owenmore river showed substantial downstream migration, which was not shown by native and ranched parr. This appears to have been an active migration rather than competitive displacement and may reflect an adaptation to environmental or physiographic conditions within the Owenmore River catchment, where the main nursery habitat is downstream of the spawning area. There were no differences between native and ranched in smolt output or adult return. Both of these measures, however, were significantly lower for the non-native group. A greater proportion of the non-native Atlantic salmon was taken in the coastal drift nets compared to the return to the Burrishoole system, probably as a result of the greater size of the non-native fish. The overall lifetime success of the non-native group, from fertilized egg to returning adult, was some 35% of native and ranched. The ranched group showed a significantly greater male parr maturity, a greater proportion of 1-year-old smolts, and differences in sex ratio and timing of freshwater entry of returning adults compared to natives, which may have fitness implications under specific conditions.

4.7 Methods to Reduce Such Effects

4.7.1 Induction of Sterility/Triploidy

Induction of sterility in reared strains of fish and invertebrates has been extensively investigated in conjunction with the aquaculture industry, because in many species both appearance and flesh quality deteriorates greatly at sexual maturity. In certain species, such as gadoid and pleuronectid fish, males tend to mature younger and at a much smaller size, leading to an interest in all female production. Sterility would also be desirable in animals that might escape from rearing facilities or in commercial ranching (although, obviously where the object of stocking is to produce or enhance self-sustaining populations, sterility would not be desirable). Sterility can be achieved directly by hormonal treatment although this is recommended against in certain regions, and triploidy, induced by temperature or pressure shock in fish or by chemical means in shellfish, is an alternative. In certain fish species (e.g., salmonids) male triploids have a weak reproductive capability and sex reversal has to be achieved in the female parent (converted to a so-called pseudo male) by hormonal means. All-female production is possible in salmonids because the female is the homogametic sex. Objections have been raised to sterile fish production in salmon ranching since it is not clear whether the process will inhibit freshwater migration. Furthermore, induction of triploidy seems to inhibit some aspects of physiological performance, such as reduced tolerance of low oxygen conditions. Since these conditions pertain in cage farming during sea louse treatments or at excessively high water temperatures, there is considerable industry objection to the process.

Sex determination mechanisms are poorly known in other finfish species used in aquaculture (but see Mank et al. 2006) and need further study before triploidy should be attempted with all-female strains. Where used, the incorporation of triploid induction into existing and future breeding programmes is also recommended, so as to avoid lowered genetic variability in triploids. Early attempts to induce triploidy in bivalve shellfish using chemicals like cytochalasin-B were only partially successful and rarely produced 100% triploids. These methods are not therefore reliable as a means of inducing sterility. However, using a proprietary, patented, process, a USA company can now produce tetraploid brood stock. Males are identified from select brood stock lines and sacrificed for their sperm. The sperm from these tetraploid males is naturally diploid (2N), instead of haploid (1N) as it would be from typical diploid males. The diploid sperm is then added to haploid eggs from the customer's normal diploid brood stock. No chemical or pressure induction treatment is used. The resulting zygotes are genetically triploid, with two sets of chromosomes contributed by the sperm and one set contributed by the egg (Benoit et al. 2000).

4.8 Risk Analysis

Important questions must be raised with respect to species in which wild populations are most at risk from incursions of reared conspecifics. Table 4.2 addresses some of these questions. At present the numbers of marine and anadromous fish and invertebrate species cultured in Europe, other than for salmon, flat oyster and mussel is much lower than for wild populations. However, this is completely opposite to the situation in salmon in the eastern North Atlantic, where cultured fish outnumber wild conspecifics by more than two orders of magnitude.

The marine species that are currently farmed in Europe (such as cod, halibut, turbot, sea bass, sea bream, lobster, scallop, mussels, native and Pacific oysters, abalone and Manila clams) differ from Atlantic salmon in many important respects. As noted above, census population sizes of native wild marine fish and invertebrates are several orders of magnitude greater than salmonids and genetic population (stock) structure appears to be much less well defined (Table 4.2). In addition, as mentioned above, the extent of local adaptation for the majority of marine species has not been established. Thus, from a genetic viewpoint the risk to wild populations might be considered to be far less (but see Bekkevold et al. 2006). However, as farmed production of some or all of these species is predicted to increase greatly, with extensive ocean ranching being promoted in several quarters (Leber et al. 2004), and wild populations are steadily decreasing due to overfishing (Worm et al. 2006), the situation may change in the future. Furthermore, diseases, which are likely to be much more of a problem under high density rearing conditions and yet are usually controllable by medication in captivity, might become a severe problem

Table 4.2 Major marine and anadromous aquaculture species produced in Europe with consideration of risk to wild conspecifics and other species. In compiling potential risk, level of production, whether hatchery reared or collected from the wild, the degree of population differentiation and the amount of research, which has been conducted on genetic effects of interactions are considered

Species	Production	*Hatchery*	Farmed (F) Stocked (S)	Native/ exotic	Wild population Differentiation	Direct problems	Indirect problems	Research status	Potential risk
Atlantic salmon	Very high	Yes	F/S	Native	High	Yes	Yes	Much	Proven to be high
Mussels	Very high	No	F(in wild)	Native	Low (intra-specifically)	Yes	Yes	Little	Low?
Pacific oyster	Very high	Yes	F(in wild)	Exotic	NA	No	Yes	Little	Medium
Sea bass	High	Yes	F	Native	Medium	Yes	Yes	Little	Medium
Sea bream	High	Yes	F/S	Native	Low	Yes	Yes	Little	Medium
Cod	Medium	Yes	F/S	Native	Medium	Yes	Yes	Some	Medium
Flat oyster	Medium	No	F(in wild)	Native	Low	Yes	Yes	Little	Low
Manila clam	Medium	Yes	F(in wild)	Exotic	NA	No	Yes	Little	Medium
Turbot	Medium	Yes	F/S	Native	Low	Yes	Yes	Little	Low
Halibut	Low	Yes	F	Native	Low	Yes	Yes	Little	Low
Lobster	Low	Yes	S	Native	Low	Yes	Yes	Little	Low (except locally)
Scallop	Low	Yes	F/S	Native	Low	Yes	Yes	Little	Low

for wild populations in areas where interactions occur. In this context the sea louse *Caligus elongatus* may become a major threat because it can infect both wild and farmed salmon and cod. A recently completed EU project (GENIMPACT (www.genimpact.imr.no)) is considering the genetic aspects of cultured/wild interactions in European cultured aquatic species.

It is concluded that the likely severity of detrimental effects on individual native species depends on size and status of native populations, and extent and frequency of introductions (and on genetic composition of cultured strains). Effects are most severe where wild populations sizes were historically small and have been further reduced by overfishing and increased natural mortality (e.g., salmon throughout their range), and also where introductions are large and regular (annual) and where cultured strains have undergone several generations of hatchery rearing and strong artificial selection.

4.9 Recommendations

1. Governmental regulations in different countries should be reviewed to determine whether these give adequate protection to wild populations potentially threatened by direct and/or indirect interactions with reared fish or invertebrates.
2. For many species currently reared in Europe and elsewhere there is inadequate knowledge of the genetic population structure. Detailed investigations are urgently needed and also some earlier studies will need repeating as more discriminatory molecular markers become available
3. It should be recognised that entirely different genetic principles are involved in producing strains for farming under conditions of confinement throughout the lifecycle, compared with strains for stocking/ranching (or indeed for farming in the wild without confinement, e.g., for bivalve molluscs).
4. It is recognised that traditional genetic methods of strain improvement by selection will continue, assisted by Marker Assisted Selection and other novel molecular techniques. In contrast minimum genetic manipulation should be applied to organisms to be used for stocking/ ranching.
5. Experiments are urgently required to learn more about domestication selection.
6. Cage/ tank outflow design should be matched to particular site conditions so as to minimise escapes and potential for interactions.
7. While beyond the defined scope of the current review, it is recognised that there is a vast and expanding production of marine algae, chiefly in Asia. Because there is no containment involved in the culture methods, the potential for introgression with wild populations must be considerable. It is recommended that the effects of such interaction events be investigated.
8. The production of sterile strains should be considered (note that the production of single-sex animals of several species is also desired by the aquaculture industry). While this will ameliorate the genetic effects of direct interaction, it will do almost nothing to avoid indirect effects.

9. Genetically Modified (transgenic) animals should be prohibited from "open" culture until further experiments are conducted.
10. Movement of animals for rearing between documented major population groupings should be prohibited, e.g., eastern North America, western Europe and rivers around the Baltic Sea for Atlantic salmon; western and eastern North Atlantic and Baltic for cod; and Mediterranean and east Atlantic for sea bass.
11. Most experimental investigations of interactions to date have involved Atlantic salmon. This species appears to be very different both in life history and genetics from all of the other species cultured in Europe, so studies are needed with some of these other species to investigate whether general principles can be developed.

4.10 Conclusions

As noted at the beginning of this chapter it is anticipated that aquaculture will continue to grow both for closed-cycle farming and for stocking/ranching. This expanded production will mean that there is the potential for greater numbers of reared fish and invertebrates to be inadvertently or deliberately introduced to the wild, with the latter being potentially as detrimental as the former. Thus, the problem of genetic interactions may increase, unless (i) measures are taken to avoid escapes or ameliorate their effects, and (ii) more research is undertaken to quantify and reduce the effect of deliberately released reared species. Another incompletely researched area is whether it is possible to extrapolate from previous results, mostly with Atlantic salmon, to other marine fish and invertebrate species. Most previous studies have concentrated on direct genetic effects and used relative survival as a surrogate for reproductive fitness. It is now clear from some of the case studies reported here that indirect effects, mediated either pathologically (by diseases) or ecologically, may have severe detrimental influence on wild conspecifics or other native species in the natural environment. Since these indirect effects may be masked by direct effects, it is important to devise experiments to unravel these influences. Advances in molecular genetics, particularly in the area of genomics and the study of adaptive genes, e.g., associated with disease resistance, should greatly assist future investigations.

References

Agnalt A-L, Jørstad KE, Kristiansen T, Nøstvold E, Farestveit E, Næss H, Paulsen OI, Svåsand T (2004) Enhancing the European lobster (*Homarus gammarus*) stock at Kvitsøy Islands: Perspectives of rebuilding Norwegian stocks. In: Leber KM, Kitada S, Blankenship HL, Svåsand T (eds.), Stock enhancement and sea ranching: developments, pitfalls and opportunities. Blackwell, Oxford, pp. 415–426

Agnalt AL, van der Meeren GI, Jorstad KE, Naess H, Farestveit E, Nostvold E, Svasand T, Korsoen E, Ydstebo L (1999) Stock enhancement of European lobster (*Homarus gammarus*):

a large-scale experiment off South Western Norway. In: Howell BR, Moksness E, Svasand T (eds.), Stock enhancement and sea ranching. Fishing News Books, pp. 401–419

Andrews JD (1988) Epizootiology of the disease caused by the oyster pathogen *Perkinsus marinus* and its effect on the oyster industry. American Fisheries Society Special Publications, 18: 47–63

Aoyagi K, Dijkstra JM, Xia C, Denda I, Ototake M, Hashimoto K, Nakanishi T (2002) Classical MHC class I genes composed of highly divergent sequence lineages share a single locus in rainbow trout (*Oncorhynchus mykiss*). Journal of Immunology, 168: 260–273

Arkush KD, Giese AR, Mendonca HL, McBride AM, Marty GD, Hedrick PW (2002) Resistance to three pathogens in the endangered winter-run chinook salmon (*Oncorhynchus tshawytscha*): effects of inbreeding and major histocompatibility complex genotypes. Canadian Journal of Fisheries and Aquatic Sciences 59: 966–975

Armitage D (2006) Unpublished M.Sc. thesis, University College Cork, Ireland

Bakke TA, Jansen PA, Hansen LP (1990) Differences in the host resistance of Atlantic salmon, *Salmo salar* L., stocks to the monogenean, *Gyrodactylus salaris*, Malmberg, 1957. Journal of Fish Biology 30: 713–721

Beebee T, Rowe G (2004) An introduction to molecular ecology. Oxford University Press, New York, 346 pp

Bekkevold D, Hansen MM, Nielsen EE (2006) Genetic impact of gadoid culture on wild fish populations: predictions, lessons from salmonids, and possibilities for minimising adverse effects. ICES Journal of Marine Sciences 63: 198–208

Bekkevold D, Andre C, Dahlgren TG, Clausen LAW, Torstensen E, Mosegaard H, Carvahlo GR, Christensen TB, Norlinder E, Ruzzante DE (2005) Environmental correlates of population differentiation in Atlantic herring. Evolution 59: 2656–2668

Benoit E, Allen SK, Guo X (2000) Delayed meiosis and polar body release in eggs of triploid Pacific oysters, *Crassostrea gigas*, in relation to tetraploid production. Journal of Experimental Marine Biology and Ecology 248: 151–161

Brown BL, Butt AJ, Shelton SW, Meritt D, Paynter KT (2005a) Resistance of *Dermo* in eastern oysters *Crassostrea virginica* (Gmelin), of North Carolina but not Chesapeake Bay heritage. Aquaculture Research 36: 1391–1399

Brown BL, Butt AJ, Meritt D, Paynter KT (2005b) Evaluation of resistance to *Dermo* in eastern oyster strains tested in Chesapeake Bay. Aquaculture Research 36: 1544–1554

Boudry P, Heurtebise S, Collet B, Cornette F, Gérard A (1998) Differentiation between populations of the Portuguese oyster *Crassostrea angulata* (Lamark) and the Pacific oyster, *Crassostrea gigas* (Thunberg), revealed by mtDNA RFLP analysis. Journal of Experimental Marine Biology and Ecology 226: 279–291

Clifford S, McGinnity P, Ferguson A (1998) Genetic changes in Atlantic salmon (*Salmo salar* L.) in NW Irish rivers resulting from escapes of adult farm salmon. Canadian Journal of Fisheries and Aquatic Sciences 55: 358–363

Conover DO, Clarke LM, Munch SB, Wagner GN (2006) Spatial and temporal scales of adaptive divergence in marine fishes and the implications for conservation. Journal of Fish Biology 69: 21–47

Coughlan JP, Imsland AK, Galvin PT, FitzGerald RD, Naevdal G, Cross TF (1998) Microsatellite DNA variation in wild populations and farmed strains of turbot from Ireland and Norway: a preliminary study. Journal of Fish Biology 52: 916–922

Coughlan J, McGinnity P, O'Farrell B, Dillane E, Diserud O, de Eyto E, O'Farrell K, Whelan K, Stet RJM, Cross TF (2006) Temporal variation in an immune response gene (MHC I) in anadromous *Salmo trutta* in an Irish river before and during aquaculture activities. ICES Journal of Marine Science 63: 1248–1255

Cross TF, niChallanain DG (1991) Genetic characterisation of Atlantic salmon *(Salmo salar)* lines farmed in Ireland. Aquaculture 98: 209–216

Cross TF, King J (1983) Genetic effects of hatchery rearing in Atlantic salmon. Aquaculture 33: 33–40

Cross T, Bailey J, Friars G, O'Flynn F (1993) Maintenance of genetic variability in reared Atlantic salmon (*Salmo salar*) stocks. In: Mills DH (ed.), Salmon in the sea, pp. 356–366

Cross TF, McGinnity P, Coughlan J, Dillane E, Ferguson A, Koljonen M-L, Milner N, O'Reilly P, Vasemagi A (2007) Genetic considerations for Stocking and Ranching of Atlantic salmon. In: Verspoor E, Stradmeyer L, Nielsen JL (eds.), The Atlantic salmon: genetics, conservation and management. Blackwell, Oxford, pp. 325–356

Culloty SC, Novoa B, Pernas M, Longshaw M, Mulcahy MF, Feist SW, Figueras A (1999) Susceptibility of a number of bivalve species to the protozoan parasite *Bonamia ostreae* and their ability to act as vectors for this parasite. Diseases of Aquatic Organisms 37: 73–80

Culloty SC, Cronin MA, Mulcahy MF (2004) Potential resistance of a number of populations of the oyster *Ostrea edulis* to the parasite *Bonamia ostreae*. Aquaculture 237: 41–58

Devlin RH, D'Andrade M, Uh M, Biagi CA (2004) Population effects of GH transgenic salmon are dependant upon food availability and genotype by environment interactions. Proceedings of National Academy of Sciences USA 101: 9303–9308

Devlin RH, Yesaki TY, Biagi CA, Donaldson EM, Swanson P, Chan WK (1994) Extraordinary salmon growth. Nature 371: 209–210

DeWoody JA, Avise JC (2000) Microsatellite variation in marine, freshwater and anadromous fishes compared with other animals. Journal of Fish Biology 56: 461–473

Duchesne P, Godbout MH, Bernatchez L (2002) PAPA (package for the analysis of parental allocation): a computer program for simulated and real parental allocation. Molecular Ecology Notes 2:191–193

de Eyto E, McGinnity P, Consuegra S, Coughlan J, Tufto J, Farrell K, Jordan WC, Cross T, Megens H-J, Stet R (2007) Natural selection acts on Atlantic salmon MHC variability in the wild. Proceedings of the Royal Society B 274: 861–869

Einum S, Fleming I (1997) Genetic divergence and interactions in the wild among native, farmed and hybrid Atlantic salmon. Journal of Fish Biology 50: 634–651

Ferguson A, Fleming IA, Hindar K, Skaala Ø, McGinnity P, Cross T, Prodöhl P (2007) Farm Escapes. In: Verspoor E, Stradmeyer L, Nielsen JL (eds.), The Atlantic salmon: genetics, conservation and management. Blackwell, Oxford, pp. 357–398

FAO (2007) State of World Fisheries and Aquaculture, Food and Agriculture Organization of the United Nations, Rome 2007

Fevolden SE, Pogson GH (1997) Genetic divergence at the synaptophysin (Syp I) locus among Norwegian coastal and north-east Arctic populations of Atlantic cod. Journal of Fish Biology 51: 895–908

Fischer W (1995) Ecological genetics of mussels (*Mytilus edulis* and *M. galloprovincialis*) farmed in suspended culture in south-west Ireland. M.Sc. thesis, University College Cork, Ireland, 125 pp

Fleming IA, Einum S (1997) Experimental tests of genetic divergence of farmed from wild Atlantic salmon due to domestication. ICES Journal of Marine Science 54: 1051–1063

Fleming IA, Hindar K, Mjølnerød IB, Jonsson B, Balstad T, Lamberg A (2000) Lifetime success and interactions of farm salmon invading a native population. Proceedings of the Royal Society B 267: 1517–1523

Ford SE, Haskin HH (1987) Infection and mortality patterns in strains of oysters *Crassostrea virginica* selected for resistance to the parasite *Haplosporidium nelsoni* (MSX). Journal of Protozoology 73: 368–376

Gjedrem T (1999) Genetic improvement of cold water fish species. Aquaculture Research 31: 25–33

Gosling EM (2003) Bivalve molluscs: biology, ecology and culture. Fishing News Books, 443 pp

Grimholt U, Drablos F, Jorgensen SM, Hoyheim B, Stet RJM (2002) The major histocompatibility class I locus in Atlantic salmon (*Salmo salar* L.): polymorphism, linkage analysis and protein modelling. Immunogenetics 54: 570–581

Grimholt U, Larsen S, Nordmo R, Midtlyng P, Kjoeglum S, Storset A, Saebø S, Stet RJM (2003) MHC polymorphism and disease resistance in Atlantic salmon (*Salmo salar*); facing pathogens with single expressed major histocompatibility class I and class II loci. Immunogenetics 55(4):210–219

Guenguen Y, Cadoret JP, Flament D, Barreau-Roumiguiere C, Girardot AL, Garnier J, Hoareau A, Bachere E, Escoubas JM (2003) Immune gene discovery by expressed sequence tags generated from hemocytes of the bacteria challenged oyster *Crassostrea virginica*. Gene 303: 139–145

Hansen LP (2006) Migration and survival of farmed Atlantic salmon (*Salmo salas* L.) released from two Norwegian fish farms. ICES Journal of Marine Science 63: 1211–1217

Hindar K, Balstad T (1994) Salmonid culture and interspecific hybridization. Conservation Biology 8: 881–882

Hindar K, Fleming IA, McGinnity P, Diserud O (2006) Genetic and ecological effects of salmon farming on wild salmon: modelling from experimental results. ICES Journal of Marine Science 63: 1234–1247

Hislop JRG, Webb JH (1992) Escaped farmed Atlantic salmon (*Salmo salar*) feeding in Scottish coastal waters. Aquaculture and Fisheries Management 23: 721–723

Hutchinson WF, Carvalho GR, Rogers SI (2001) Marked genetic structuring in localised spawning populations of cod *Gadus morhua* in the North Sea and adjoining waters, as revealed by microsatellites. Marine Ecology Progress Series 223: 251–260

Hutchinson WF, Oosterhout CV, Rogers SI, Carvalho GR (2003) Temporal analysis of archived samples indicates marked genetic changes in declining North Sea cod (*Gadus morhua*). Proceedings of the Royal Society of London, Series B, Biological Sciences 270: 2125–2132

Imsland AK, J6nsd6ttir ÓDB (2002) Is there a genetic basis to growth in Atlantic cod? Fish and Fisheries 3: 36–52

Jonasson J, Gjerde B, Gjedrem T (1997) Genetic parameters for return rate and body weight in sea ranched Atlantic salmon. Aquaculture 154: 219–231

Jones AG, Ardren WR (2003) Methods of parentage analysis in natural populations. Molecular Ecology 12: 2511–2523

Jonsson B, Jonsson N (2006) Cultured Atlantic salmon in nature: a review of their ecology and interaction with wild fish. ICES Journal of Marine Science 63: 1162–1181

Jorstad KJ, van der Meeren T, Paulsen OI, Thomsen T, Svasand T (2006) "Escapement" of eggs from farmed cod spawning in net pens and offspring intermingling with natural spawned larvae. Sea ranching and stock enhancement symposium. Seattle, USA, September 2006

Jorstad KE, Paulsen OI, Naevdal G, Thorkildsen S (1994) Genetic studies of cod, *Gadus morhua* L. in Masfjorden, western Norway: comparison between the local stock and released, artificially reared cod. Aquaculture and Fisheries Management 25:77–91

Jorstad KE, Prodohl PA, Kristiansen TS, Hughes M, Farestveit E, Taggart JB, Agnalt A-L, Ferguson A (2005) Communal larval rearing of European lobster (*Homarus gammarus*): family identification by microsatellite DNA profiling and offspring fitness comparisons. Aquaculture 247: 275–285

Langefors Å, von Schantz T, Widegren B (1998) Allelic variation of Mhc class II in Atlantic salmon: a population genetic analysis. Heredity 80:568–575

Launey S, Barre M, Gerard A, Naciri-Graven Y (2001) Population bottleneck and effective size in *Bonamia ostreae* resistant populations of *Ostreae edulis* as inferred by microsatellite markers. Genetic Research 78 : 259–270

Leber KM, Kitada S, Blankenship HL, Svasand T (2004) Stock enhancement and sea ranching: developments, pitfalls and opportunities. Blackwell, Oxford. 2nd edn, 562 pp

Lohm J, Grahn M, Langefors Å, Andersen O, Storset A, von Schantz T (2002) Experimental evidence for major histocompatibility complex–allele-specific resistance to a bacterial infection. Proceedings of the Royal Society B 269: 2029–2034

Mank JE, Promislow DEL, Avise JC (2006) Evolution of alternative sex-determining mechanisms in teleost fishes. Biological Journal of the Linnean Society 87: 83–93

Mantovani S, Castaldelli G, Rossi R, Fano EA (2006) The infaunal community in experimentally seeded low and high density Manila clam (*Tapes philippinarum*) beds in a Po River Delta lagoon, Italy. ICES Journal of Marine Science 63: 860 – 866

Marteil L (1976) La conchyliculture française. 2. Biologie de l'huître et de la moule. Rev. Trav. Inst. Pêches marit. 40: 149–345

Marty GD, Bower SM, Clarke KR, Meyer G, Lowe G, Osborn AL, Chow EP, Hannah H, Byrne S, Sojonky K, Robinson JH (2006) Histopathology and a real-time PCR assay for detection of *Bonamia ostreae* in *Ostrea edulis* cultured in western Canada. Aquaculture 261: 33–42

McElligott EA, Maguire TMF, Cross TF (1987) The amount and nature of electrophoretically-detectable genetic polymorphism in hatchery reared Atlantic salmon (*Salmo salar* L.) in Ireland. Conseils Internationale pour l'Exploration de la Mer CM Doc. 1987/M:13, Ref E, 10 pp

McGinnity P, de Eyto E, Cross T, Coughlan J, Ferguson A (in press) Population specific smolt development, migration and maturity schedules in Atlantic salmon in a natural river environment. Proceedings of the 7th International Workshop on Salmonid Smoltification held in Tono, Japan, 2005. Aquaculture (Special Issue)-in press

McGinnity P, Prodöhl P, O'Maoiléidigh N, Hynes R, Cotter D, Baker N, O'Hea B, Ferguson A (2004) Differential lifetime success and performance of native and non-native Atlantic salmon examined under communal natural conditions. Journal of Fish Biology 65 (Supplement A): 173–187

McGinnity P, Stone C, Taggart JB, Cooke D, Cotter D, Hynes R, McCamley C, Cross T, Ferguson A (1997) Genetic impact of escaped farmed salmon (*Salmo salar* L.) on native populations: use of DNA profiling to assess freshwater performance of wild, farmed, and hybrid progeny in a natural river environment. ICES Journal of Marine Science 54: 998–1008

McGinnity P, Prodöhl P, Ferguson A, Hynes R, O'Maoiléidigh N, Baker N, Cotter D, O'Hea B, Cooke D, Rogan G, Taggart J, Cross T (2003) Fitness reduction and potential extinction of wild populations of Atlantic salmon, *Salmo salar*, as a result of interactions with escaped farm salmon. Proceedings of the Royal Society B 270: 2443–2450

McMeel OM, Hoey EM, Ferguson A (2001) Partial nucleotide sequences, and routine typing by polymerase chain reaction-restriction fragment length polymorphism, of the brown trout (*Salmo trutta*) lactate dehydrogenase, LDH-C1*90 and *100 alleles. Molecular Ecology 10: 29–34

Miller KM, Shaorang L, Ming TJ, Kaukinen KH, Schulze AD (2006) The salmonid MHC class I: more ancient loci uncovered. Immunogenetics 58: 571–589

Naciri-Graven Y, Martin AG, Baud JP, Renault T, Gerard A (1998) Selecting the flat oyster *Ostreae edulis* (L.) for survival when infected with the parasite *Bonamia ostreae*. Journal of Experimental Marine Biology and Ecology 224: 91–107

Naciri-Graven Y, Haure J, Gerard A, Baud JP (1999) Comparative growth of *Bonamia ostreae* resistant and wild flat oyster *Ostrea edulis* in an intensive system II Second year of the experiment. Aquaculture 171: 195–208

Nell JA, Perkins B (2006) Evaluation of the progeny of third generation Sydney rock oyster *Saccostrea glomerata* (Gould 1850) breeding lines for resistance to QX disease *Marteilia sydneyi* and winter mortality *Bonamia roughleyi*. Aquaculture Research 37: 693–700

Norris AT, Bradley DG, Cunningham EP (1999) Microsatellite genetic variation between and within farmed and wild Atlantic salmon (*Salmo salar*) populations. Aquaculture 180: 247–264

O'Leary DB, Coughlan J, Dillane E, McCarthy TV, Cross TF (2007) Microsatellite variation in cod *Gadus morhua* throughout its geographic range. Journal of Fish Biology 70: 310–335

Rajagopal S, van der Velde G, Jansen J, van der Gaag M, Atsma G, Janssen-Mommen J, Polma H, Jenner HA (2005) Thermal tolerance of the invasive oyster *Crassostrea gigas*: feasibility of heat treatment as an antifouling option. Water Research 39: 4335 – 4342

Reece KS, Bushek D, Hudson KL, Graves JE (2001) Geographic distribution of *Perkinsus marinus* genetic strains along the Atlantic and Gulf coasts of the USA. Marine Biology 139: 1047–1055

Reisenbichler RR, Rubin SP, Wetzel L, Phelps SR (2004) Natural selection after release from a hatchery leads to domestication in steelhead, *Oncorhynchus mykiss*. In: KM Leber, HL Blankenship, S Kitada, T Svåsand (Editor), Stock enhancement and sea ranching: developments, pitfalls and opportunities, 2nd edn. Blackwell, Oxford, pp. 371–383

Roberge C, Einum S, Guderly H, Bernatchez L (2006) Rapid parallel evolutionary changes of gene transcription profiles in farmed Atlantic salmon. Molecular Ecology 15: 9–20

Ruzzante DE, Taggart CT, Cook D (1999) A review of evidence for genetic structure of cod (*Gadus morhua*) populations in the NW Atlantic and population affinities of larval cod off Newfoundland and the Gulf of St. Lawrence. Fisheries Research 43: 79–97

Sanger F, Nicklen S, Coulson AR (1977) DNA sequencing with chain-terminating inhibitors. Proceedings of the National Academy of Sciences (USA) 74: 5463–5467

Sato A, Figueroa F, Murray BW, Malaga-Trillo E, Zaleska-Rutczynska Z, Sultmann H, Toyosawa S, Wedekind C, Steck N, Klein J (2000) Nonlinkage of major histocompatibility complex class I and class II loci in bony fishes. Immunogenetics 51:108–116

Shum BP, Guethlein L, Flodin LR, Adkison MD, Hedrick RP, Nehring RB, Stet RJ, Secombes C, Parham P (2001) Modes of salmonid MHC Class I and II evolution differ from the primate paradigm. Journal of Immunology 166: 3297–3308

Shumway SE, Parsons GJ (eds.) (2006) Scallops: Biology, ecology and aquaculture (2nd edn). In: Developments in aquaculture and fisheries science, 35, Elsevier, 1406 pp

Stefansson MO, Coughlan J, Cross TF (2001) Microsatellite DNA variation in reared strains of turbot (*Scophthalmus maximus*) and Atlantic halibut (*Hippoglossus hippoglossus*) compared with wild samples. Conseils Internationale pour l'Exploration de la Mer CM Doc. 2001/L:16, 29 pp

Stet RJM, deVries B, Mudde K, Hermsen T, van Heerwaarden J, Shum B P, Grimholt U (2002) Unique haplotypes of co-segregating major histocompatibility class II A and class II B alleles in Atlantic salmon (*Salmo salar*) give rise to diverse class II genotypes. Immunogenetics 54: 320 – 331

Tanguy A, Guo X, Ford SE (2004) Discovery of genes expressed in response to *Perkinsus marinus* challenge in Eastern (*Crassostrea virginica*) and Pacific (*C. gigas*) oysters. Gene 338: 121–131

Triantaphyllidis A, Apostolidis A, Katsares V, Kelly E, Hughes M, Jorstad K, Tsolou R, Hynes R, Triantaphyllidis C (2005) Mitochondrial DNA variation in the European lobster throughout the range. Marine Biology 146:223–235

Uki N (2006) Stock enhancement of the Japanese scallop, *Patinopecten yessoensis* in Hokkaido. Fisheries Research 80: 62 – 66

Van Banning P (1987) Further results on the *Bonamia ostreae* challenge tests in Dutch oyster culture. Aquaculture 67: 191–194

Verspoor E (1988) Reduced genetic variability in first generation hatchery populations of Atlantic salmon (*Salmo salar*). Canadian Journal of Fisheries and Aquatic Sciences 45:1686–1690

Verspoor E, Stradmyer L, Nielsen JL (eds.) (2007) The Atlantic salmon: genetics, conservation and management. Blackwell, Oxford, 500 pp

Waples R (1998) Separating the wheat from the chaff: patterns of genetic differentiation in high gene flow species. Journal of Heredity 89: 438–450

Wegner KM, Reusch TBH, Kalbe M (2003) Multiple parasites are driving major histocompatibility complex polymorphism in the wild. Journal of Evolutionary Biology 16: 224–232

Worm B, Barbier E B, Beaumont N, Duffy JE, Folke C, Halpern BS, Jackson JBC, Lotze HK, Micheli F, Palumbi SR, Sala E, Selkoe KA, Stachowicz JJ, Watson R (2006) Impacts of biodiversity loss on ocean ecosystem services. Science 314: 787–790

Youngson AF, Webb JH, Thompson CE, Knox D (1993) Spawning of escaped farmed Atlantic salmon (*Salmo salar*): hybridization of females with brown trout (*Salmo trutta*). Canadian Journal of Fisheries and Aquatic Sciences 50: 1986–1990

Youngson AF, Dosdat A, Saroglia M, Jordan WC (2001) Genetic interactions between marine finfish species in European aquaculture and wild conspecifics. Journal of Applied Ichthyology 17:153–162

Glossary

(terms listed in bold in explanations are themselves explained):

Adaptive loci loci influenced by natural selection, usually coding for proteins

Alleles DNA sequences occurring at a **locus** in **diploid** systems

Allozymes allelic forms of enzymes (determined by **functional genes**) and detected by protein **electrophoresis**

Assortative mating non random mating, choosing a particular sub-set of a population for mating

Breeding programme programme involving rearing and artificial selection for desirable traits such as growth rate

Common garden experiment where two groups are tested in a common environment so focussing on genetic differences

Dam female parent

Diploid nuclear systems where one **allele** derives from the mother and one from the father

Direct interactions involving interbreeding

DNA sequences nucleic acid (adenine, thymine, cytosine, guanine) arrangement

Electrophoresis protein or DNA separation method based on rate of migration in an electric field

F1 and F2 hybrids first and second generation **hybrids**

Fitness Darwinian fitness, reproductive output

Functional genes DNA sequences coding for proteins or protein regulation

Genetic composition frequencies of alleles at each **locus**

Genetic drift chance alteration of allele frequencies between generations; much greater in small populations

Genetic variability measures of numbers of allelic genes at each **locus**

Genome all the DNA of an individual including the **functional genes**

Haploid mitochondrial systems with only one maternally-derived DNA **sequence**

Heterozygote individual with two different **alleles** at a **locus**

Heterozygosity proportion of **diploid** individuals in a sample that are **heterozygotes**

Homozygote individual with two identical **alleles** at a **locus**

Hybrid cross between two forms or species

Indirect interactions not involving interbreeding

Individual assignment statistical technique using genotype to assess likely population membership

Interspecific between species

Intraspecific within a species

Local adaptation where a population is optimally adapted to its particular area

Locus place on a pair of homologous chromosomes where two **alleles** occur in diploid systems

Mass spawning where several females and males are placed in a tank and allowed to spawn randomly (opposite to single pair mating)
Marker assisted selection using molecular markers to enhance traditional breeding programmes
MHC major histocompatibility complex; immune-response genes
Microarray plate with sequences of large numbers of functional genes applied, so that the level of activity of each can be assessed
Microsatellite loci tandemly arrayed DNA repeats of two or more nucleotides as used in human forensics
Mitochondria organelles in each cell with **haploid genomes**
Natal homing returning to the area of birth to reproduce (=philopatry)
Neutral loci loci unaffected by natural selection
Outbreeding reduction of fitness due to crossing of distinct populations or species, due to break down of co-adapted genes complexes
Progenitor population ancestral population
Quantitative trait factor such as growth rate which is normally-distributed amongst individuals of a population
Sex ratio number of females relative to males
Sire male parent
Stock fisheries term for a management unit, which may be a genetic population
Subtractive hybridisation technique to concentrate on genes with different levels of activity in microarrays
Transgenic GM organism where there has been the introduction of a gene from another species
Translocated movement of species or population outside its native area
Triploid organism with three sets of chromosomes instead of the normal two, usually with two female sets
VNTR variable number tandem repeat, microsatellite and minisatellite DNA sequences

Chapter 5
Non-Native Aquaculture Species Releases: Implications for Aquatic Ecosystems

Elizabeth J. Cook[1,*], Gail Ashton[1], Marnie Campbell[4], Ashley Coutts[2], Stephan Gollasch[3], Chad Hewitt[4], Hui Liu[5], Dan Minchin[6], Gregory Ruiz[7], and Richard Shucksmith[1]

Abstract Aquaculture is undergoing a rapid worldwide expansion. Of significant concern is the increasing use of non-native species, with subsequent escapes of these species and their associated pathogens and parasites posing a serious threat to native biodiversity, economic value and ecosystem function, particularly in regions rich in endemic species. The contribution of non-native species to the growth of the global aquaculture industry and the economic benefits that it has brought to many developing countries cannot be underestimated. However, minimizing the escapes of non-native aquaculture species must be a high priority for resource managers, conservationists and the aquaculture industry. This paper reviews intentional and unintentional non-native aquaculture introductions and the environmental consequences that escapes can have on the aquatic environment and presents a potential system of risk evaluation, management and funding mechanisms to assist in the long term sustainable development of the aquaculture industry.

Keywords Non-native species; aquaculture; introductions; aquatic ecosystems; biodiversity hotspots; risk evaluation and management

---

[1] Present Address: Scottish Association for Marine Science, Dunstaffnage Marine Laboratory, Oban, PA37 1QA, UK. Tel. +44 1631 559 243; Fax +44 1631 559 001; E-mail: ejc@sams.ac.uk

[2] Australian Government Department of Agriculture, Fisheries and Forestry GPO Box 858, Canberra Act 2601, Australia

[3] GoConsult, Bahrenfelder Str. 73a, Hamburg 22765, Germany

[4] National Centre for Marine and Coastal Conservation, Australian Maritime College, Private Mail Bag 10, Rosebud, Victoria 3939, Australia

[5] Yellow Sea Fisheries Research Institute, Chinese Academy of Fishery Science, 106 Nanjing Road, Qingdao, China 266071

[6] Marine Organism Investigations, Marina Village, Ballina, Killaloe, Co. Clare, Ireland

[7] Smithsonian Environmental Research Centre, 647 Contees Wharf Road, P.O. Box 28, Edgewater MD 21037, USA

* Corresponding author

M. Holmer et al. (eds.), *Aquaculture in the Ecosystem.*

5.1 Intentional Introduction of Non-Native Species for Aquaculture

Global aquaculture production has reached 59.4 million tonnes per year, worth US$70.3 billion accounting for almost 50% of world seafood production (FAO 2006a). It has experienced average annual growth rates of 8.8% from 1950 to 2004 (FAO 2006b) and it exceeded wild capture fisheries in Asia in 2002 (FAO 2006a) (Fig. 5.1). In many regions of the world, non-native species have been intentionally introduced for aquaculture purposes and have contributed significantly to the expansion of the industry (Welcomme 1992; Dextrase and Cocarelli 2000) (Fig. 5.2). These species provide considerable economic and social benefits, particularly in developing countries and are typically selected for production based upon: (a) the perceived poor performance of available native species relative to non-natives, including their slow growth rates, lower yield, reduced resistance to disease, tolerance to overcrowding and hardiness to environmental fluctuations; (b) proven production techniques that are readily transferred to new locations; and (c) new commercial opportunities, specifically in developing regions, utilising pre-established global markets (FAO 2006b).

The majority of intentional introductions have occurred in the last century for stocking and aquaculture purposes (Holick 1984; Welcomme 1991; Minchin and Rosenthal 2002; Goren and Galil 2005) and, with the current pace of technological development, it is highly likely that further non-native species and their hybrids will be trialled in countries outside their native range (Minchin and Rosenthal 2002). At present, four non-native species are the focus of intensive aquaculture efforts on

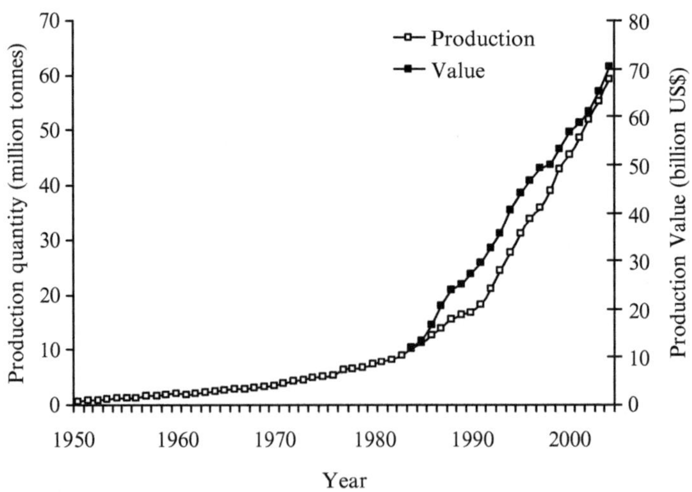

Fig. 5.1 World aquaculture production (including plants) (million tonnes) and value (billion US$) from 1950 to 2004 (FAO 2006a)

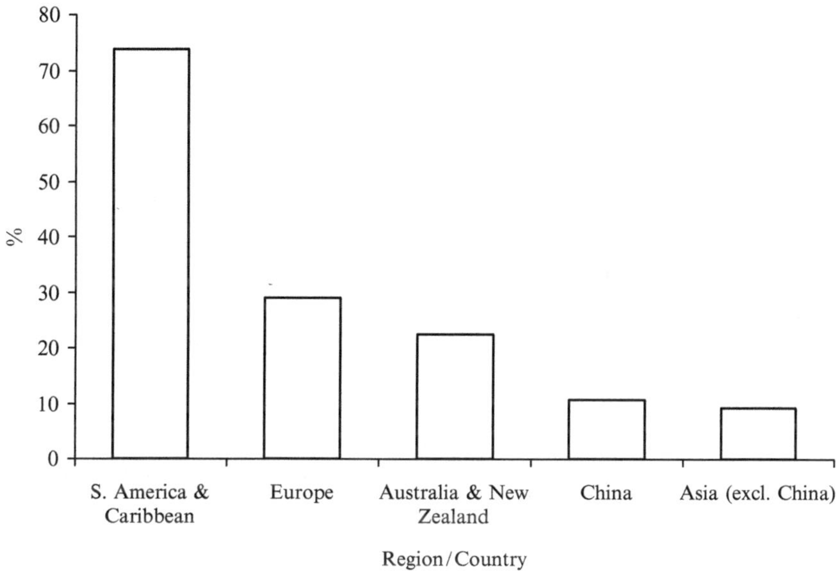

Fig. 5.2 Percentage of aquaculture production by weight (excluding plants) in 2005 based on non-native species in regions/countries with intensive aquaculture activity (FAO 2006a)

multiple continents; the Pacific white shrimp *Penaeus vannemei*, the Nile tilapia *Oreochromis niloticus*, the Atlantic salmon *Salmo salar*, and the Pacific cupped oyster *Crassostrea gigas* (FAO 2006b).

5.1.1 Non-Native Aquaculture Production in China

China has by far the greatest aquaculture industry, producing over 41.3 million tonnes in 2004 and approximately 10% of the total production in 2005 consisted of non-native species (FAO 2006a; Fig. 5.2). The importance of non-native species to the rapid increase in China's aquaculture production in the latter half of the 20th century can not be underestimated. In 1959, China introduced rainbow trout *Oncorhynchus mykiss*, an indigenous species of America, from North Korea: the first of many aquatic species introductions. Over the next half a century, more than one hundred aquatic species were introduced to China and over 20% of them have been widely cultivated (Zhu 2000). These species include 83 finfish species, such as *Oreochromis niloticus*, *Scophthalmus maximus*, *Colossoma brachypomum* and *Micropterus salmoides*; six crustacean species such as *Litopenaeus vannamei* and *Cherax quadricarinatus*; fourteen mollusc species such as *Argopecten irradians* and *Ampullaria gigas*; and nine species of turtle and tortoise (Li 2005). Currently, over 10,000 t are produced annually for each of thirteen introduced aquaculture species in China (Li 2005).

Tilapia was first introduced into China, either from Vietnam in the 1950s (Liu et al. 2000) or Africa in the 1970s (Zhu 2000), dependent on the source; this freshwater finfish is now cultivated in all 29 provinces in China with an annual production of 805,000t (2003), which accounts for 60% of the world's total production (Li 2005). The bay scallop *Argopecten irradians* introduced from the United States in 1982 has increased in production from less than 100,000t per year to an annual production of more than 600,000t (Liu and Zhu 2006). The Yesso scallop *Patinopecten* (*Mizuhopecten*) *yessoensis,* an indigenous species of Japan, Korea and Pacific Russia was introduced to China in 1981 and in 15 years has become a major off-bottom mariculture species at a shell-on production of 910,000t in 2004 (FAO 2006c). This production is three times the amount produced by Japan, the only other major producer of this species (FAO 2006c) and generates a total value of more than US$1 billion per year. The Zhangzidao Fishery Cooperation Group in Dalian is the largest producer and supplies 90% of total Yesso scallop products in China. About 800 million spat of Yesso scallop are produced annually by the hatcheries of the company, which seed the company's 400 km^2 seabed culture area.

Introductions of non-native species have also helped the industry in the face of serious problems. The outbreak of virus disease in Chinese shrimp *Penaeus (Fennerpopenaeus) orientalis* affected mariculture shrimp production in the early 1990s; however, the expanded farming of the Pacific white shrimp *Penaeus vannamei,* the Japanese shrimp *Penaeus japonicus* and the Giant tiger shrimp *Penaeus monodon* rapidly reversed the decline in shrimp production. The first commercial shipment of disease resistant *P. vannamei* broodstock from the Americas to Asia was from Hawaii to Taiwan Province of China in 1996, and from Hawaii to mainland China in 1998 (Wyban 2002). In 2004, over 735,000t of *P. vannamei* were produced in China, more than the rest of the world combined (Chen 2006; FAO 2006a).

5.1.2 Non-Native Aquaculture Production in Europe

Several non-native species have been in various forms of culture for over 2,000 years in Europe. Perhaps the earliest species cultured was the Common carp *Cyprinus carpio* in ponds in Eastern Europe, which originated from the Manchurian region of China. However, it has only been since Victorian times that aquaculture in Europe evolved and this was mainly out of concern over the depletion of existing fisheries (Wilkins 1989). Early experiments on rearing native oysters *Ostrea edulis* to produce settlements in ponds during the first few decades of the 1900s were occasionally successful. However, it was stock movements of the native oyster from continental Europe that were used to increase production. These were supplemented with imports of the American oyster *Crassostrea virginica* to both Britain and Ireland. This became a regular trade over about 40 years from the ~1880s. Such long distance movements became possible with reduced journey times owing to the development of steam transport (Minchin 2006). Intercontinental trade soon led to the movement of other species including fertilised salmonid eggs, easily transported and managed in hatchery flow trays from the 1880s. Several species later became exchanged or spread to different world regions which led to introductions of the Rainbow trout *Oncorhynchus mykiss,* subsequently cultured in freshwater as well as in sea cages, and Brook trout *Salvelinus fontinalis* and Lake trout *S. namaycush* for stocking mountain lakes in Europe.

With increases in international trade, improved biological knowledge, production of food for young stages, and increased technological developments, cultivation became practical. The hatchery techniques for bivalves developed by Loosanoff and Davis (1963) in North America soon were utilised in Britain and France from the 1960s and 1970s making it possible to raise several species. Not all of the species imported and used in experimental trials were considered useful (Utting and Spencer 1992). It was the Pacific oyster *Crassostrea gigas* and the Manila clam *Venerupis philippinarum* that became widely used in culture throughout much of northern Europe. Total production of *C. gigas* reached 122,000 t in 2004 and 29% of all aquaculture production consisted of non-native species in Europe by 2005 (FAO 2006a; Fig. 5.2). Other species that were intentionally introduced but are in cultivation at comparatively small levels of production, are, for example, the Japanese abalone *Haliotus discus hannai* in Ireland and the Japanese shrimp *Penaeus japonicus* in Spain.

Some species arrived in Europe accidentally and have subsequently been utilised. One of these, the red alga *Asparagopsis armata,* arrived in ~1940 and is now cultivated for the production of cosmetic products (Kraan and Barrington 2005).

5.1.3 Non-Native Species Production in Latin America and the Caribbean

Countries in Latin America and the Caribbean have exhibited the greatest expansion in their aquaculture industries compared to other regions, experiencing a 21.3% annual growth rate since the 1950s, when aquaculture production was minimal (<7,000 t) (FAO 2006b). Substantial growth in aquaculture production began in the late 1970s, primarily supported by shrimp and salmon production in three countries in South America: Ecuador, Brazil and Chile. The development of the world shrimp market in the 1970s and 80s saw considerable investment in these countries, particularly in Ecuador, which concentrated on the native shrimp species *Penaeus vannamei.*

Brazil also concentrated its production efforts on shrimp and imported the non-natives *P. monodon* and *P. japonicus* in the 1970s (FAO 2006a). The culture of the non-native *P. vannemei* began to increase substantially in the early 1990s in Brazil and this species is now the dominant shrimp species grown in the country with the production of 76,000 t in 2004. The non-native Common carp, *Cyprinus carpio* and the various tilapia species, including the blue *Oreochromis aureus*, Mozambique *O. mossambicus*, Nile *O. niloticus* and Wami *O. urolepis* imported to Brazil in the 1960s and 70s also comprise a large proportion of Brazil's aquaculture production with 114,248 t produced in 2004 (FAO 2006a).

In the late 1980s, Chile began to develop their salmon industry based on the Atlantic salmon *Salmo salar*, which is native to the north-east Atlantic and had been introduced to Chile in 1935 (FAO 2006a). Since 1990, this industry has exhibited one of the highest average annual growth rates (31.4%) compared to other countries' aquaculture activities. The production of non-native salmonids had reached over 550,000t by 2004 (Buschmann et al. 2006; FAO 2006a) and the Chilean government plans to double the production output of this species by 2013 (Ridler et al. 2006).

In the Caribbean, the four main aquaculture producers are Belize, Costa Rica, Cuba and Honduras. In Belize and Honduras, the non-native shrimp, *P. vannemei* is the dominant aquaculture species, comprising 97% and 80% respectively of the total aquaculture production in 2004 (FAO 2006a). In Costa Rica, 18,000t of the non-native Nile tilapia were produced in 2004, comprising 73% of the total aquaculture production for the country. In Cuba, the main aquaculture species is the non-native Silver carp *Hypophthalmichthys molitrix,* which accounts for 54% of the aquaculture production (FAO 2006a).

In 2005, over 74% of the annual production in Latin America and the Caribbean was attributed to non-native species (FAO 2006a; Fig. 5.2), with an economic value of US$3.9 billion in 2004, representing 75% of the total value of aquaculture production in the region. This production is now concentrated on non-native Pacific Whiteleg shrimp *P. vannemei* (in non-Pacific countries), Atlantic *S. salar* and Coho salmon *Oncorhynchus kisutch*, Rainbow trout *O. mykiss*, Nile tilapia *Oreochromis niloticus* and various carp species (Fig. 5.3).

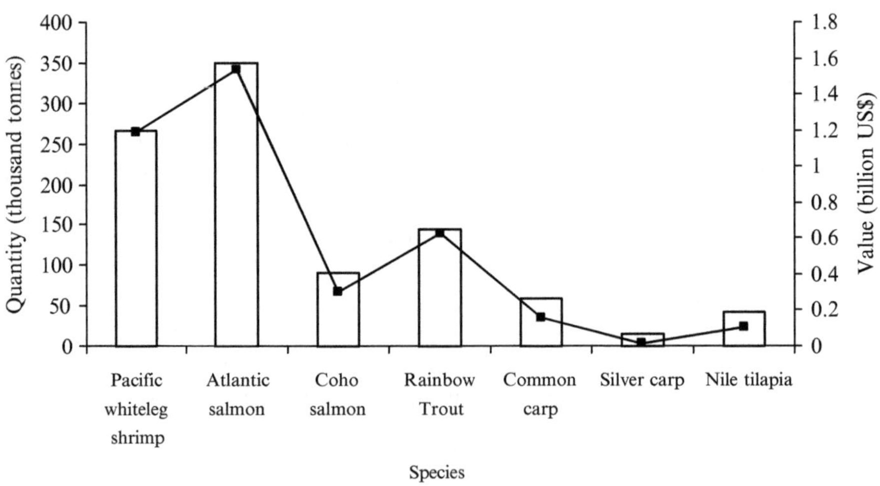

Fig. 5.3 Aquaculture production of non-native species (thousand tonnes) and value (billion US$) in Latin America and the Caribbean in 2004 (FAO 2006a)

5.1.4 New Zealand and Australia

Both New Zealand and Australia have significant and growing aquaculture industries that rely on their "clean and green" image – many of the common Northern Hemisphere diseases and parasites are absent from the aquaculture facilities in these two countries, yet almost a quarter (22%) of their aquaculture production by weight was based on non-native species in 2005 (FAO 2006a; Fig. 5.2).

5.1.4.1 New Zealand

New Zealand produces ~97,700 t of aquaculture product per year worth ~US$217 million (~NZ$315 million), equating to approximately 20% of total NZ fisheries production (NZAC 2006). Over 98% of New Zealand's aquaculture industry is based on three species: the endemic Greenshell mussel *Perna canaliculus*, and two non-native species, the Pacific oyster *Crassostrea gigas*, and the King (or Quinnat) salmon *Oncorhynchus tshawytscha*. Non-native species however, represent 33.3% of New Zealand aquaculture product by value (Table 5.1).

There are three species of salmon in New Zealand, all of which are non-native: King or Chinook salmon *O. tschawytscha* introduced from the United States in 1907, Sockeye salmon *O. nerka* introduced from Canada in 1902, and Atlantic salmon *Salmo salar* introduced in the 1960s (FAO 2006a). Only the King or Chinook salmon (also known as "Quinnat") are successfully farmed on a significant scale in New Zealand. This is in contrast to the rest of the world where salmon aquaculture is focused on the Atlantic salmon, except for some Chinook salmon in Canada and Coho salmon in Chile.

King salmon are grown in sea cages in the marine environment and in freshwater raceways throughout the South Island. There are about 29 salmon farms in New Zealand covering a total of around 128 hectares (as of December 2005) producing around 7,000 metric tonnes per annum. These 29 farms account for roughly half of the worldwide farmed king salmon production.

The main Pacific oyster farming areas are located in sheltered bays and harbours around the North Island. The farming method for Pacific oysters consists of wooden racks to which the oysters are attached. The racks are anchored in the lower intertidal region. There are about 236 Pacific oyster farms in New Zealand covering a total of ~928 hectares (as of December 2005) and producing over 2,000 t in 2004 (FAO 2006a).

5.1.4.2 Australia

Australia produces ~47.1 million tonnes of aquaculture product per year, worth ~US$480 million (~AU$610 million), equivalent to 30% of Australia's total fisheries

Table 5.1 Aquaculture production in 2004–05 for New Zealand (after NZAC 2006) and Australia (by state after (ABARE 2006). Bold names and figures represent non-native species information; figures in parentheses represent non-native species subtotals)

Value	New Zealand $'000	Australia $'000	NSW $'000	Vic $'000	Qld $'000	WA $'000	SA $'000	Tas $'000	NT $'000
FISH									
Salmon [a]	**81,000**	**112,357**	**0**	**0**	**0**	**0**	**0**	**112,357**	**NR**
Trout [b]	–	**12,862**	**1,784**	**10**	**533**	**0**	**545**	**0**	**NR**
Tuna	–	139,955	0	0	0	0	139,955	0	NR
Silver perch	–	2,821	2,431	0	390	0	0	0	NR
Barramundi	–	14,625	1,360	0	11,000	0	2,265	0	NR
Other	29,000	9,434	1,201	6,065	280	1,888	0	0	NR
CRUSTACEANS									
Prawns	–	49,864	4,464	0	45,400	0	0	0	NR
Yabbies [c]	–	1,866 **(1, 120)**	362	78	0	**1,120**	306	0	NR
Marron [d]	–	2,072 **(1,485)**	0	0	0	**1,485**	587	0	NR
Redclaw	–	1,402	2	0	1,400	0	0	0	NR
MOLLUSCS									
Edible oysters[e]	**26,000**	73,287 **(36,799)**	35,788[f]	0	700	0	**19,995**	**16,804**	NR
Pearl oysters	–	122,550	0	0	550	122,000	0	0	NR
Mussels	181,000	7,571	215	2,816	0	1,515	657	2,369	NR
Other	4,000	9,772	0	4,454	0	0	5,318	0	NR
OTHER (non food)		50,666	0	0	2,000	467	17,015	3,384	27,800
Total value	**321,000 (107,000)**	**611,103 (264,623)**	**47,605 (1,784)**	**23,946 (10)**	**61,720 (533)**	**128,475 (2,605)**	**186,643 (20,540)**	**134,914 (129,161)**	**27,800 (?)**

[a] *Salmo salar, S. trutta.*
[b] New Zealand: *Oncorhynchus tshawytschca*; Australia: *Oncorhynchus mykiss.*
[c] *Cherax albidus* (introduced to WA from Victoria; (Morrissey and Cassells 1992), *C. crassimanus, C. destructor, C. glaber, C. plebijanus, C. quinquecarinatus.*
[d] *Cherax quadricarinatus* (introduced from Qld and NSW to WA), *C. rotundus.*
[e] New Zealand: *Crassostrea gigas;* Australia: *Crassostrea gigas,* and the natives: *Ostrea angasii, Sacostrea glomerata, S. cucullata.*
[f] some component of the New South Wales production is *Crassostrea gigas*

production. Approximately 60 species are under aquaculture production, of which several are introduced from other regions of the world or from other regions of Australia (see Table 5.1). While many of these species are for human consumption (e.g., salmonids, oysters and prawns), aquaculture in Australia includes a variety of products for other purposes.

Non-native species represent 43.3% of Australian aquaculture production by value (in 2004–05). These non-native species include introduced salmonids (Atlantic salmon, *Salmo salar*; Brown trout, *Salmo trutta*; Brook trout, *Salvenlinus fontinalis* and Rainbow trout, *O. mykiss*), Pacific oysters, and freshwater crayfish (yabbies: *Cherax albidus*; *C. quadricarinatus*) translocated from one Australian state to another.

Atlantic salmon and rainbow, brown and brook trout are cultured commercially in Australia. Tasmania is the major force in Australian production. Atlantic salmon and ocean trout (rainbow trout) are grown in sea cages, trout are also grown in freshwater dams and raceways where large supplies of cold, flowing water are readily available. Sea cage culture contributes more than 60% of the total salmon and trout production in Australia.

The Pacific oyster is extensively farmed in Tasmania and South Australia, and comprises a minor component of the industry in New South Wales where several native species are grown (*Ostrea angasii*, *Sacostrea cucullata*, and *S. glomerata*). The farming method is similar to New Zealand where both wooden racks and stakes are used. The blue mussel, *Mytilus edulis*, is commercially farmed in Victoria, Tasmania, New South Wales, and Western Australia.

5.2 Unintentional Aquaculture Related Introductions of Non-Native Species

Despite the apparent success in increased aquaculture production through the use of non-native species, current practices can pose significant risks of unintentional introductions from net pens or pond systems into freshwater and marine systems. These introductions have been widely reported (Naylor et al. 2001; Nico et al. 2001; SAMS 2002) and are often associated with weather events (e.g., flooding or hurricanes) or accidents of operation. It is estimated that; up to 2 million farmed Atlantic salmon escape into the North Atlantic each year (McGinnity et al. 2003), over 500,000 Atlantic salmon escaped from cages between 1987 and 1997 on the west coast of North America (McKinnell and Thomson 1997), up to 80% of adult salmon entering rivers in Norway were escapees (Fiske and Lund 1999) and that the introduced Rainbow trout *Oncorhynchus mykiss* now occupies over 51% of Slovenian territory (Povz and Sumer 2005). Mass escapes of the Pacific white shrimp *Penaeus vannemei* have also occurred in both the United States (Balboa et al. 1991; Wenner and Knott 1992; Howells 2001) and Thailand (Barnette et al. 2006).

5.3 Ecological Consequences of Intentional and Unintentional Introduction of Non-Native Species

The intentional and unintentional introduction of non-native cultured species represents a "biological introduction", which are human mediated movements of organisms to regions where they did not evolve. Biological introductions are widely recognised as a major threat to species diversity (CBD 1992; Worm et al. 2006) arising from habitat modification, changes in ecosystem functioning, extinction of native fauna and flora, disease transfer and genetic effects such as hybridisation with native congeners (Lovei 1997; Ruiz et al. 1997; D'Antonio et al. 2001; Jonsson and Jonsson 2006). Regions supporting high levels of endemic species are particularly vulnerable to these introductions.

5.3.1 Habitat Modification

The accidental or intended introductions of exotic species can cause significant changes to ecosystems (Ruesink et al. 2006). However, the response of natural communities to the introduction of a non-native species is complex, and impacts can have positive, negative or negligible, depending on the species, location, age, or type of habitat considered (Neira et al. 2005; Gribben and Wright 2006). To highlight the potential effects of introduced species on habitat structure, two case-studies will be considered: the first in Willapa Bay, Washington USA (Ruesink et al. 2006) and the second on the South African coast (Robinson et al. 2005).

In Willapa Bay, USA the Pacific oyster *Crassostrea gigas* was introduced in 1928 for aquaculture purposes due to the overexploitation of the native oyster *Ostreola conchaphila* (Ruesink et al. 2006) and is now the main oyster species cultivated. The Pacific oyster naturally recruits to uncultivated regions of the bay and forms dense intertidal hummocks of shell and live oysters (Ruesink et al. 2006). This recruitment has also been observed in a number of other countries where *C. gigas* has been introduced (Orensanz et al. 2002; Nehring 2003; Diederich et al. 2005). The importation of *C. gigas* and the development of the oyster industry also unintentionally brought the invasive smooth cordgrass *Spartina alterniflora* to Willapa Bay in the form of packaging material for transplanted *Crassostrea virginica* in 1890 (Townsend 1893, 1896; Feist and Simenstad 2000).

Species that can change habitat structure and modify the local environment are known as ecosystem engineers (Crooks 2002). *C. gigas* and *S. alterniflora* can substantially re-engineer a habitat to provide biogenic structures which provide substrate for fish, invertebrate and macroalgal recruitment and sediment accumulation (Ruesink et al. 2006). The expansion of culture sites and biogenic reefs formed by oysters can also cause significant changes in sediment porosity, bioturbation activity and have an effect on biogeochemical cycling (Ruesink et al. 2006).

In South Africa, the Mediterranean mussel *Mytilus galloprovincialis* was accidentally introduced in ~1979 and is now grown commercially (Robinson et al. 2005) with over 6,100 t produced in 2005 (FAO 2006a). *M. galloprovincialis* has become the dominant intertidal mussel along the west coast, where it has considerably modified the natural community composition by dominating rock surfaces (Robinson et al. 2005). *M. galloprovincialis* forms dense, multi-layered structures and supports a higher biomass per m^2 than the single layered beds of the indigenous mussels *Choromytilus meridionalis* and *Aulacomya ater* (Robinson et al. 2005). The increased vertical range of *M. galloprovincialis*, due to a greater dessication tolerance, higher fecundity and faster growth rates than the native species (Van Erkom Schurink and Griffiths 1990, 1991; Hockey and van Erkom Schurink 1992; Van Erkom Schurink and Griffiths 1992), has led to a massive increase in non-native mussel biomass along the South African west coast (Griffiths et al. 1992).

The introduction of certain non-native species to a region can considerably modify the system, as shown by the introduction of *C. gigas*, *S. alterniflora* and *M. galloprovincialis*. Predicting the impact that non-native species will have on habitat structure and, as a consequence, existing food webs and community composition is inherently difficult. A greater understanding of how these non-native "ecosystem engineers" alter energy flow, ecological processes, biogeochemical cycles and ecosystem function is critical in determining the impact that these species will have on the ecosystem as a whole.

5.3.2 Changes in Ecosystem Functioning

Ecosystem services are a set of ecosystem functions that are useful to humans and many are critical to our survival (climate regulation, air purification, pollination, nutrient recycling) while others enhance it (aesthetics) (Kremen 2005). Ecosystem functioning is intrinsically linked to biodiversity and changes in biodiversity and community structure can cause drastic changes in ecosystem function and hence in the provision of ecosystem services.

The majority of studies of ecosystem function have concentrated on biodiversity loss due to extinctions; however, many biological invasions have resulted in a net gain at the local or regional level (Sax and Gaines 2003). This causes a net increase in diversity at the ecosystem level and an important consideration is how these species additions affect ecosystem functioning (Stachowicz and Tilman 2005). Few studies have been undertaken to specifically address this question, although it is clear that invasive species can affect ecosystem structure and function (Stachowicz and Tilman 2005). For example, Levin et al. (2006) showed that invasion by a *Spartina* hybrid in San Francisco Bay (USA) shifted the system from an algae based to a primarily detrital-based system. Furthermore, the *Spartina* hybrid canopy changed the hydrodynamic regime causing drastic and multiple changes in the physical, chemical and biological properties in the benthic system (Neira et al. 2006). These changes caused a reduction in survivorship of key taxa that supported higher trophic levels, such as migratory shorebirds (Neira et al. 2006).

Tilapia has been used worldwide as an aquaculture species and has escaped in many regions where they are cultured (Peterson et al. 2005). Tilapia can significantly alter the ecosystem they invade, yet the impact is often hard to predict (Figueredo and Giani 2005). For example, the Redbelly tilapia (*Tilapia zilli*) was accidentally introduced into a power plant reservoir in North Carolina, where it reduced all aquatic macrophytes through grazing, which coincided with a dramatic decline in native fishes (Crutchfield 1995). The Common carp (*Cyprinus carpio*) has been widely translocated around the world for aquaculture purposes and through unintentional introductions is now a successful invader in parts of Europe, Asia, Africa, North, Central and South America, Australia and Oceania (Lever 1996; FAO 2002). Carp can reach high densities (1000 individuals ha^{-1}) and biomass (3144 kg ha^{-1}) (Harris and Gehrke 1997) and this can result in reduced photosynthetic production and visibility for visually feeding fish (Koehn 2004) through increasing the water turbidity whilst feeding (Fletcher et al. 1985; King et al. 1997), a decline in the abundance of aquatic plants (Fletcher et al. 1985; Roberts et al. 1995) and finally, cause trophic cascades in shallow lakes (Khan et al. 2003).

5.3.3 Extinction of Native Flora and Fauna

There is no doubt that biological invasions are causing dramatic widespread changes to communities and altering many ecological systems (Parker et al. 1999; Ruiz et al. 1999; Levi and Francour 2004; Neira et al. 2005; Gribben and Wright 2006; Ruesink et al. 2006). However, many extinctions have been attributed to biological invasions when there have been many other environmental factors (eutrophication, habitat loss, land use changes, over grazing) which could have played a key role in causing the decline of the native species (Gurevitch and Padilla 2004). Of the 762 species globally documented to have become extinct as a result of human activities in the past few hundred years, < 2% list non-native species as a cause (Gurevitch and Padilla 2004).

In many cases, species do not go “extinct”, but are lost from a large part of their former range which greatly reduces and/or fragments the populations (Hobbs and Mooney 1997). Non-native species have been identified as part of the problem and, in combination with habitat loss, modification and degradation of the environment, have lead to the loss of species in a particular region. For example, Fellers and Drost (1993) resurveyed 16 historic sites and 34 other sites for the Cascade frog *Rana cascadae* and only found two frogs at one site. The population extinction was attributed to several factors, principally to the introduction of non-native predatory fish, drought and habitat loss due to management activities (Hobbs and Mooney 1997). The introduction of the Grass carp *Catenopharyngodon idella*, the Bighead carp *Aristichthys nobilis* and the Taihu Lake noodlefish *Neosalanx taihuensis* during the 1970s and 1980s to the southern provinces of Guangdong,

Guangxi and Yunan from the Yangtze River system, has severely affected and has contributed to the extinction of some local finfish species (Li and Xie 2002). The Nile perch *Lates niloticus* was introduced into Lake Victoria in the 1960s apparently causing the extinction of many cichlids species – viewed as the biggest vertebrate extinction of the 20th century (Witte et al. 2000). However, Gurevitch and Padilla (2004) suggest that development of the railroad in the 1920s caused erosion and shoreline destruction (Verschuren et al. 2002) and urbanization during the 1970s increased eutrophication and decreased lake transparency from 8 to 1.5 m (Verschuren et al. 2002; Aloo 2003). Increased nutrient loading and anoxic events resulting in fish kills are now common. The increase in nutrient loads, however, has favoured the non-native water hyacinth *Eichhornia crassipes*, which alters nursery areas for juvenile fish (Witte et al. 2000).

Species in the marine environment are typically considered to have a lower risk of extinction because of the large continuous habitats they occupy and the life history characteristics of many species that results in extensive dispersal potential enabling the recolonisation and repopulation of impoverished areas (Gurevitch and Padilla 2004). Caution should be taken, however, as this perception was derived from experiences when marine populations were much larger than they are today (Dulvy et al. 2003) and when the current rate of exploitation of marine species and the level of associated by-catch of non-target species was significantly lower (Worm et al. 2006). Unintentional introductions of non-native aquaculture species are likely to increase with the rapid expansion of the aquaculture industry on a global scale and there is an urgent need for more research into the role of non-native species in pushing native species towards extinction and to evaluate their impact relative to that of other factors (Gurevitch and Padilla 2004).

5.3.4 Disease Transfer

To be economically viable, cultivation of a species must normally take place at a high density either within contained units (on account of the capital costs of the equipment) or as bottom culture on shores (where space may be limited). Under these conditions introduced pests, parasites and diseases are provided with increased opportunities to thrive (Minchin and Rosenthal 2002).

There are many cases of stock movements introducing unwanted pests, parasites and diseases and some have had serious economic impacts on aquaculture production, for example, oysters (Heral 1990), shrimp (Kinne 1984; Sindermann 1993) and fishes (Kinne 1984). For example, the trematode *Gyrodactylus salaris* was carried with Atlantic salmon *Salmo salar* from Swedish hatcheries to Norway (Johnsen and Jensen 1991) and resulted in serious salmon mortalities in the recipient region.

Movements of harmful biota over larger distances, however, are more common. For example, consignments of half-grown Pacific oysters have resulted in a large suite of invertebrates being spread throughout the world (Gruet et al. 1976), with their uneven shells providing a large surface area for the attachment of cryptic species. Biota may also reside in the mantle cavity, the gut or in various tissues. During the early large-scale movements of oysters these associated species were tolerated as a nuisance. However, with present knowledge and management such releases are unlikely to be repeated on account of the wide range of microbiota and syndromes that have been associated with such movements (Cheyney et al. 2000).

The movement of stock in seemingly small quantities can also have serious consequences for native species. For example, the importation of Japanese eels *Anguilla japonica* for cultivation trials in Europe released a rotund nematode that in its final stage lodges in the visceral cavity near the air bladder and has caused significant internal damage in other eel species such as the native freshwater eel *Anguilla anguilla* (Kennedy and Fitch 1990). This nematode is easily dispersed by copepods, and a wide range of paratenic hosts that include other fishes and insects. The species has now become widely spread in Europe and the consequences for the stock of the North Atlantic eel, already in decline, are unknown.

The spread of viral diseases through stock movements has been particularly prevalent in Penaeid shrimp and has caused significant declines in production (Subasinghe et al. 2000). Viruses may also be spread *via* other crustaceans, and barnacles may even be capable of transmitting these to different countries as hull fouling on ships. Pathogenic species may also be carried in the water and sediments in the ballast tanks of ships and many species in commercial culture have been found associated with hull fouling (Minchin and Gollasch 2002). No studies have been undertaken on the potentially harmful biota carried on ships' hulls although it is suspected that the oyster disease *Bonamia osteae* was carried to different bays on the hull of a barge (Howard 1994).

5.3.5 Genetic Impacts

Marine aquaculture species are increasingly being selected or modified with respcct to genetic traits linked to performance. Cross (2000) described the genetic improvement of aquaculture species as an economic imperative and without it, the industry would find it impossible to compete. For example, Coho salmon *Oncorhynchus kisutch* with introduced growth hormone genes from Chinook salmon *Oncorhynchus tshawytscha*, demonstrated much faster growth compared to the control group (Devlin et al. 1994). Hybridization between the Yesso scallop *Patinopecten* (*Mizuhopecten*) *yessoensis* and a local species *Chlamys farreri* have also been undertaken to improve growth performance (Yang et al. 2004; Yu et al. 2006). In addition, Chinese researchers have recently introduced a new batch of Yesso scallop broodstock from Russia (Meng 2006) in an effort to reconstruct their genetic diversity (Li and Xue 2005). These experiments have produced new strains of scallops and some individuals have already been put out to sea for a pilot grow-out.

As a result, a substantial fraction of genetic variation in aquaculture species resides at a higher organisational level (among populations) than in natural populations where all variation resides below the family level (Youngson et al. 2001). Within the population, genetic complexes will develop, often relating to the environment in which the population has developed, constituting spatial, behavioural or temporal isolating mechanisms. Aquaculture practices of both inbreeding and selection of individuals for specific traits magnifies the development of genetic complexes in a population.

When aquaculture escapes breed with natural populations, hybridisation and subsequent introgression can lead to a breakdown of the genetic complexes which have developed, forcing a reduced fitness in the hybrid individuals (Skaala et al. 2006). This can lead to a decline in fitness and increased threat of extinction in the now hybridised natural population (Mooney and Cleland 2001). Outbred largemouth bass *Micropterus salmoides* crossed from two distinct populations suffered a reduction in fitness of approximately 14% relative to parental stocks (Goldberg et al. 2005). F2 generation hybrids suffered higher mortality rates and increased susceptibility to infectious disease. Collection and translocation between previously isolated stocks can have similar effects, which have been shown in stocks of black-lipped pearl oyster, *Pinctada margaritifera cumingii*, in French Polynesia (Arnaud-Haond et al. 2004). When large populations of invading species are introduced, the threat to native species is unavoidable, however evidence suggests that even when small populations of an invader are introduced (for example, escaping aquaculture individuals) the native population is still threatened (Mooney and Cleland 2001).

Hybridisation can be either inter- or intraspecific. Hybridisation between native brown trout, *Salmo trutta* and Atlantic salmon, *Salmo salar*, in Europe is an example of the former and female salmon escapees have been shown to hybridize relatively freely with the brown trout (Youngson et al. 1993). Intraspecific hybridisation would involve escape of a different strain of the species into a native population (Cross 2000). This is likely, due to the modification of aquaculture species with traits chosen for performance. Spawning success is lower for cultured salmon than for wild fish (Fleming et al. 1996, 2000), even when released to the wild as smolts (Jonsson et al. 1990). In Spain, where rivers have been highly stocked with non-native trout *S. trutta*, 25% of native populations had evidence of introgression by genes of hatchery origin (Almodóvar et al. 2001). Evidence has also been found for the introgression of the Mediterranean mussel, *Mytilus galloprovincialis* genes into native Australian populations (Sanjuan et al. 1997).

5.3.6 Trans-Boundary Effects

Biological invasions, whether intentional or accidental, are by their very nature not limited by geo-political boundaries. This is even more the case for marine bioinvasions where oceanic currents and natural dispersal mechanisms can lead to significant range expansions, following initial establishment, that transcend state and national boundaries. Examples include the escape and spread of the macroalga *Undaria pinnatifida* from aquaculture facilities in Brittany, Atlantic France (Pérez et al. 1984) across the English Channel to southern England and along the coasts northwards to the Netherlands and southwards to Spain (Fletcher and Manfredi 1995; Wallentinus 1999). Similarly, the expansion of the European green crab, *Carcinus maenas*, along the West Coast of North America following its initial accidental establishment in San Francisco Bay resulted in an expansion from the state of California, to Oregon and Washington (Grosholz and Ruiz 1995).

The Conference of Parties of the Convention on Biological Diversity (CBD 1992) in Decision VII/5, identified the need for regional and international collaboration to address trans-boundary impacts of mariculture on biodiversity, such as spread of disease and invasive alien species (paragraph 51), particularly where non-native species are grown for mariculture purposes. Similarly, the FAO through the Code of Conduct for Responsible Fishing (CCRF) (FAO 1995) and Technical Guideline Number 5 (FAO 1997) has explicitly addressed aquaculture development in relation to trans-boundary obligations. Article 9.1.2 of the CCRF identifies the potential genetic impacts of introduced (alien) species through introgression and competition with native stocks and Article 9.2.3 explicitly discusses the need for consultation with neighbouring states when considering the introduction of alien species into a trans-boundary system.

From an aquaculture perspective, trans-boundary effects include both the intentional release of a species that has the ability to disperse across geo-political boundaries and cause harm to a neighbouring coastal state, as well as, the operational or regulatory management failure to prevent or mitigate non-native species escapes that may cause harm to a neighbouring coastal state.

5.3.7 Implications for Biodiversity Hotspots

Biodiversity hotspots are defined as those areas where "exceptional concentrations of endemic species are experiencing exceptional loss of habitat" (Myers et al. 2000; Orme et al. 2005). Of the top five regions identified as major global hotspots for marine biodiversity (Roberts et al. 2002), two regions are major aquaculture producers; the Philippines and Indonesia with an annual production of over 1.4 million tonnes. The Caribbean is ranked ninth (Roberts et al. 2002) and this region has experienced an annual growth rate in aquaculture production of 21.3%, almost three times higher than the global production average of 8.8% since the 1950s. Over 65% of the production in the Caribbean is due to introduced species (FAO 2006b). Chile is identified in the top 25 terrestrial hotspots and has experienced an annual increase in aquaculture production of 40.0% from 1980 to 2004 (FAO 2006a). From a conservation perspective, it could be argued that concerns about aquaculture effects related to non-native species need to be primarily focused on these areas.

5.4 Future Directions

As the landings from capture fisheries stagnate (SOFIA 2004; Hilborn 2007), aquaculture is critical to the provision of global resources. The industry provides full time employment for over 3.3 million people in China alone (De Silva 2000) and many millions more could be employed either directly or indirectly in aquaculture worldwide – provided there is wise environmental management. A sustainable approach to coastal aquaculture is especially key given that 65% of humanity, 3.6 billion people, live within 150 km of the coast and are dependent on ecosystem based services (Cohen 1995; Sachs and Reid 2006) and that a number of major aquaculture regions support biodiversity hotspots. Much of the future aquatic production will be dependant on good water quality and how developments evolve that might otherwise conflict with the space required for cultivation. The present ease of transportation will allow for the movement of aquaculture species over large distances rapidly enabling a wide range of species to become transferred. Legislation and risk management in the movement of species is becoming recognised as an important area in order to prevent undesired impacts, as a result of an intended introduction.

5.4.1 Legislation for the Introduction of Non-Native Species for Aquaculture Purposes and Management Strategies

The International Council for the Exploration of the Sea (ICES) Code of Practice on Introductions and Transfers of Marine Organisms (ICES 2005a) and the European Inland Fisheries Advisory Commission (EIFAC) Code of Practice for Consideration of Introductions and Transfers of Marine and Freshwater Organisms (Turner 1988), provided guidelines for the intentional introduction of non-native species for aquaculture purposes. Furthermore, it has been recommended that the new IUCN code of practice should be incorporated into national development strategies (Hewitt et al. 2006). These codes aim to minimise negative impacts of non-natives used in aquaculture on the recipient environments. Australia and New Zealand are well advanced in the development of their national strategies; however, it is recognised that these procedures take time to implement and there are circumstances where there is an urgent requirement to provide food for the vast population, as in China or where there has been serious environmental degradation, as in the case of deforestation in the Indo-Pacific (Coates 1995).

5.4.1.1 Australia and New Zealand

Australia has experienced a number of high profile invasions from a variety of sectors resulting in serious environmental and economic impacts (Hewitt et al. in press). As a consequence of these invasions, Australia identified the need for a coordinated approach across national and state agencies through the development of a National System for the Prevention and Management of Marine Pest Incursions (National System) to address all potential marine pest vectors underpinned by a risk assessment framework and to specifically establish arrangements for prevention, emergency preparedness and response, and ongoing management and control (Hewitt et al. in press).

The National System is coordinated by the Department of Agriculture, Fisheries and Forestry with all Australian States and the Northern Territory, marine industries (shipping, ports, fishing, aquaculture), conservation groups and researchers. Australia's biosecurity system is largely managed under the Quarantine Act (1903). At present, biosecurity management of aquaculture is partitioned into: quarantine activities associated with import standards, established by Biosecurity Australia and implemented by the Australian Quarantine Inspection Service (AQIS); and operational management at State and Territory levels. The importation of a new species for use as an aquaculture product must be assessed and approved by Biosecurity Australia, with appropriate approvals by AQIS. Once these approvals are in place, importation can proceed once approvals from the State or Territory are provided. Under the current National System, it would be unlikely that approvals for a new importation of a species for open water culture would proceed due to the obligations to prevent and minimise impacts of non-native species in the marine environment. If approvals were given, the operator would be required to submit and have approved an Emergency Marine Pest Plan that outlines options for action in the event of escape or other problems such as a disease outbreak. Similarly, it is likely that ongoing monitoring would be required with mandatory reporting to State and Territory authorities.

For the purposes of New Zealand's regulatory requirements, non-native fish, aquatic life or seaweeds approved for use in New Zealand must be in the exclusive and continuous possession or control of the person undertaking the activity AND must be able to be distinguished or kept separate from naturally occurring fish, aquatic life or seaweeds.

Importation of plants and animals, including aquatic organisms for aquaculture, is rigorously controlled by the New Zealand Environmental Risk Management Authority (ERMA). Biosecurity arrangements restricting the importation and quarantine of new species (that is species not occurring in the wild prior to 1996) involves a thorough investigation of the potential risk of introducing this species into New Zealand including the disease risk it presents. The ERMA makes decisions on applications to introduce hazardous substances or new organisms, including genetically modified organisms.

5.4.1.2 China

In China, before the issue of the Quarantine Act of Import and Export Organisms (2004, People's Congress) and the Aquaculture Seedling Management Procedures

(2005, Chinese Ministry of Agriculture, MOA), the lack of management had resulted in a somewhat chaotic situation in non-native species introductions, and some species were introduced repeatedly. The Aquaculture Seedling Management Procedures was enacted to deal with this situation and, for the first time the introduction of broodstock, juveniles, larvae and fertilized eggs for aquaculture (research or production) purposes is under government control. All aquaculture seedlings are categorized by the MOA in collaboration with relevant branches of the State Council, as (i) whose import and export are forbidden; (ii) whose import and export rely on the approval of MOA; or (iii) whose import and export rely on the approval of Provincial Fisheries Administrations. Among other requirements, all applications for the import of aquaculture seedlings should contain a Safety Impact Report (including environmental and biological impact and possible disease transfer) and a Certificate of Origin. These measures are inadequate, in that there is still no integrated risk assessment system in China to prevent aquatic bio-invasion, no legislation governing the early-warning, removal and control of introduced species, and no ecological remediation and compensation liability measures to combat bio-invasion.

In recent years, however, Chinese central government has strengthened legislative and administrative measures supervising aquatic species introduction, and encouraging research efforts on risk assessment and the control of bio-invasions. Guided by the above mentioned acts and a number of regulations, the National Biosafety Office, affiliated with the State Environmental Protection Agency (SEPA), the MOA along with its provincial level agencies and the General Administration of Quality Supervision, Inspection and Quarantine of the People's Republic of China (AQSIQ) undertake the management of species introduction and the inspection of pests, parasites and diseases carried by any imported organisms. Strict inspection and risk assessment procedures have also been implemented on import and export of genetically modified organisms (GMO).

5.4.2 Risk Evaluation and Management

Risk evaluation has become a useful management tool to assess the biological and ecological aspects of ecosystems when using limited available data (i.e., managing under uncertainty). For example, ecologically sustainable development seeks a balance between the benefits and the costs (environment, economic, social) of an activity. In many instances, the information necessary to determine benefits and costs will be unknown and risk evaluation can aid the decision-making process. In simple terms, risk analysis is used to determine how often an event may occur (frequency) and what the consequences of such an event would be. Risk evaluations can inform decisions before allowing the import of a new species (pre-border) or before allowing release of a new species into the environment (post-border).

A standardised risk management process can be summarized in four steps: (1) establishing the context; (2) identifying the risk; (3) assessing the risks (risk analysis and risk evaluation); and (4) and treating the risks (e.g., Australian and New Zealand Standard Risk Management AS/NZ4360:2004). This is readily applicable to assessing pre-border biosecurity risk (e.g., microalgae import decision-tree; Campbell 2006b) and post-border biosecurity risk in the form of organism impact assessments (Campbell 2005, 2006a).

In an aquaculture context, risk evaluation must assess: (1) the introduced species being imported for commercial purposes (e.g., use of abalone, *Haliotis rufescens* and *H. discus hannai*, in Chile); (2) the mechanism of transfer to determine hitch-hikers including pathogens and parasites; and (3) the feed species (e.g., *Thalassiosira wiessfloggi* is fed to rotifers that are then used as aquaculture feed) imported to sustain both native and introduced aquaculture species.

Management of imported introduced species is typically controlled with the aid of Import Health Standards (IHS), that operate as codified rule structures that identify how, when and where a specific "risk good" can be imported, and adhere to the World Trade Organizations (WTO) related standards (Hewitt and Campbell 2007). IHS seeks to minimise the risk and identify appropriate management options (Orensanz et al. 2002; DAFF 2003; Pheloung 2003).

IHS's are often underpinned by species specific risk analyses. A decision tree model is one example of risk analysis where a series of simple yes/no questions progresses the assessment through the process, indicating where importation should be rejected, approved with or without stipulations (Fig. 5.4). The model can be qualitative, semi-quantitative or quantitative (data input dependent). Each step is assessed against a risk mitigation context (such as a management procedure) with the endpoint derived by the questions asked at each step in the process. The decision-tree applies the same set of criteria to all species, ensuring a consistent, objective and verifiable manner to assess all import requests and invariably considers specific national and international obligations.

Countries can also apply the risk evaluation embedded in the ICES Code of Practice (ICES 2005b), or develop more individualised importation processes (e.g., Hewitt et al. 2006). For example, a generic importation model for aquaculture species identifies risk as an integral component, followed by an economic assessment of cost: benefit (Fig. 5.5). The model is initiated when a request to import a non-indigenous species or non-indigenous genome occurs. Decision makers undertake a risk evaluation that defines: unacceptable impacts, risk methods used and *a-priori* states the acceptable level of risk. The process is supervised by a scientific review committee and produces contingency and action plans or guidelines that deal with the accidental release of a non-indigenous species.

5.5 Concluding Remarks

Rapid increases in the production of non-native species and the associated risks of unintentional introductions and pathogen and parasite transfers to native populations underscores the urgent need for concerted global action in advancing

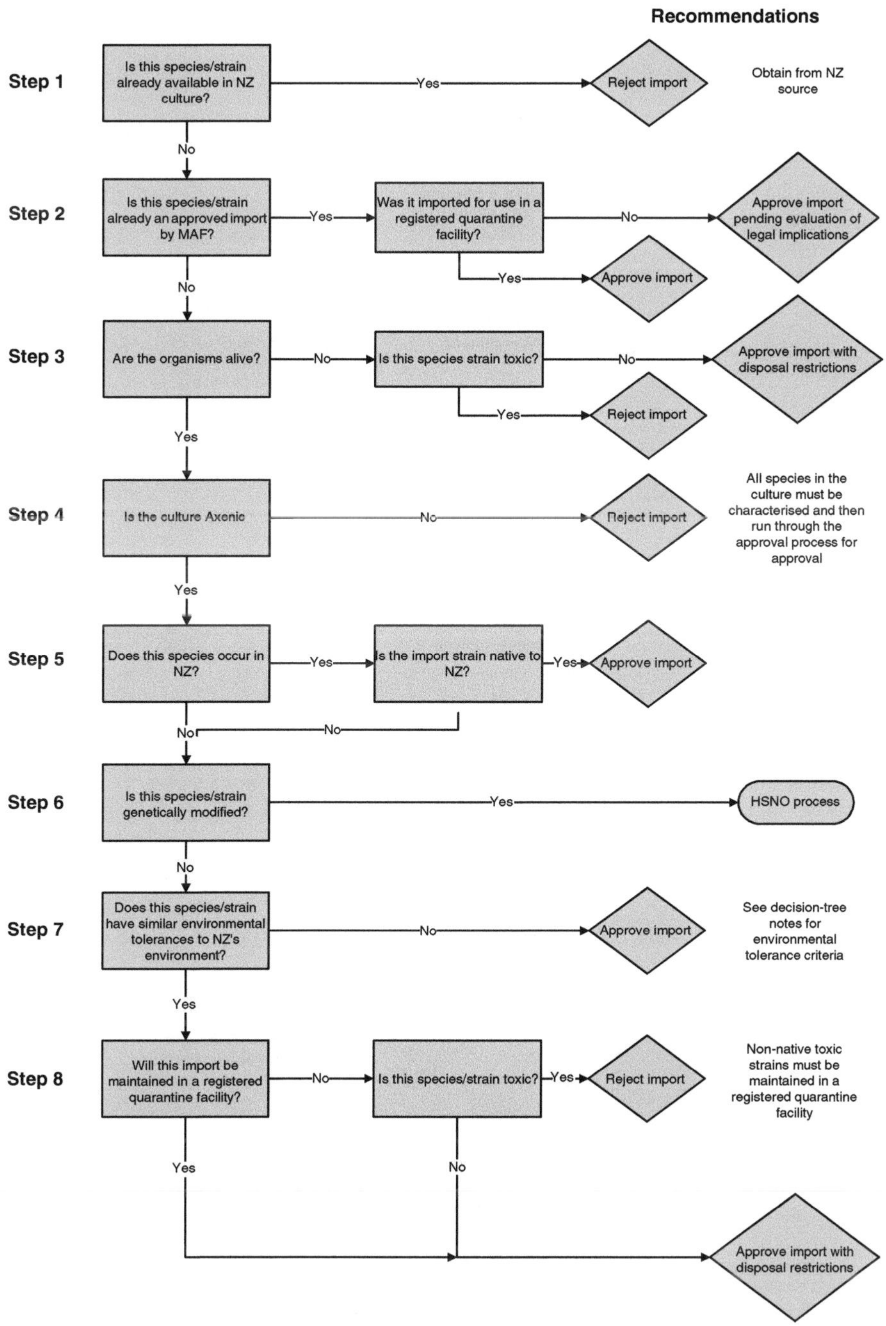

Fig. 5.4 Microalgae decision-tree developed for assessing the risk of importation of microalgae to New Zealand

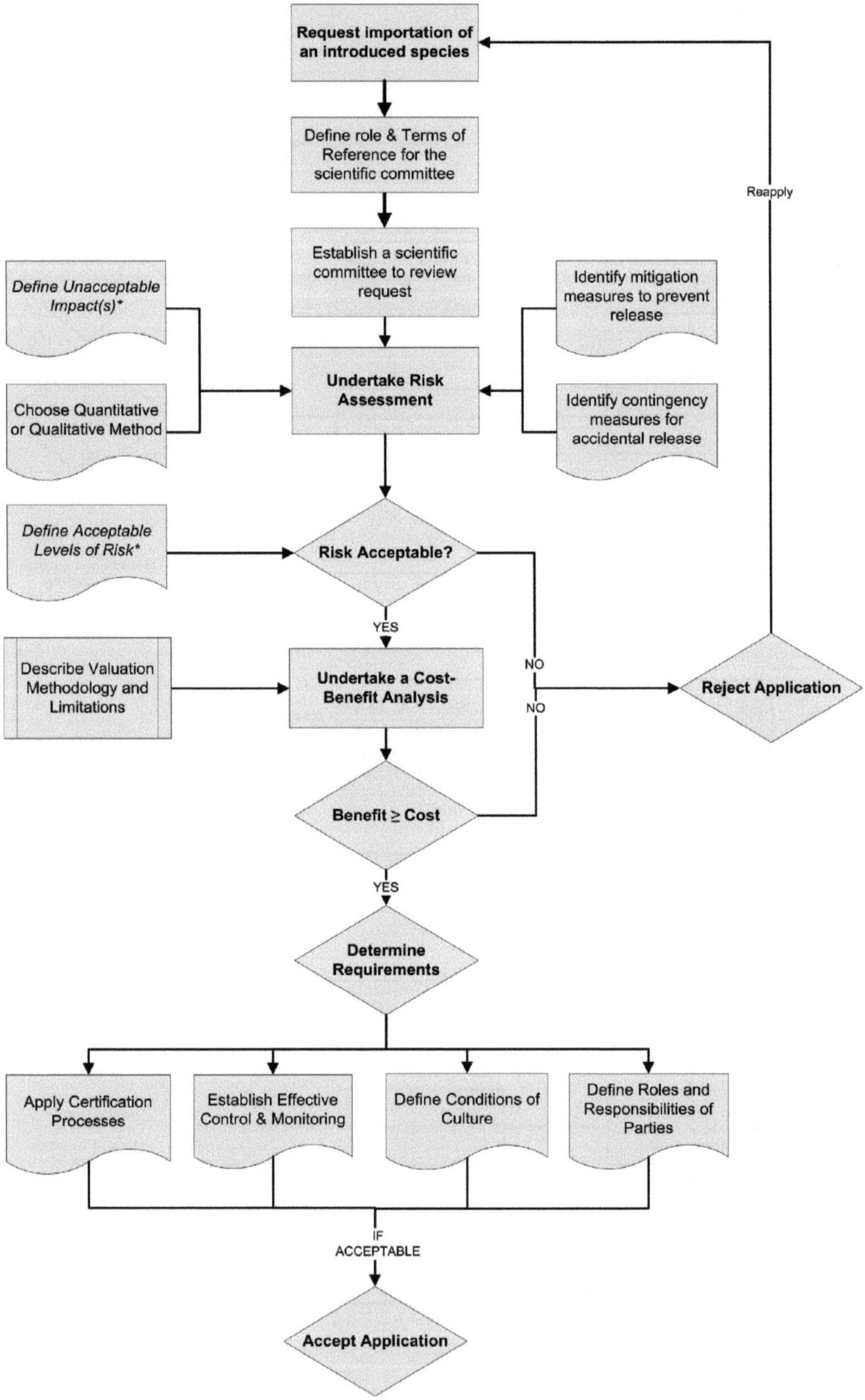

Fig. 5.5 A conceptual risk framework used for the importation of non-indigenous species for aquaculture purposes

environmentally sound aquaculture practices. If aquaculture is to be sustainable and the ecosystem safeguarded, particularly in regions of high endemism (i.e., biodiversity hotspots), effective controls on the introduction of non-native species associated with production are needed. Species diversity has been linked to increased robustness of systems to exploitation (Worm et al. 2006), making protection of biodiversity hotspots a clear priority (Webster et al. 2005). Existing codes of practice and risk evaluation models serve as important guidelines and should be carefully considered and actively promoted in planning non-native aquaculture. The 10-year Global Conservation Fund or the 5-year Critical Ecosystem Partnership Fund, which are aimed exclusively at hotspots (Brooks et al. 2006) should be used to assist the regions of highest risk to adopt international regulations and risk assessments for the introduction of non-native species for aquaculture purposes.

It is evident that a multi-disciplinary approach is needed to draw together experts, particularly from aquaculture, invasion biology, sociology and economics, which till now have had relatively limited interaction. In addition, efforts should be directed towards joint partnerships between countries and experts that have pioneered aquaculture research and those which possess the greatest biodiversity to improve growth rates, immunological resistance, product quality and market availability for native cultured species and/ or to design more robust aquaculture systems. Such action would either reduce the need to introduce non-native species for aquaculture purposes or minimise the risk of escape.

Finally, efforts should be advanced to increase the profile of concerns surrounding non-natives, in order to educate and involve a broad cross-section (scientists, industry, managers and the public) and promote sustainable aquaculture practices. This should include an international forum of experts and countries prepared to aid development in developing countries, symposia and workshops that engage a diverse community, ready access to the above codes of practice and related information, and explore market identity for environmentally sound products (Bartley and Minchin 1996).

References

ABARE (2006) Australian fisheries statistics 2005, Canberra, Australia

Almodóvar A, Suárez J, Nicola GG, NuevoAlmod M (2001) Genetic introgression between wild and stocked brown trout in the Douro River basin, Spain. Journal of Fish Biology 59: 68–74

Aloo PA (2003) Biological diversity of the Yala Swamp lakes, with special emphasis on fish species composition, in relation to changes in the Lake Victoria Basin (Kenya): threats and conservation measures. Biodiversity and Conservation 12: 905–920

Arnaud-Haond S, Vonau V, Bonhomme F, Boudry P, Blanc F, Prou J, Seaman T, Goyard E (2004 Spatio-temporal variation in the genetic composition of wild populations of pearl oyster (*Pinctada margaritifera* cumingii) in French Polynesia following 10 years of juvenile translocation. Molecular Ecology 13: 2001–2007

Balboa WA, King TL, Hammerschmidt PC (1991) Occurrence of Pacific white shrimp in Lower Laguna Madre, Texas, Annual Conference Southeast Association Fish and Wildlife Agencies, pp. 288–292

Barnette P, Sananan W, Khururat O, Tandkrad-Olan N, Upatham S, Pantuwatana S, C W (2006) The prevalence of taura syndrome virus, white spot syndrome virus and yellow head virus in wild shrimp species in Thailand, 14th International conference on aquatic invasive species, Florida, pp. 107–107

Bartley D, Minchin D (1996) Precautionary approach to the introduction and transfer of aquatic species. Precautionary approach to fisheries. Part 2. Scientific papers. FAO, Rome, pp. 159–189

Brooks TM, Mittermeier RA, da Fonseca GAB, Gerlach J, Hoffmann M, Lamoreux JF, Mittermeier CG, Pilgrim JD, Rodrigues ASL (2006) Global biodiversity conservation priorities. Science 313: 58–61

Buschmann AH, Riquelme VA, Hernandez-Gonzalez MC, Varela D, Jimenez JE, Henriquez LA, Vergara PA, Guinez R, Filun L (2006) A review of the impacts of salmonid farming on marine coastal ecosystems in the southeast Pacific. ICES Journal of Marine Science 63: 1338–1345

Campbell ML (2005) Organism impact assessment (OIA) for potential impacts of *Didymosphenia geminata*. In: All Oceans Ecology (ed.), Biosecurity, New Zealand; Melbourne, Australia, 64 pp

Campbell ML (2006a) Risk Assessment (modified Organism Impact Assessment) to update information on *Undaria pinnatifida*. In: All Oceans Ecology. (ed.), Biosecurity, New Zealand; Melbourne, Australia, 78 pp

Campbell ML (2006b) Risk analysis of microalgae import health standard listed marine species. In: All Oceans Ecology, (ed.), Biosecurity, New Zealand; Melbourne, Australia, 88 pp

CBD (1992) Convention on Biological diversity. Rio de Janeiro, Brazil, Montreal, Canada.

Chen J (2006) China's coastal aquaculture after strategic transfer of fishery model, World Aquaculture Society, Florence, Italy, 477 pp

Cheyney DP, MacDonald BF, Elston RA (2000) Summer mortality of Pacific oysters, *Crassostrea gigas* (Thunberg): initial findings on multiple environmental stressors in Puget Sound, Washington, 1998 Journal of Shellfish Research 19: 353–359

Coates D (1995) Implementation of the EIFAC/ICES Code of Practice: Experiences with the evaluation of international fish transfers into the Sepik River Basin, Papua New Guinea. Theme 3 – Protection of the aquatic environment. In: Voiglander, CW (ed.), World Fisheries Congress. Oxford and IBH, New Delhi, pp. 159–173

Cohen JE (1995) Population-growth and Earth's human carrying-capacity. Science 269: 341–346

Crooks AJ (2002) Characterizing ecosystem-level consequences of biological invasions: the role of ecosystem engineers. Oikos 97: 153–166

Cross TF (2000) Genetic implications of translocation and stocking of fish species, with particular reference to Western Australia. Aquaculture Research 31: 83

Crutchfield JU (1995) Establishment and expansion of redbelly tilapia and blue tilapia in a power plant cooling reservoir. American Fisheries Society Symposium 15: 452–461

D'Antonio C, Meyerson LA, Denslow J (2001) In: Soulé MS, Orians GH (eds.), Conservation biology research priorities for the next decade, Island, Washington, DC, pp. 59–80

DAFF (2003) Import Risk Analysis Handbook. Department of agriculture, fisheries, and forestry, Canberra, Australia, 45 pp

De Silva SS (2000) A global perspective of aquaculture in the new millennium. In: Subasinghe RP, Bueno P, Phillips MJ, Hough C, McGladdery SE, Arthur JR (eds.), Aquaculture in the Third Millennium. NACA, Bangkok and FAO, Rome, Thailand, pp. 431–459

Devlin RH, Yesaki TY, Biagi CA, Donaldson EM, Swanson P, Chan WK (1994) Extraordinary salmon growth. Nature 371: 209–210

Dextrase AJ, Cocarelli MA (2000) Intentional introductions of nonindigenous freshwater organisms in North America. In: Claudi R, Leach JH (eds.), Nonindigenous freshwater organisms: vectors, biology and impacts. Lewis, Boca Raton, Florida, pp. 61–98

Diederich S, Nehls G, van Beusekom JEE, Reise K (2005) Introduced Pacific oysters (*Crassostrea gigas*) in the northern Wadden Sea: invasion accelerated by warm summers? Helgoland Marine Research 59: 97–106

Dulvy KN, Sadovy Y, Reynolds DJ (2003) Extinction vulnerability in marine populations. Fish and Fisheries 4: 25–64

FAO (1995) Code of conduct for responsible fisheries. In: FAO (ed.), Rome, Italy

FAO (1997) Aquaculture development. In: FAO (ed.), Technical guidelines for responsible fisheries, Rome, Italy, 40 pp

FAO (2002) Evaluation of introductions

FAO (2006a) Fisheries Global Information System

FAO (2006b) State of world aquaculture: FAO Fisheries technical paper. No. 500, Rome, Italy, 134 pp

FAO (2006c) Cultured aquatic species information programme – *Patinopecten yessoensis*. In: (FIRI), I.W.R.a.A.S. (ed.), Cultured aquatic species fact sheets. FAO, Rome

Feist BE, Simenstad CA (2000) Expansion rates and recruitment frequency of exotic smooth cordgrass, *Spartina alterniflora* (Loisel), colonizing unvegetated littoral flats in Willapa Bay, Washington. Estuaries 23: 267–274

Fellers GM, Drost CA (1993) Disappearance of the cascades frog *Rana-cascadae* at the southern end of its range, California, USA. Biology Conservation 65: 177–181

Figueredo CC, Giani A (2005) Ecological interactions between Nile tilapia (*Oreochromis niloticus*, L) and the phytoplanktonic community of the Furnas Reservoir (Brazil). Freshwater Biology 50: 1391–1403

Fiske P, Lund RA (1999) Escapes of reared salmon in coastal and riverine fisheries in the period 1989–1998. NINA Oppdragsmelding 603: 1–23

Fleming IA, Jonsson B, Gross MR, Lamberg A (1996) An experimental study of the reproductive behaviour and success of farmed and wild Atlantic salmon (*Salmo salar*). Journal of Applied Ecology 33: 893–905

Fleming IA, Hindar K, Mjølnerød IB, Jonsson B, Dal- stad T, Lamberg A (2000) Lifetime success and interactions of farm salmon invading a native population. Proceedings of the Royal Society of London B 267: 1517–1523

Fletcher AR, Morison AK, Hume DJ (1985) Effects of carp (*Cyprinus carpio* L.) on aquatic vegetation and turbidity of waterbodies in the lower Goulburn River Basin. Australian Journal of Marine & Freshwater Research 36: 311–327

Fletcher RL, Manfredi C (1995) The occurrence of *Undaria pinnatifida* (Phaeophyceae, Laminariales) on the south coast of England. Botanica Marina 38: 355–358

Goldberg TL, Grant EC, Inendino KR, Kassler TW, Claussen JE, Philipp DP (2005) Increased infectious disease susceptibility resulting from outbreeding depression. Conservation Biology 19: 455–462

Goren M, Galil BS (2005) A review of changes in the fish assemblages of Levantine inland and marine ecosystems following the introduction of non-native fishes. Journal of Applied Ichthyology 21: 364–370

Gribben PE, Wright JT (2006) Invasive seaweed enhances recruitment of a native bivalve: roles of refuge from predation and the habitat choice of recruits. Marine Ecology Progress Series 318: 177–185

Griffiths CL, Hockey PAR, van Erkom Schurink C, Le Roux PJ (1992) Marine invasive aliens on South African shores: implications for community structure and trophic functioning. In: Payne ALL, Brink KH, Mann KH, and Hilborn R (eds.), Benguela Trophic Functioning. South African Journal of Marine Science, pp. 713–722

Grosholz ED, Ruiz GM (1995) Spread and potential impact of the recently introduced European green crab, *Carcinus maenas*, in Central California. Marine Biology 122: 239–247

Gruet Y, Heral M, Robert J-M (1976) Premières observations sur l'introduction de la faune associée au naissain d'huîtres Japonaises *Crassostrea gigas* (Thunberg), importé sur la côte Atlantique Française. Cahiers de Biologie Marine 17: 173–184

Gurevitch J, Padilla DK (2004) Are invasive species a major cause of extinctions? Trends in Ecology and Evolution 19: 470–474

Harris JH, Gehrke PC (1997) Fish and rivers in stress – the NSW rivers survey. NSW fisheries office of conservation & the cooperative research centre for freshwater ecology, Cronulla & Canberra

Héral M (1990) Traditional oyster culture in France. In: Barnabé (ed.), Aquaculture. Ellis-Horwood, London, pp. 342–387

Hewitt CL, Campbell ML (2007) Mechanisms for the prevention of marine bioinvasions for better biosecurity. Marine Pollution Bulletin 55: 395–401
Hewitt CL, Campbell M, Gollasch S (2006) Alien species in aquaculture: considerations for responsible use. In: IUCN (ed.), Gland, Switzerland.
Hewitt CL, Everett RA, Parker N (in press) Examples of current international, regional and national regulatory frameworks for preventing and managing marine bioinvasions. In: Rilov G, Crooks J (eds.), Marine bioinvasions: ecology, conservation and management perspectives. Kluwer, Dordrecht, The Netherlands
Hilborn R (2007) Defining success in fisheries and conflicts in objectives. Marine Policy 31: 153–158
Hobbs RJ, Mooney HA (1997) Broadening the extinction debate: population deletions and additions in California and Western Australia. Conservation Biology 2: 271–283
Hockey PAR, Van Erkom Schurink C (1992) The invasive biology of the mussel *Mytilus galloprovincialis* on the southern African coast. Transactions of the Royal Society of South Africa 48: 123–139
Holick J (1984) Some considerations on the role of introduced species of fish in the management of inland fisheries. In: EIFAC (ed.), pp. 488–495
Howard AE (1994) The possibility of long distance transmission of *Bonamia* by fouling on boat hulls. Bulletin of the European Association of Fish Pathologists 14: 211–212
Howells R (2001) Introduced non-native fishes and shellfishes in Texas waters: an updated list and discussion, Management Data Series. Texas Parks and Wildlife Department, 188 pp
ICES (2005a) Report of the Working Group on Introductions and Transfers of Marine Organisms (Available at: http://www.ices.dk/reports/ACME/2005/WGITMO05.pdf)
ICES (2005b) ICES Code of Practice on the introductions and transfers of marine organisms. International Council for the Exploration of the Seas, Copenhagen, Denmark, 30 pp
Johnsen BO, Jensen AJ (1991) The *Gyrodactylus* story in Norway. Aquaculture 98: 289–302
Jonsson B, Jonsson N (2006) Cultured Atlantic salmon in nature: a review of their ecology and interaction with wild fish. ICES Journal of Marine Science 63: 1162–1181
Jonsson B, Jonsson N, Hansen LP (1990) Does juvenile experience affect migration and spawning of adult Atlantic salmon? Behavioural Ecology and Sociobiology 26: 225–230
Kennedy CR, Fitch DJ (1990) Colonisation, larval survival and epidemiology of the nematode *Anguillicola crassus*, parasite in the eel *Anguilla anguilla* in Britain. Journal of Fish Biology 36: 117–131
Khan TA, Wilson ME, Khan MT (2003) Evidence for invasive carp mediated trophic cascade in shallow lakes of western Victoria, Australia. Hydrobiologia 506–509: 465–472
King AJ, Robertson AI, Healey MR (1997) Experimental manipulations of the biomass of introduced carp (*Cyprinus carp*io) in billabongs, I. Impacts on water-column properties. Marine and Freshwater Research 48: 435–443
Kinne O (1984) Diseases of marine animals: pices. Biologische Anstalt Helgoland, Hamburg, 541 pp.
Koehn JD (2004) Carp (*Cyprinus carpio*) as a powerful invader in Australian waterways. Freshwater Biology 49: 882–894
Kraan S, Barrington KA (2005) Commercial farming of *Asparagopsis armata* (Bonnemaisoniaceae, Rhodophyta) in Ireland, maintenance of an introduced species? Journal of Applied Phycology 17: 103–110
Kremen C (2005) Managing ecosystem services: what do we need to know about their ecology? Ecology Letters 8: 468–479
Lever C (1996) Naturalized fishes of the world. Academic Press, London, UK
Levi F, Francour P (2004) Behavioural response of *Mullus surmuletus* to habitat modification by the invasive macroalga *Caulerpa taxifolia*. Journal of Fish Biology 64: 55–64
Levin AL, Neira C, Grosholz ED (2006) Invasive cordgrass modifies wetland trophic function. Ecology 87: 419–432
Li SF (2005) Reflection on the positive and negative situation of aquatic species introduction 7: 1–3

Li WJ, Xue ZF (2005) Healthy sustainable proliferation and cultivation of scallop *Patinopecten yessoensis*. Fisheries Sciences 24: 49–51

Li ZY, Xie Y (2002) Bio-invasion in China. China Forestry, Beijing

Liu H, Zhu JX (2006) Bivalve cultivation in China. Shellfish News 21: 14–17

Liu SL, Li BS, Li J (2000) Study on present status and development strategy of introduction of new technology and new species of fishery in China. Marine Fisheries Research 21: 72–76

Loosanoff VL, Davis HC (1963) Rearing of bivalve molluscs. Advances in marine Biology 1: 1–136

Lovei GL (1997) Biodiversity – global change through invasion. Nature 388: 627–628

McGinnity P, Prodohl P, Ferguson K, Hynes R, O'Maoileidigh N, Baker N, Cotter D, O'Hea B, Cooke D, Rogan G, Taggart J, Cross T (2003) Fitness reduction and potential extinction of wild populations of Atlantic salmon, *Salmo salar*, as a result of interactions with escaped farm salmon. Proceedings of the Royal Society of London Series B-Biological Sciences 270: 2443–2450

McKinnell S, Thomson AJ (1997) Recent events concerning Atlantic salmon escapees in the Pacific. ICES Journal of Marine Science 54: 1221–1225

Meng QL (2006) Shandong Province plans to introduce *Patinopecten yessoensis* broodstock from Russia. Hebei Fisheries 9: 62

Minchin D (2006) The transport and the spread of living aquatic species. In: Davenport J, Davenport JL (eds.), The ecology of transportation: managing mobility for the environment. Springer, Dordrecht, The Netherlands, pp. 77–97

Minchin D, Gollasch S (2002) Vectors – how exotics get around. In: Leppäkoski, E, Gollasch S, Olenin S (eds.), Invasive aquatic species of Europe. Distribution, impacts and management. Kluwer, Dordrecht, The Netherlands, pp. 183–192

Minchin D, Rosenthal H (2002) Exotics for stocking and aquaculture, making correct decisions. In: Leppäkoski E, Gollasch S, Olenin S (eds.), Invasive aquatic species of Europe: distribution, impact and management. Kluwer, Dordrecht, the Netherlands, pp. 206–216

Mooney HA, Cleland EE (2001) The evolutionary impact of invasive species. Proceedings of the National Academy of Science 98: 5446–5451

Morrissey NM, Cassells G (1992) Spread of the introduced yabbie *Cherax albidus* Clark, 1936 in Western Australia, Fisheries Research Report No. 92, 27 pp

Myers N, Mittermeier RA, Mittermeier CG, da Fonseca GAB, Kent J (2000) Biodiversity hotspots for conservation priorities. Nature 403: 853–858

Naylor RL, Williams SL, Strong DR (2001) Aquaculture– a gateway for exotic species. Science 294: 1655–1656

Nehring S (2003) Pacific oysters in the European Wadden Sea: an irreversible impact in a highly protected ecosystem, Invasive Species Specialist Group IUCN Species Survival Commission, pp. 20–21

Neira C, Levin LA, Grosholz ED (2005) Benthic macrofaunal communities of three sites in San Francisco Bay invaded by hybrid *Spartina*, with comparison to uninvaded habitats. Marine Ecology Progress Series 292: 111–126

Neira C, Grosholz ED, Levin AL, Blake R (2006) Mechanisms generating modification of benthos following tidal flat invasion by a *Spartina* hybrid. Ecological Applications 16: 1391–1404

Nico LG, Williams JD, Herod JJ (2001) Black carp (*Mylopharyngadas piccus*): a biological synopsis and updated risk assessment. Risk Assessment and Management Committee, US Geological Survey, Gainsville, FL

NZAC (2006) The New Zealand Aquaculture Council Annual Report 2004–2005, Nelson, New Zealand.

Orensanz JM, Schwindt E, Pastorino G, Bortolus A, Casas G, Darrigan G, Elias R, Lopez Gappa JJ, Benat S, Pascual M, Penchaszadeh P, Piriz ML, Scarabino F, Spivak ED, Vallarino EA (2002) No longer the pristine confines of the world ocean: a survey of exotic marine species in the southwestern Atlantic. Biological Invasions 4: 115–143

Orme CDL, Davies RG, Burgess M, Eigenbrod F, Pickup N, Olson VA, Webster AJ, Ding TS, Rasmussen PC, Ridgely RS, Stattersfield AJ, Bennett PM, Blackburn TM, Gaston KJ, Owens

IPF (2005) Global hotspots of species richness are not congruent with endemism or threat. Nature 436: 1016–1019
Parker IM, Simberloff D, Lonsdale WM, Goodell K, Wonham M, Kareiva PM, Williamson MH, Von Holle B, Moyle PB, Byers JE, Goldwasser L (1999) Impact: toward a framework for understanding the ecological effects of invaders. Biological Invasions 1: 3–19
Pérez R, Kaas R, Barbaroux O (1984) Experimental culture of the brown seaweed *Undaria pinnatifida* along the coasts of France. Science et Peche 343: 3–15
Peterson MS, Slack WT, Woodley CM (2005) The occurrence of non-indigenous Nile Tilapia, *Oreochromis niloticus* (Linnaeus) in coastal Mississippi, USA: ties to aquaculture and thermal effluent. Wetlands 25: 112–121
Pheloung P (2003) An Australian perspective on the management of pathways for invasive species. In: Ruiz GM, Carlton JT (eds.), Invasive species: vectors and management strategies. Island Press, Boca Raton, pp. 249–269
Povz M, Sumer S (2005) A brief review of non-native freshwater fishes in Slovenia. Journal of Applied Ichthyology 21: 316–318
Ridler N, Robinson B, Chopin T, Robinson S, Page F (2006) Development of integrated multi-trophic aquaculture in the Bay of Fundy, Canada: a socio-economic case study. World Aquaculture 37: 43–48
Roberts CM, McClean CJ, Veron JEN, Hawkins JP, Allen GR, McAllister DE, Mittermeier CG, Schueler FW, Spalding M, Wells F, Vynne C, Werner TB (2002) Marine biodiversity hotspots and conservation priorities for tropical reefs. Science 295: 1280–1284
Roberts J, Chick LO, Thompson P (1995) Effects of carp, *Cyprinus carpio* L., an exotic benthivorous fish, on aquatic plants and water quality in experimental ponds. Marine and Freshwater Research 46: 1171–1180
Robinson TB, Griffiths CL, McQuaid C, Rius M (2005) Marine alien species of South Africa – status and impacts. African Journal of Marine Science 27: 297–306
Ruesink JL, Feist BE, Harvey CJ, Hong JS, Trimble AC, Wisehart LM (2006) Changes in productivity associated with four introduced species: ecosystem transformation of a 'pristine' estuary. Marine Ecology Progress Series 311: 203–215
Ruiz GM, Carlton JT, Grosholz ED, Hines AH (1997) Global invasions of marine and estuarine habitats by non-indigenous species: mechanisms extent, and consequences. American Zoologist 37: 621–632
Ruiz GM, Fofonoff P, Hines AH, Grosholz ED (1999) Non-indigenous species as stressors in estuarine and marine communities: Assessing invasion impacts and interactions. Limnology and Oceanography 44: 950–972
Sachs JD, Reid WV (2006) Environment – investments toward sustainable development. Science 312: 1002–1002
SAMS (2002) Review and synthesis of the environmental impacts of aquaculture. Scottish Association for Marine Science and Napier University, Edinburgh, Scotland, 42 pp
Sanjuan A, Zapata C, Alvarez G (1997) Genetic differentiation in *Mytilus galloprovincialis* Lmk throughout the world. Ophelia 47: 13–31
Sax DF, Gaines SD (2003) Species diversity: from global decreases to local increases. Trends in Ecology and Evolution 18: 561–566
Sindermann CL (1993) Disease risks associated with importation of non-indigenous marine animals. Fisheries Review 54: 1–10
Skaala O, Wennevik V, Glover KA (2006) Evidence of temporal genetic change in wild Atlantic salmon, *Salmo salar* L., populations affected by farm escapees. ICES Journal of Marine Science 63: 1224–1233
SOFIA (2004) The state of the world fisheries and aquaculture (SOFIA). Available online: www.fao.org/sof/sofia/index_en.htm
Stachowicz JC, Tilman D (2005) Species invasions and the relationship between species diversity, community saturation, and ecosystem functioning. In: Sax DF, Gaines SD, Stachowicz JJ (eds.), Species Invasions. Insights into ecology, evolution, and biogeography. Sinauer Associates, Sunderland, Massachusetts

Subasinghe RP, Arthur JR, Phillips MJ, Reantoso MB (2000) Thematic review on management strategies for major diseases in shrimp aquaculture. FAO, UN, Cebu, Philippines.
Townsend CH (1893) Report of observations respecting the oyster resources and oyster fishery of the Pacific coast of the United State, U.S. Fisheries Commission Report. U.S. Fisheries Commission, Government Printing Office, Washington, DC, pp. 343–372
Townsend CH (1896) The transplanting of eastern oysters to Willapa Bay, Washington, with notes on the native oyster industry. U.S. Fisheries Commission Report. U.S. Fisheries Commission, Government Printing Office, Washington, DC, pp. 193–202
Turner GE (1988) Codes of Practice and Manual of Procedures for Consideration of Introductions and Transfers of Marine and Freshwater Organisms. European Inland Fisheries Advisory Commission. Food and Agriculture Organization of the United Nations
Utting SD, Spencer BE (1992) Introductions of marine bivalve molluscs into the United Kingdom for commercial culture – case histories. ICES Marine Science Symposium 194: 84–91
Van Erkom Schurink C, Griffiths CL (1990) Marine mussels of southern Africa – their distribution patterns, standing stocks, exploitation and culture. Journal of Shellfish Research 9: 75–85
Van Erkom Schurink C, Griffiths CL (1991) A comparison of reproductive cycles and reproductive output in four southern African mussel species. Marine Ecology Progress Series 76: 123–134
Van Erkom Schurink C, Griffiths CL (1992) Physiological energetics of four South African mussel species in relation to size, ration and temperature. Comparative Biochemistry and Physiology 101: 779–789
Verschuren D, Kling HJ, Edgington DN, Leavitt PR, Brown ET, Talbot MR, Hecky RE (2002) History and timing of human impact on Lake Victoria, East Africa. Proceedings of the Royal Society London Series B 269: 289–294
Wallentinus I (1999) *Undaria pinnatifida* (Harvey) Suringar. In: Gollasch S, Minchin D, Rosenthal H, Voigt M (eds.), Case histories on introduced species: their general biology, distribution, range expansion and impact. Department of Fishery Biology, Institute for Marine Science, University of Kiel, Germany, pp. 13–19
Webster PJ, Holland GJ, Curry JA, Chang HR (2005) Changes in tropical cyclone number, duration, and intensity in a warming environment. Science 309: 1844–1846
Welcomme RL (1991) International introductions of freshwater fish species into Europe. Finnish Fisheries Research 12: 11–18
Welcomme RL (1992) A history of international introductions of inland aquatic species. ICES Marine Science Symposia 194: 3–14
Wenner EC, Knott DM (1992) Occurrence of Pacific white shrimp, *Penaeus vannamei*, in coastal waters of South Carolina. In: DeVoe MR (ed.), Introductions and transfers of marine species: achieving a balance between economic development and resource protection, pp. 173–181
Wilkins NP (1989) Ponds, passes and parks: aquaculture in Victorian Ireland, Glendale, Dublin.
Witte E, Msuku BS, Wanink JH, Seehausen O, Katunzi EFB, Goudswaard PC, Goldschmidt T (2000) Recovery of cichlid species in Lake Victoria: an examination of factors leading to differential extinction. Reviews of Fish Biology and Fisheries 10: 233–241
Worm B, Barbier EB, Beaumont N, Duffy JE, Folke C, Halpern BS, Jackson JBC, Lotze HK, Micheli F, Palumbi SR, Sala E, Selkoe KA, Stachowicz JJ, Watson R (2006) Impacts of biodiversity loss on ocean ecosystem services. Science 314: 787–790
Wyban J (2002) White shrimp boom continues. Global Aquaculture Advocate: 18–19
Yang AG, Wang QY, Liu ZH, Zhou LQ (2004) The hybrid between the scallops *Chlamys farreri* and *Patinopecten yessoensis* and the inheritance characteristics of its first filial generation. Marine Fisheries Research 25: 1–5
Youngson AF, Webb JH, MacLean JC, Whyte BM (1993) Frequency of occurrence of reared salmon in Scottish salmon fisheries. ICES Journal of Marine Science 54: 1216–1220
Youngson AF, Dosdatz A, Saroglia M, Jorda WC (2001) Genetic interactions between marine finfish species in European aquaculture and wild conspecifics. Journal of Applied Ichthyology 17: 153–162

Yu RH, Bao ZM, Wang ZP Kong LF, Tong W, Li JQ (2006) A study on hybridization between scallops *Clamys farreri* and *Patinopecten yessoensis*. Periodical of Ocean University of China 36: 85–88

Zhu PR (2000) China actively introduces resource of foreign species. World Agriculture 253: 17–20

Chapter 6
Safe and Nutritious Aquaculture Produce: Benefits and Risks of Alternative Sustainable Aquafeeds

J. Gordon Bell[1]* and Rune Waagbø[2]

Abstract It is estimated that by 2010 >85% of globally available fish oil (FO) and ~50% of fish meal (FM) will be consumed by aquaculture so, it is vital that reliance on marine raw materials is reduced and that sustainable aquafeeds are developed using more terrestrial plant products. In addition, levels of persistent organic pollutants (POPs), principally dioxins/furans and polychlorinated biphenyls (PCBs), in some European FO may breach new EU limits and prevent their use in aquafeeds. Current evidence suggests that salmonids can be grown on diets where 100% of the FO is replaced by vegetable oils (VO), and that bass and bream fed up to 60% VO showed no detrimental effects on growth. However, use of VO can result in reductions of the n-3 highly unsaturated fatty acids, DHA and EPA, of between 50% and 65%, although these values can be restored to 70–100% of the values in fish fed FO by the use of FO-containing finishing diets. Such high levels of FO replacement can only be used if essential fatty acid levels are maintained by inclusion of adequate FM levels. Simultaneous reductions in FM *and* FO will require considerable care if fish health and welfare, as well as product quality, are to be maintained. The efficacy of n-3 highly unsaturated fatty acids (HUFA), principally EPA and DHA, in the prevention or modulation of many of the inflammatory conditions prevalent in the developed world is well established. However, there is concern that the levels of POPs (dioxins, PCBs and PBDEs), as well as the presence of toxic metals, (e.g., Pb, As, Cd and Hg), present a potential risk to human health. The nutrients, as well as contaminants, found in fish flesh are derived largely from the feed and, thus, farmed fish can be tailored to provide optimal levels of fatty acids, and selected vitamins and minerals for human consumption.

Keywords Sustainable aquafeeds, vegetable oils, plant proteins, micronutrients, n-3 fatty acids, organic contaminants

[1]Institute of Aquaculture, University of Stirling, Stirling FK9 4LA, Scotland, U.K, Tel. + 44 1786 467997, Fax. + 44 1786 472133, E-mail: g.j.bell@stir.ac.uk

[2]National Institute of Nutrition and Seafood Research (NIFES), PO Box 2029, Nordnes, 5817 Bergen, Norway

* Corresponding author

M. Holmer et al. (eds.), *Aquaculture in the Ecosystem.*

6.1 Introduction

The rapid growth of aquaculture production worldwide, estimated to be around 10.5% per annum over the last 10 years, (Tacon 2003), has meant that demand for extruded aquafeeds has increased in parallel with production. The culture of carnivorous species in Europe, principally Atlantic salmon (*Salmo salar*), rainbow trout (*Oncorhynchus mykiss*), European sea bass (*Dicentrarchus labrax*) and gilthead sea bream (*Sparus aurata*), has meant that aquafeed formulations have contained relatively large amounts of both fish meal (FM) and fish oil (FO) derived from marine feed grade fisheries. Until recently this has been regarded as sound practice as fish meal is rich in essential amino acids, is readily accepted and digested by fish and has been readily available and relatively cheap. In addition, FO meets the high requirements of fish for n-3 essential fatty acids and, while freshwater fish and salmonids can convert 18:3n-3 to 20:5n-3 and 22:6n-3, but thrive on 20:5n-3 and 22:6n-3, marine fish cannot perform this conversion and have an absolute requirement for 20:5n-3 and 22:6n-3 (Sargent et al. 2002).

Fish meal and oil are obtained from feed grade fisheries that have effectively reached their sustainable limits and production of FM and FO has remained relatively stable for the last 20 years (Pike 2005). Previous estimates that global FO demand would outstrip production by 2010 or earlier (Barlow and Pike 2001) have now been revised but current estimates suggest that by 2010 more than 85% of the fish oil will be consumed in aquafeed production (Tacon 2004). Therefore, it is vital to reduce dependence on marine raw materials and that sustainable aquafeeds are developed using more terrestrial plant products. An additional reason to include more plant products, especially vegetable oils (VO), is because levels of persistent organic pollutants, principally dioxins/furans and polychlorinated biphenyls (PCBs), in some Northern hemisphere FO may breach new EU limits and prevent their use in aquafeeds (Easton et al. 2002; Foran et al. 2005; EC, 2006b).

Over the last 5–10 years considerable research has been conducted on testing replacements for FO and FM in a range of cultured freshwater, anadromous and marine finfish species. A large amount of data on FO and FM replacement has arisen from the EU Framework 5 projects, Researching Alternatives to Fish Oil in Aquaculture (RAFOA, www.rafoa.stir.ac.uk) and Perspectives on Plant Protein Alternatives (PEPPA, www.st-pee.inra.fr/ici/stpee/nut/peppa/peppa.htm). RAFOA established that salmonids could be cultured on diets where 100% of the FO is replaced by single VO, or a VO blend and that for bass and bream replacement of up to 60% of FO showed no detrimental effects on growth. PEPPA established that trout and bream can be cultured with 75% replacement of FM, by plant products with no loss of growth performance. However, such high levels of FO replacement can only be achieved provided essential fatty acid (EFA) levels are maintained by the inclusion of adequate FM levels. Likewise, high levels of FM replacement were achieved in feeds using FO as oil source and, thus, future studies seeking to replace both FM and FO may prove more challenging.

Fish have proved fairly tolerant of changes in lipid and protein sources, in terms of growth and survival, provided EFA and essential amino acid requirements are met. However, there are potential detrimental effects, in the edible flesh, on biologically important fatty acid concentrations, due to replacement of FO with VO, and also in micronutrient concentrations, when FO and FM are replaced. High dietary VO inclusion can result in reductions of flesh DHA and EPA of ~65% in salmon fed 100% VO or up to 50% in bass and bream fed 60% VO. However, flesh EPA and DHA values can be restored to 70–100% of the values in fish fed FO, for the whole grow out period, by the use of FO-containing finishing diets in the pre-harvest period (Bell et al. 2004a; Torstensen et al. 2005; Mourente et al. 2005; Izquierdo et al. 2005).

Significant replacement of FO and FM also leads to changes in dietary supply, bioavailability and requirement of micronutrients for the farmed fish. Greatest focus has been on mineral bioavailability aspects related to the inherent anti-nutrient factors (ANF) in plant derived raw materials (Francis et al. 2001). Besides the risks for suboptimal micronutrient nutrition for the fish, by lower gross nutrient concentrations and bioavailability, this also implies subsequent alteration in product composition and quality, since several vitamins and minerals in fish flesh are tailored through diet (Baker 2001; Lie 2001). There has been less focus on these secondary consequences of changes in feed ingredients, which especially includes nutrients with antioxidant properties and those normally associated with the benefits of seafood consumption, such as vitamins B_{12}, D, E, carotenoids (astaxanthin and canthaxanthin), iodine and selenium. Changes in product composition will, like lipid retention, also depend on fish species (lean or fat), as well as their feed intake and growth rate.

The importance of n-3 highly unsaturated fatty acids (HUFA), principally EPA and DHA, in human nutrition was first recognised in the 1970s by Dyerberg and Bang who suggested that the diet of Greenland Inuit populations resulted in reduction, or absence, of disease conditions which were prevalent in developed societies (Dyerberg et al. 1975; Bang et al. 1980). Over the last 50 years the prevalence of diseases with an inflammatory pathology has increased dramatically especially pathologies of the cardiovascular system (Kris-Etherton et al. 2002; Wang et al. 2003). However, more recent research has implicated many more disease conditions with an inflammatory pathology that may respond to n-3 HUFA supplementation. These include asthma (Broughton et al. 1997), rheumatoid arthritis (Calder and Zurier 2001), Alzheimer's disease (Morris et al. 2003), Crohn's disease (Belluzzi and Miglio 1998), lupus (Kelley et al. 1985), cancer (Hardman 2002), diabetes (Lombardo and Chicco 2006), psoriasis (Ziboh 1998), schizophrenia (Peet et al. 2001), bipolar disorder (Noaghiul and Hibbeln 2003) and autism (Bell et al. 2004b). In recent times, many countries in the developed world, as well as the World Health Organisation and NATO, have produced recommendations on combined EPA and DHA intake for improved human health which are generally in the range of 0.3–0.5 g day^{-1}. In 2004, ISSFAL (www.issfal.org.uk) updated their recommendation for n-3 HUFA intake suggesting that consuming 500 mg of EPA + DHA day^{-1}, or 3.5 g $week^{-1}$ should provide for good cardiac health in adults.

Nowadays, the efficacy of n-3 HUFA in the prevention or modulation of many of the inflammatory conditions prevalent in the developed world is well established (Connor 2000). However, while the increased consumption of fish is widely recommended, there is currently concern that the levels of persistent organic pollutants (POPs) including dioxins, furans, polychlorinated biphenyls (PCB) and polybrominated dephenylether (PBDE) flame retardants, as well as the presence of toxic metals, including Pb, As, Cd and Hg, present a potential risk to human health (Jacobs et al. 2002a; Hites et al. 2004a,b; MAFF 1999). In 2001, the EU introduced new limits on dioxins and furans in fish feeds and fish for human consumption (SCAN 2000; SCF 2001). These values were 2.25 ng dioxin toxic equivalents (TEQ) kg^{-1} in feed and 4.0 ng TEQ kg^{-1} in fish. Currently, the EU has set limits for the 12 dioxin-like (DL) PCBs, in addition to the values for the 17 dioxins/furans assigned toxic equivalency factors, with combined values of 8 ng TEQ kg^{-1} in fish (EC, 2006a).

In this chapter we will discuss why there is a need to develop aquafeeds that are less reliant on marine raw materials including the effects of including plant meals and oils on the nutritional quality of farmed fish in terms of n-3 HUFA content, as well as vitamin and mineral content. In addition, the importance of n-3 HUFA for human health, current recommended intake levels of n-3 HUFA and the concentrations of n-3 HUFA and other nutrients provided by farmed fish will be discussed. Finally, the concentrations of POPs in farmed fish, as well as mechanisms by which these concentrations can be reduced by manipulation of feed ingredients will be discussed.

6.2 Why Do We Need Alternatives to Fishmeal and Fish Oil?

Food-grade capture fisheries have maintained a fairly stable supply over the last 20 years, although increasing demand has meant that the shortfall in sea food availability for human consumption has been met by increasing aquaculture production (Tidwell and Allan 2002; FAO 2005). In 2003, production of Atlantic salmon and marine-cultured rainbow trout in Europe was ~824,000 t with bass and bream contributing ~150,000 t (FAO 2005). Salmonids are the biggest consumers of extruded aquafeeds in Europe and to produce this tonnage of fish required over 1 million tonnes of extruded feed (Tacon 2005). In 2003, aquaculture utilised 79% of global FO production and 42% of FM production and, of this, salmonid culture used 52 and 27% of FO and FM production, respectively (Tacon 2004).

Therefore, as the supplies of FO and FM have remained static at ~6 million tonnes and ~1 million tonnes per annum, respectively, for around 20 years, the continued increased demand for these products for aquaculture is unsustainable. It is therefore vital that reliance on marine raw materials is reduced and that sustainable aquafeeds are developed using more terrestrial plant products.

In addition to a potential shortfall in the supply of FM and FO, concerns have recently been raised regarding the sustainability and ethical arguments for utilising fish species that could be used directly as food for humans, in animal feeds (Tacon 1997; Goldburg and Naylor 2005). Furthermore, there is also heightened public and non-governmental organisations (NGO) awareness over the management of feed grade fisheries and the potential impact on marine ecosystems particularly effects on sea birds and marine mammals (Huntington et al. 2004).

A further reason to substitute FO for VO in fish diets is that carnivorous fish have evolved to use oil for the energy required to fuel growth and reproduction. Thus, in effect, most of the oil consumed by fish is effectively "burned" to supply energy and fish do not seem to discriminate greatly between which fatty acids are retained and which are metabolised. Therefore, once essential fatty acid requirements have been met the excess n-3 PUFA in FO are stored as triacylglycerols in body stores and then used for energy production (Sargent et al. 2002). For that reason, it might be more productive to feed VO for energy production, in place of FO, for most of the production cycle and not waste valuable n-3 HUFA in the production of energy for growth. Further, the use of FO finishing diets can still be used to increase n-3 HUFA in fish body oils at a point where the fish has undergone most of its growth increment but has not yet reached the reproductive phase.

A final reason for seeking substitutes for FM and FO relates to the levels of POPs present in some geographical sources of marine raw materials. It is generally established that the lowest levels of contaminants are found in pelagic fish species from South America, while the highest levels are found in those caught in Northern Europe (Easton et al. 2002; FIN 2004). In addition, due to tightening of EU legislation in 2001, a significant proportion of Northern European FO and FM, particularly some of those originating from the Baltic and Barent seas, would no longer be eligible for inclusion in animal and aquafeeds (SCAN 2000; FIN 2004; EC, 2006b).

6.3 Fish Oil and Fish Meal Substitution

6.3.1 Effects of Fish Oil Replacement on Growth Performance

Over the last five years a large number of studies have investigated a range of FO substitutes in a wide range of species. However, all of these earlier studies fed the experimental diets for only a small part of the growth cycle for each species, usually only for 8–20 weeks. By contrast, the studies undertaken as part of the EU project, RAFOA, were unique in that they were long term trials from juvenile stages or covered the whole cycle from first feeding to harvest (www.rafoa.stir.ac.uk).

In salmon smolts, replacement of FO with increasing quantities (25–100%) of linseed oil (LO), rapeseed oil (RO) and olive oil (OO) had no effect on growth or survival (Torstensen et al. 2004a; Bell et al. 2004a) and confirms reports of similar studies using either low (Hardy et al. 1987; Bell et al. 1991; Waagbø et al. 1991, 1993) or high energy diets in Atlantic salmon (Bell et al. 2001a, 2002; Rosenlund et al. 2001; Torstensen et al. 2000, 2004b). These studies show similar growth responses to other studies that have used RO (Bell et al. 2001a, 2003a,b; Rosenlund et al. 2001) or LO (Bell et al. 1997, 2003a; Tocher et al. 2000) as full, or partial, replacement of FO and suggest that the energy requirements of salmon can be satisfied by VO with variable fatty acid compositions. Furthermore, the lack of any negative growth response suggests that the contribution of EPA and DHA from dietary fishmeal is sufficient to satisfy EFA requirements of salmon up to 100% replacement of FO with VO.

In a full production cycle trial in salmon, a dietary fatty acid composition was formulated by mixing RO, LO and palm oil (PO) to provide similar levels of the different fatty acid classes (saturated, monounsaturated and polyunsaturated n-3 fatty acids) to capelin oil. It was hoped this balance might be better physiologically for fish health and welfare. Thus, at high levels of FO replacement (75 or 100%) a balanced fatty acid composition should be less stressful physiologically compared to the more extreme fatty acid compositions obtained by replacing FO with a single VO. Thus, when salmon were fed either 75 or 100% of the VO blend for the whole production cycle, at two different geographical locations, growth was high in all treatments. However, for the 100% VO group in Norway, significantly higher final mean weight was found compared to the FO group (Torstensen et al. 2005). The higher mean weight after 22 months post-first feeding (PFF) correlated with higher protein sparing in the 100% VO group compared to the FO group indicating that, during the late autumn and winter period of the sea water growth phase, the fatty acid composition of the 100% VO diet favoured protein growth and spared dietary protein from energy production (Torstensen et al. 2005). Previously, dietary lipid content, but not dietary oil source (Torstensen et al. 2000; Bendiksen et al. 2003), has been shown to affect protein utilisation, growth rate, muscle lipid level and feed conversion (Watanabe 1977; Arzel et al. 1993, 1994).

In rainbow trout fed the same single VOs as described for salmon, for 12 weeks, or the 75% and 100% VO blend, for 62 weeks, there were no significant effects of diet on final weight, SGR, TGC or FCR (Kaushik and Corraze 2004, Richard et al. 2006). This data supports earlier studies with rainbow trout and other salmonids where no detrimental effects on growth were observed with different FO substitutes, including soybean oil (SO), RO, OO, PO, LO and lard (Dosanjh et al. 1988; Greene and Selivonchick 1990; Guillou et al. 1995; Caballero et al. 2002; Figueiredo-Silva et al. 2005; Fonseca-Madrigal et al. 2005). No increase in final weight of rainbow trout fed the 100% VO blend was observed. In salmon, the beneficial growth effect during the winter period in Norway may have been due to increased fatty acid digestibility, and thereby increased protein sparing, which lead to improved growth at low water temperatures of less than 5°C. By contrast, the rainbow trout trial was conducted at a constant 17°C where any differences in digestibility at low temperature would not be apparent (Ng et al. 2004a).

In sea bass and sea bream, replacement of up to 60% of FO with VO had no detrimental effects on growth or feed conversion (Izquierdo et al. 2003, 2005; Mourente et al. 2005). However, replacement with 80% linseed oil or 100% of a VO blend in sea bream did reduce growth rates (Izquierdo et al. 2003, 2005) although, with the VO blend, growth reduction was not seen in fish over 250 g suggesting that EFA requirements in larger fish were less stringent than in smaller fish.

6.3.2 Effects of Fish Meal Replacement on Growth Performance

The investigation of fish meal substitutes to supply dietary protein in aquafeeds has been conducted for many years, and although often driven by the desire for more cost effective raw materials, more recently the focus has been to introduce more sustainable aquaculture practices. Generally, despite lower protein content, lower levels of some essential amino acids and the presence of anti-nutritional factors, replacement of around 30% of fish meal can be achieved without loss of growth performance, depending on the degree of product refinement (Teskeredzic et al. 1995; Medale et al. 1998; Glencross et al. 2004; Morris et al. 2005). Some trials have been conducted with fish meal-free diets but generally these have resulted in loss of growth performance (Kaushik et al. 1995; Watanabe et al. 1998). In recent studies in rainbow trout, replacement of FM above 75% resulted in growth reduction as well as some changes in sensory properties, even though amino acid contents were optimized by addition of crystalline amino acids (de Francesco et al. 2004). However, this was a long term trial of almost 6 months and growth rates only became depressed in the high plant protein group after 12 weeks, which is similar to the length of many trials conducted on FM replacement and emphasizes the need to conduct longer term trials (de Francesco et al. 2004).

Generally, Atlantic salmon appear less able to cope with high levels of plant proteins than rainbow trout which may be related to different digestive capacity as well as sensitivity to ANFs (Refstie et al. 2000; Glencross et al. 2004; Mundheim et al. 2004). Salmon fed a range of plant protein concentrations, provided by full-fat soya meal and maize gluten (2:1 w:w), from 15% to 65% showed a linear decrease in growth with each addition of plant protein (Mundheim et al. 2004), although there were no significant differences in SGR, TGC or FCR.

Diets replacing all fish meal with maize gluten and soy protein showed significant growth reduction in European sea bass (Dias 1999) compared to studies utilizing lower levels of replacement (Tibaldi et al. 1999; Tulli et al. 1999). A more recent study, in which sea bass were fed up to 98% of dietary protein as plant meals, showed no reduction of growth over a 12 week period (Kaushik et al. 2004), and, unlike previous studies (Gomes et al. 1995; Dias 1999), there was no reduction of voluntary feed intake in the study by Kaushik et al. (2004). In gilthead sea bream, previous studies have indicated that about one third of the fish meal could be replaced without reducing the levels of indispensable amino acids or reduction in growth rate (Pereira and Oliva-Teles 2002; Gomez-Requeni et al. 2003). However, in a more recent study over 12 weeks, sea bream showed a slight growth reduction when 50 and 75% of protein was provided by plant sources but a 30% growth reduction was seen at 100% plant protein inclusion and this was associated with a marked reduction in feed intake (Gomez-Requeni et al. 2004). By contrast, a recent longer term trial showed that 75% plant protein inclusion did not result in growth reduction (Sitja-Bobadilla et al. 2005).

In turbot (*Psetta maxima*), studies with either maize gluten or lupin have shown promising results as potential fish meal substitutes although reduced feed intake was observed (Burel et al. 2000a,b). In an attempt to provide a more balanced amino acid composition Fournier et al. (2004) fed a mixture of lupin, wheat and maize gluten with supplementary crystalline amino acids and showed that growth rate in turbot was only compromised when fed 90 or 100% replacement of fish meal over 12 weeks.

The inclusion of high levels of plant proteins can be limited by the presence of ANFs including protease inhibitors, phytates, glucosinolates, tannins, lectins, phytoestrogens and antivitamins among others (Francis et al. 2001). At levels of individual product inclusion in fish feeds many of these factors should not affect growth performance and some can be reduced or eliminated by solvent extraction, steam extrusion or enzymatic treatment. These anti-nutritional factors can reduce growth by affecting palatability and reduction of feed intake or by limited digestibility. Besides, these direct effects on nutrient supply and utilisation, indirect toxic effects with organ damage and endocrine disruption are evident for some ANFs. Increased use of plant proteins in aquafeeds requires more information on which factors are present in specific plant meals so that measures to limit their effects can be achieved by appropriate processing techniques.

A further concern regarding plant proteins is the presence of genetically modified (GM) products, currently used in terrestrial animal production, especially those derived from soya, canola and maize (Pusztai and Bardocz 2006). However, studies conducted with Atlantic salmon suggest that while short transgenic sequences (~120 bp) can be detected in gut tissues, no transgenic fragments have been found in liver, muscle or brain (Sanden et al. 2004). For this reason, there should be no danger of transgenic plant material entering the human food chain from consumption of farmed salmon flesh.

6.3.3 Flesh Fatty Acid Compositions Including Success of Finishing Diets

Numerous studies, in a wide range of fish species, have shown that flesh fatty acid compositions are closely correlated to dietary fatty acid compositions and that feeding high levels of VO will strongly influence flesh fatty acid compositions (Bell et al. 2004a; Izquierdo et al. 2003; Caballero et al. 2002; Mourente et al. 2005; Visentainer et al. 2005; Glencross et al. 2003). However, the influence of dietary lipid on flesh fatty acids is also related to the lipid content of the flesh and the ratio of neutral to polar lipid present, since the correlation with diet is closest in lipid rich flesh, which is high in neutral lipid, especially triacylglycerols (Sargent et al. 2002). In this regard, the rank order of flesh lipid content would be salmon > trout > sea bream > sea bass > cod and flesh lipid deposition tends to increase with fish weight, especially in salmonids (Hemre and Sandnes 1999; Torstensen et al. 2001). Several studies with salmon have shown a clear linear relationship between dietary and flesh fatty acid compositions where a number of VO including RO, PO, SO and blends of RO and LO have been used, along with FO, in diet formulations (Rosenlund et al. 2001; Torstensen et al. 2001, 2004a; Bell et al. 2001a, 2002, 2003a).

The data from studies with salmon, and similar studies with other species, confirm that individual fatty acids, within a blend of fatty acids, are selectively retained or metabolised depending on their concentration in the diet and the biological function of each specific fatty acid. One of the most striking effects, in all species, is the preferential deposition and retention of DHA in flesh lipids, regardless of the concentration present in the diet. This selectivity presumably reflects the specificity of the fatty acyl transferase enzymes that incorporate the individual fatty acids into flesh triacylglycerols and phospholipids, a phenomenon that has been observed in previous studies with salmon fed different combinations of VO (Torstensen et al. 2000; Bell et al. 2001a, 2002, 2003a; Rosenlund et al. 2001) as shown in Table 6.1.

In comparison to DHA, the other PUFA and HUFA seem to be directed more towards metabolism, presumably being largely catabolised for energy production rather than deposition, especially when present at high concentrations. When present at lower concentrations, only EPA appeared to be selectively retained as demonstrated by higher flesh values compared to diet values, specifically in fish fed 100% LO (Table 6.1). In contrast, both 18:2n-6 and especially 18:3n-3 were selected against in terms of deposition in flesh. The tendency towards preferential metabolism of C_{18} PUFA by β-oxidation has been observed not only in fish (Bell et al. 2001b; Bell et al. 2003c) but also in humans, in whom 18:3n-3 was preferred over 18:2n-6 as an oxidative substrate (DeLany et al. 2000). However, it should also be noted that both 18:2n-6 and 18:3n-3 are substrates for Δ6-desaturase, and salmon hepatocytes reportedly favour desaturation and elongation of 18:3n-3 over 18:2n-6 (Bell et al.,1997; Ruyter et al. 2003). In addition to PUFA, the long chain monoene fatty acids (20:1 & 22:1), found in high latitude FO, are thought to be important catabolic substrates (Sargent et al. 2002). This appears to be confirmed in the present studies, particularly so in the salmonids fed capelin oil, as 20:1 and especially 22:1, were selected against in terms of flesh deposition. The literature suggests that 22:1n-11 and 18:2n-6 are preferred substrates for β-oxidation, along

Table 6.1 The differences (Δ) between diet and flesh total lipid fatty acid values for salmon fed 100% fish oil, 50% linseed oil (LO), 100% LO, 33% rapeseed oil (RO) and 100% RO diets

Fatty acid	Δ 100% FO	Δ 50% LO	Δ 100% LO	Δ 33% RO	Δ̃100% RO
16:0	0.8	1.4	2.2	0.8	1.7
18:1n-9	4.1	2.5	1.6	−1.8	−6.8
18:2n-6	−0.3	−1.2	−2.0	−1.0	−3.3
18:3n-3	−0.1	−5.5	−11.7	−0.9	−2.1
20:1n-9	−1.3	−0.3	0.5	−0.2	0.9
22:1n-11	−3.4	−1.7	0.0	−1.2	−0.7
20:5n-3	−1.6	−1.0	0.3	−1.7	−0.4
22:6n-3	3.1	1.9	1.6	1.3	2.1

Data from Bell et al. 2003, 2004. Fatty acid concentrations are g/100 g fatty acid in flesh and diet. Negative Δ values indicate lower values in flesh compared with diet whereas positive values indicate accumulation in flesh relative to diet.

with 16:0, 16:1 and 18:1n-9 (Henderson 1996; Kiessling and Keissling 1993; Frøyland et al. 2000), although it should be noted that much of this work was done using tissue homogenates, isolated cells or mitochondria. In our recent studies, 16:0 was selectively deposited in flesh, suggesting that it may not be readily used as a catabolic substrate (Bell et al. 2003a, 2004a; Table 6.1). By comparison, 18:1n-9 was selectively deposited except when present in high concentrations, e.g., where salmonids were fed > 33% RO or OO (Torstensen et al. 2004a,b) and in sea bass and bream fed 60% RO and 60% OO (Izquierdo et al. 2005; Mourente et al. 2005). It is generally accepted that DHA is selectively retained due to the biological importance of this fatty acid in cell membrane functional integrity, especially in neural, reproductive and immune tissues (Sargent et al. 2002). The selective deposition of 16:0 and 18:1n-9, rather than mobilisation, may reflect the structural importance of both these fatty acids in membrane phospholipids, where they are often located in the *sn*-1 position, especially in PC and PE, with PUFA and HUFA being favoured in the *sn*-2 position (Bell and Dick 1991; Sargent et al. 2002).

The health benefits of fish consumption, related to the n-3 HUFA content, are now widely recognised (Simopoulos 1999; Connor 2000) and it is important that aquaculture maintains a healthy product image by producing seafood that is comparable with those from capture fisheries. However, evidence from several studies suggests that when fish are cultured on diets containing VO, especially at levels over 50%, then there are significant reductions in flesh EPA and DHA (Table 6.2, Bell et al. 2004a; Torstensen et al. 2004a,b; Menoyo et al. 2004; Mourente et al. 2005). To overcome this, fish can be placed on a FO finishing diet for a period prior to harvest to restore n-3 HUFA levels. In general, the ability to restore EPA and DHA concentrations was more easily achieved than the dilution or wash out of the 18:2n-6 and 18:3n-3 (Bell et al. 2004a; Torstensen et al. 2004b, 2005; Mourente et al. 2005; Izquierdo et al. 2005). In salmon and trout, although DHA and EPA levels were still significantly lower, after 24 and 12 weeks on a FO finishing diet, respectively, the values attained were at least 80% of the values in fish fed FO, in fish previously fed VO compared to fish fed FO throughout. In salmon, the DHA and EPA were largely restored after 16 weeks with only further small increases up to 24 weeks (Bell et al. 2004a) while for sea bass and bream restoration of EPA and DHA could be largely achieved in 14 weeks (Mourente et al. 2005; Izquierdo et al. 2005).

In trials where fish were fed what were considered maximal levels of VO (60–100% of total added oil) for the whole production cycle, fish were exposed to high dietary VO for an extended period of time of 50–100 weeks. However, the blend of RO, LO and PO used was selected to balance the saturated, monounsaturated and polyunsaturated fatty acids with the same levels as found in either capelin oil or anchovy oil. This generally resulted in lower levels of 18:2n-6 and 18:3n-3 than were found when either LO or RO were used as single FO substitutes. In salmon, flesh DHA and EPA levels were restored to ~90%, in fish fed 75% VO, and ~65%, in fish fed 100% VO, of values in fish fed FO throughout, after 24 weeks on a FO finishing diet (Torstensen et al. 2005). The difference between the two VO treatments can be explained, in part, by the seasonal differences in the finishing diet period between the 75% VO trial in Scotland (March–September when water

Table 6.2 Selected flesh fatty acid compositions (weight % of total fatty acids) in salmon, sea bass and sea bream fed diets containing different levels of rapeseed (RO) or linseed oils (LO) relative to fish oil (FO) from juvenile to commercial harvest weight

Species & Diet/Fatty acid	18:2n-6	18:3n-3	20:5n-3	22:6n-3
Salmon				
100% FO	3.9	0.8	4.3	8.1
100% LO	13.1	38.7	1.3	3.1
100% RO	15.0	5.1	1.7	4.9
Sea bass				
100% FO	3.0	1.0	9.6	20.2
60% LO	5.7	8.4	5.7	14.4
60% RO	8.5	2.7	5.0	9.4
Sea bream				
100% FO	5.5	0.8	9.1	7.3
60% LO	10.4	17.7	3.5	4.6
60% RO	12.9	3.4	3.6	4.9

Data from Bell et al. 2004; Torstensen et al. 2004; Mourente et al. 2005; Izquierdo et al. 2005.

temperatures and growth were high) and the 100% VO trial in Norway (January–May when water temperatures and growth were low; Torstensen et al. 2005).

While restoration of flesh DHA and EPA levels could be reasonably easily achieved, for all species, in around 14–24 weeks, the reduction of 18:2n-6 and 18:3n-3 was less easy (Bell et al. 2004a; Torstensen et al. 2005; Mourente et al. 2005; Montero et al. 2005). In salmon previously fed 100% LO or RO 18:2n-6 levels were still 75% higher compared to fish fed FO, after 24 weeks on the finishing diet. By comparison, the 18:3n-3 remaining in flesh of salmon previously fed 100% LO was 1344% higher after 24 weeks on the FO finishing diet (Bell et al. 2004a; Torstensen et al. 2004a,b). Comparable values for sea bream flesh were 18:2n-6 120% higher, in fish previously fed 60% RO and 18:3n-3 1525% higher in fish fed 80% LO, compared to fish fed FO throughout (Izquierdo et al. 2005). By contrast, for salmon in the whole life cycle trials, 18:3n-3 was only 160% or 360% higher, for fish fed the 75% or 100% VO diet, respectively, while 18:2n-6 was 60% higher than the FO fish after 24 weeks on the FO finishing diet (Torstensen et al. 2005). Thus, residual C_{18} PUFA levels were much lower when using a VO blend with lower 18:2n-6 and 18:3n-3 levels than when single VO high in C_{18} PUFA were used. A recent study with Atlantic salmon introduced the concept of a dilution model that allowed changes in flesh fatty acids to be accurately predicted when fish were switched from a VO diet to a FO diet. The model described by Jobling (2003) supports the findings of the RAFOA studies such that relatively constant levels of DHA and EPA were found after feeding salmonids a FO finishing diet for 16–24 weeks.

6.4 n-3 Highly Unsaturated Fatty Acids (HUFA) and Human Health

The efficacy of EPA and DHA in preventing or attenuating inflammatory disease in humans was first recognized in the early 1970s when epidemiological studies indicated a low incidence of cardiovascular disease in Inuit populations in Greenland and that coastal populations had different disease patterns from inland dwellers (Bang and Dyerberg 1972; Dewailly et al. 2001a,b). The reason for the differences in disease patterns were attributed to higher fish and n-3 HUFA intake in coastal populations. Historically, the human genome has changed little since Paleolithic times when humans were hunter-gatherers and consumed a diet where the ratio of n-6/n-3 PUFA was estimated to be around 1:1 (Leaf and Weber 1987; Simopoulos 1999). Thus, over the past 10,000 years the human genome will have changed little such that the nutritional input in the developed world in the 21st century will be very different to that which our genetic composition is best suited. The changes in our lipid intake from Paleolithic times to the present day are shown in Fig. 6.1 (Leaf and Weber 1987). This demonstrates the increase in total fat intake towards the end of the Agricultural revolution and similar increases in saturated fat and n-6 PUFA, with decreased n-3 PUFA during the Industrial revolution. It is also noteworthy that increased consumption of cereal grains at this time resulted in a greatly elevated starch intake that resulted in increased lipogenesis. The n-6/n-3 PUFA ratio increased steadily from around 1:1 in the early 19th century such that the ratio in the developed world now ranges from 5:1 to 25:1. The most dramatic increases have been due to increased production and use of n-6-rich seed oils which became established following the First World War and have subsequently dominated agricultural production. They gained popularity in human nutrition due to the improvement in serum lipid and cholesterol profile induced by n-6 PUFA compared to saturated fat (Keys et al. 1957). Unfortunately, the dominance of n-6 PUFA in the human food chain, due to direct consumption of vegetable oils as well as the use of oilseeds for the production of farm animals, has seen a steady decline in n-3 PUFA and HUFA in the food chain in the 20th century (Simopoulos 1999).

Over the last 50 years the prevalence of diseases with an inflammatory pathology has increased dramatically especially pathologies of the cardiovascular system (Simopoulos 1991; Zheng et al. 2001). Recent evidence suggests that supplementation with EPA and DHA can reduce death from coronary heart disease (CHD) by 25% and of sudden cardiac death by 45% (Marchioli et al. 2002), and that the risk of CHD can be predicted by a so called "Omega-3 index" based on combined blood fatty concentrations of DHA + EPA (Harris and von Schacky 2004). Using a dose response study, the authors studied the effectiveness of increasing DHA + EPA supplementation on red blood cell DHA + EPA content and thereby correlating Omega-3 index with CHD risk factors identified in earlier epidemiological studies and randomised controlled trials (Harris and von Schacky 2004). However, when considering a healthy intake of n-3 HUFA it is also vital to consider n-6 PUFA and HUFA intake as the n-3 and n-6 fatty acids compete during metabolic conversions

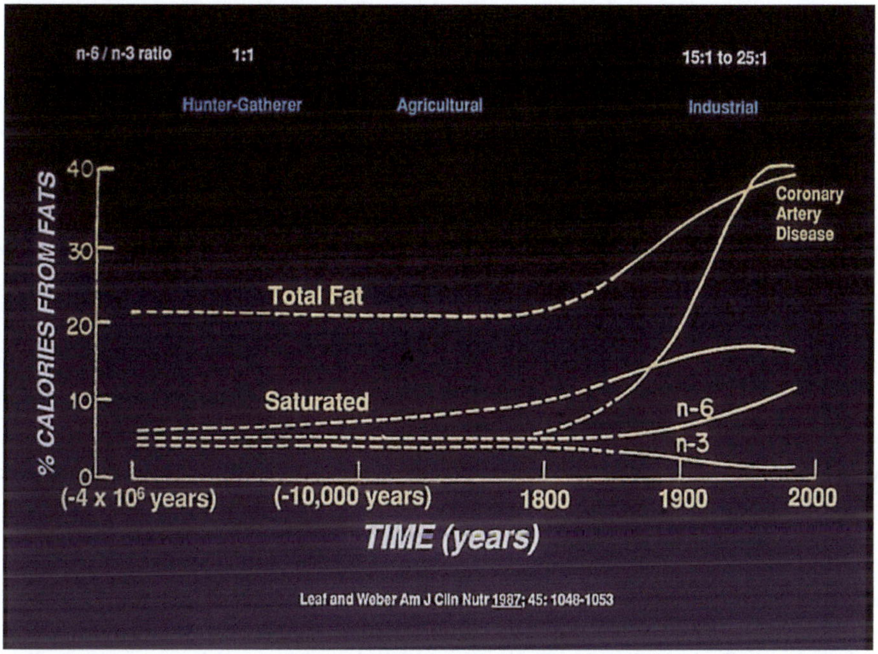

Fig. 6.1 Changes in fat intake from the Paleolithic era to the present day

including desaturation and elongation, acylation into phospholipids and production of eicosanoids (Lands et al. 1992; Lands 1992). The competitive interaction between ARA and EPA for incorporation into membrane phospholipids has been demonstrated in fish (Bell et al. 1989) and in mammals (Lands 2003) and is the basis for the anti-inflammatory activity of EPA and marine fish oils. The importance of maintaining low levels of tissue n-6 HUFA, especially ARA, is shown in the data of Lands (2003) where the positive correlation between tissue n-6 HUFA and CHD mortality is clearly demonstrated (Fig. 6.2).

However, while there has been a great deal of research conducted on the benefits of fish and fish oils on cardiovascular disease, there is also more recent research which has investigated many more disease conditions with an inflammatory pathology that may respond to n-3 HUFA supplementation. The fatty acid composition of inflammatory and immune cells is closely linked to the dietary fatty acid composition and thus provides a link between diet, inflammation and immune function (Calder 2001; Yaqoob 2004). In conditions with an inflammatory or auto-immune component, results following supplementation with marine fish oils have generally been positive. In the case of rheumatoid arthritis improvements observed included reduced stiffness and joint pain, increased grip strength and reduced reliance on non-steroidal anti-inflammatory drugs (James and Cleland 1997; Calder and Zurier 2001). Dietary supplementation with n-3 fatty acids has also shown benefits for dyslipidemia, insulin resistance, glucose homeostasis, diabetes and obesity in

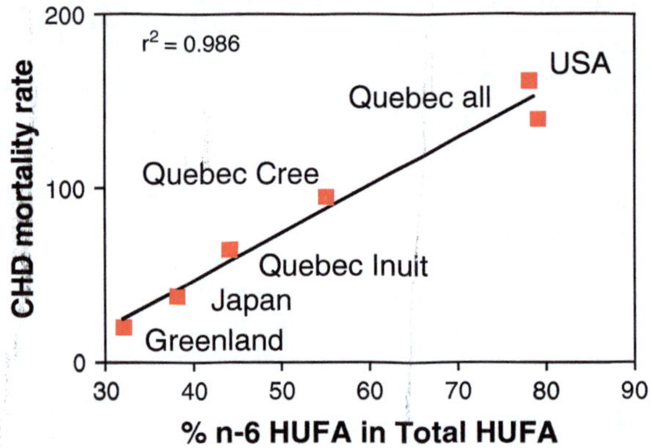

Fig. 6.2 Coronary heart disease (CHD) mortality rates versus tissue n-6 HUFA concentrations in subjects from the USA, Japan, Greenland, Quebec total population, Quebec Inuit and Quebec Cree populations (Lands, 2002)

animal and human studies (Storlien et al. 1998; Lombardo and Chicco 2006). In inflammatory conditions of the bowel, consumption of oily fish reduced relapses in patients with Crohn's disease (Belluzzi and Miglio 1998) and in ulcerative colitis, dependence on steroids was reduced as well as improved colon histology following supplementation with n-3 HUFA (Rogers 1998). In a study with patients suffering from Crohn's or ulcerative colitis, reduced disease activity, serum cholesterol and joint pain index were observed when patients were given an infusion of seal oil (Arslan et al. 2002). In patients with the inflammatory skin condition, psoriasis, improvements in itching, scaling, lesion thickness and erythema were observed following supplementation with EPA + DHA (Ziboh 1998). In childhood asthma, increased dietary n-3 HUFA tended to reduce the asthma severity, while increased n-6 PUFA had the opposite effect (Hodge et al. 1996; Haby et al. 2001). However, supplementation studies gave mixed results with some individuals reporting no effects, while other reported reduced symptoms (Broughton et al. 1997; Hodge et al. 1998). Another inflammatory condition that shows potential for response to n-3 HUFA supplementation is cystic fibrosis, where low blood DHA levels have been found (Roulet et al. 1997).

The proliferation of certain tumour cells is known to be increased by ARA, while this effect can be reversed by inhibitors of eicosanoid production (Horrocks and Yeo 1999). Subsequently, administration of n-3 HUFA has been shown to promote programmed cell death (apoptosis) in leukaemia and lymphoma cell cultures as well as in animal models (Fernandes et al. 1996; Heimli et al. 2002). In addition, animal studies have demonstrated that consumption of n-3 HUFA can reduce cancer growth rates, increase the efficacy of chemotherapy and reduce associated side effects of treatment and of the cancer (Hardman 2002; Hardman et al. 2002). The proposed mechanisms involve suppression of nuclear factor-κB activation and alteration in expression of specific genes including suppression of cyclooxygenase-2 expression and that of other genes that are implicated in tumour promotion (Hardman 2002). In a recent study, proliferation and gene expression in a B-lymphocyte cell line were investigated when supplemented with EPA or DHA (Verlengia et al. 2004). Cell proliferation was enhanced by both n-3 HUFA while production of key immunomodulatory cytokines, including IL-10, TNF-α and INF-γ was reduced. In addition, altered expression of specific genes including those involved with cytokines, signal transduction, transcription, cell cycle, defence and repair, apoptosis, cell adhesion, cytoskeleton and hormones was observed with EPA almost 3 times more active than DHA at the same concentration (Verlengia et al. 2004).

Although the health benefits of increased fish, and thereby n-3 HUFA, intake have been regarded as beneficial for CHD for over 35 years, more recent focus has turned to the role of essential HUFA in normal neural function and the prevention and treatment of neuropsychiatric disorders (Young and Conquer 2005). It should be no surprise that both ARA and DHA are vital for neural function as these two HUFAs comprise around 30% of the dry weight of brain and retinal tissue (Sastry 1985). By contrast, EPA is not particularly enriched in neural tissues and its role is more likely involved in an anti-inflammatory capacity, as an inhibitor of ARA-derived eicosanoid production and activity as well as inhibiting phospholipase A_2 (PLA_2) activity (Finnen and Lovell 1991). Reduced blood concentrations of n-3 HUFA have been observed in numerous neurodevelopmental and neurodegenerative disorders including Attention Deficit/Hyperactivity Disorder (ADHD) (Stevens et al. 1995), Alzheimer's disease and dementia (Coorigan et al. 1998), schizophrenia (Richardson et al. 2003), bipolar, unipolar and post-natal depression (Peet et al. 1998; Frasure-Smith et al. 2004; Hibbeln 2002). More recently other neurological disorders have also been implicated with abnormal blood n-3 HUFA levels, although the evidence is less well founded at the present time. These include dyslexia (Taylor et al. 2000), autism spectrum disorders (Bell et al. 2004b), dyspraxia (Richardson 2004), obsessive compulsive disorder (Fox et al. 2004) and aggression (Iribarren et al. 2004).

6.5 18:3n-3 and Human Health

While the data from section 6.3.3 above suggests that fish cultured using diets containing a significant proportion of VO have reduced levels of n-3 HUFA in their flesh lipids, they also contain significant levels of α-linolenic acid (18:3n-3; ALA) as well as linoleic acid (18:2n-6; LA). However, while the results of clinical trials with 18:3n-3 have been less clear than those with n-3 HUFA, there is still good evidence that diets that provide increased tissue levels of 18:3n-3 may also be beneficial to human health (Sanderson et al. 2002). Benefits of increased 18:3n-3 intake for various cardiovascular disorders, as well as for both breast and prostate cancer has been reported in the literature (Billman et al. 1999; Singh et al. 1997; Ferreti and Flanagan 1996; Maillard et al. 2002; Newcomer et al. 2001). Therefore, producing fish that provide moderate doses of EPA, DHA and 18:3n-3, but with low

18:2n-6 can be of significant value in human nutrition. In addition, even when salmon were cultured on 100% LO for the whole marine grow-out phase, flesh EPA and DHA concentrations were 0.12 g and 0.28 g, respectively, while ALA was 3.48 g and LA 1.12 g per 100 g of salmon flesh (Bell et al. 2004a). These values are not very different to those suggested by Simopoulos et al. (1999, 2000), who recommended 0.22 g each/day of DHA and EPA, 2.22 g/d for ALA and <6.67 g/day of LA. The studies conducted as part of the EU RAFOA project, as well as numerous other studies on FO replacement, have allowed us to advance our knowledge on how farmed fish flesh can be "tailored" to deliver ratios of DHA/EPA/18:3n-3/18:2n-6 that are beneficial to human health.

6.6 Recommended Intake of EPA and DHA for Human Health and Concentrations Provided by Farmed Fish

There is now a considerable weight of scientific evidence to support the widely recognised belief that n-3 HUFA intake, especially EPA and DHA, have wide ranging benefits for human health (SACN/COT 2004). But how much fish and n-3 HUFA do we need to consume to provide realistic benefits? Over the past 20 years studies in many countries have sought to recommend beneficial intake values for n-3 HUFA. One of the earliest, established by the Committee on Medical Aspects of Food Policy (COMA), recommended a daily intake of n-3 HUFA of 200 mg/day against an estimated daily intake, in the UK in 1994, of 100 mg/day (DH 1994). By comparison, the recommendation of the Scientific Advisory Committee on Nutrition and the Committee on Toxicity recommended increasing the intake up to 450 mg/day (SACN/COT 2004), against national intake values of 282 mg/day of which 244 mg were from EPA + DHA (Givens and Gibbs 2006). In terms of protection against CVD, Singh et al. (1997) suggested that 2 g n-3 HUFA/day reduced mortality in patients who had suffered a previous myocardial infarction. However, ISSFAL suggested in 2004 that an intake of 500 mg/day or 3.5 g/week of EPA + DHA should provide optimal cardiac health in humans (www.issfal.org.uk).

So how much fish produced by European aquaculture, whether cultured using predominantly marine raw materials or alternative feeds with increased vegetable oil inclusion, requires to be consumed to meet these recommended intake values? A moderate intake (~200 g/week) of Atlantic salmon, grown on FO containing diets, can provide the ISSFAL weekly recommended intake (3.5 g) of EPA + DHA and around 460 g/week of a salmon grown on 75% VO or 750 g/week of salmon grown on 100% VO would meet the ISSFAL intake value for EPA + DHA (Fig. 6.3). However, following the FO finishing diet period 200 g of salmon fed FO exceeds the weekly intake value or 260 g of salmon previously fed 75% VO or 290 g of salmon previously fed 100% VO would meet the recommended EPA + DHA intake value (Torstensen et al. 2005; Fig. 6.3). It is also worth noting that the capelin oil used in the RAFOA trials, during both the grow out and finishing diet phases, contains relatively low levels of EPA + DHA and using other oils with higher n-3

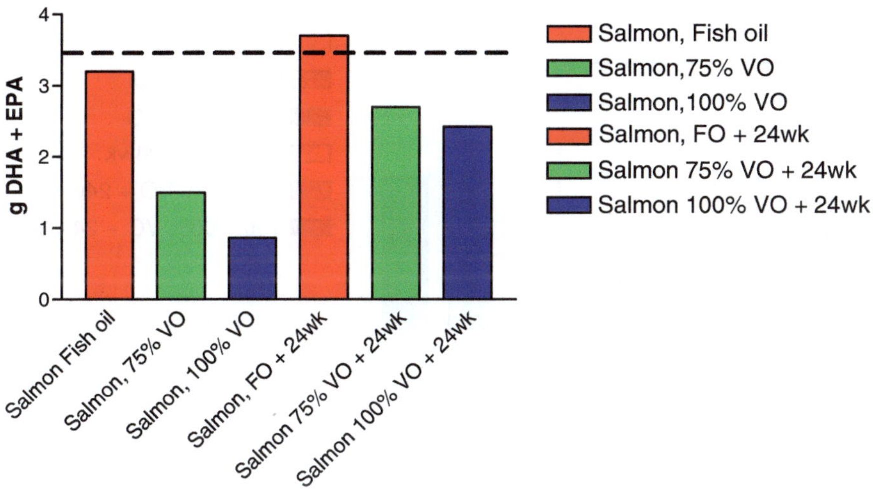

Fig. 6.3 Flesh EPA + DHA (g) in 200 g of salmon fed fish oil (FO), 75% vegetable oil (VO) or 100% VO for the whole production cycle (25 months) and following 24 weeks on a fish oil finishing diet. The dotted line shows the ISSFAL recommended intake of EPA + DHA of 3.5 g/week

HUFA levels would produce higher n-3 HUFA levels in market sized fish. The values found in the RAFOA trials were slightly higher than those reported for wild salmon but lower than those reported for farmed salmon in recent published reports (SACN/COT 2004; EFSA 2005).

By comparison, 200 g of rainbow trout, grown on FO diets, can provide ~60% of the ISSFAL weekly recommended intake (3.5 g) of EPA + DHA and around 340 g/week would be needed to fully meet the ISSFAL intake (Richard et al. 2006). Five hundred and sixty grams of trout grown on 75% VO or 650 g of trout grown on 100% VO would meet the ISSFAL weekly intake value for EPA + DHA (Fig. 6.4). However, following the FO finishing diet period 300 g of trout fed FO, 340 g of trout previously fed 75% VO or 350 g of trout previously fed 100% VO would meet the recommended EPA + DHA intake value (Fig. 6.4). The values of ~1.0 g EPA + DHA/100 g wet flesh reported in the RAFOA studies are similar to recently reported literature values (SACN/COT 2004; EFSA 2005).

For the marine species, 200 g of sea bream, grown on FO diets, can provide 35% of the ISSFAL weekly recommended intake (3.5 g) of EPA + DHA and around 580 g/week would be needed to fully meet the ISSFAL recommendation. Slightly more than 1 kg of sea bream grown on 60% VO or 2.3 kg of sea bream grown on 100% VO would need to be consumed to meet the ISSFAL weekly intake value for EPA + DHA (Fig. 6.5). Following the FO finishing diet period 800 g of sea bream fed FO would supply the weekly intake value or 950 g of sea bream fed 60% VO or 930 g of sea bream fed 100% VO would meet the recommended EPA + DHA intake value (Fig. 6.5). The increase in the required consumption of fish fed FO in the post-finishing diet period is due to the higher lipid deposition in larger bream where more of the lipid deposited is saturates and monoenes rather than HUFA (Izquierdo et al. 2005).

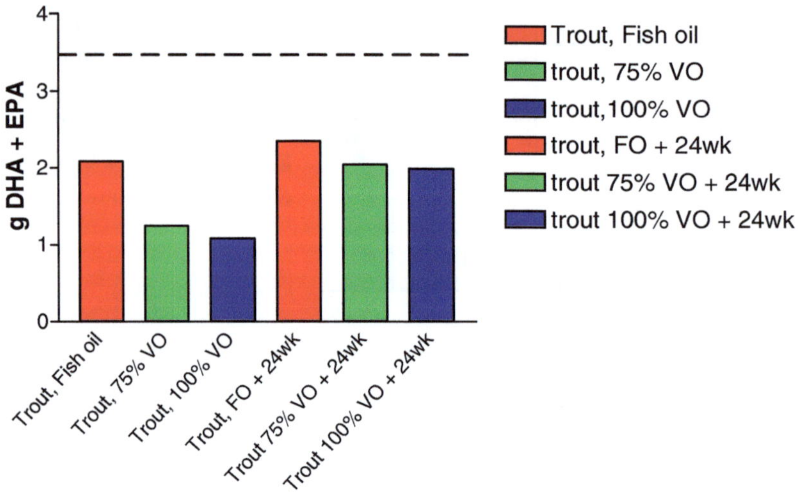

Fig. 6.4 Flesh EPA + DHA (g) in 200 g of rainbow trout fed fish oil (FO), 75% vegetable oil (VO) or 100% VO for the whole production cycle (62 weeks) and following 23 weeks on a fish oil finishing diet. The dotted line shows the ISSFAL recommended intake of EPA + DHA of 3.5 g/week

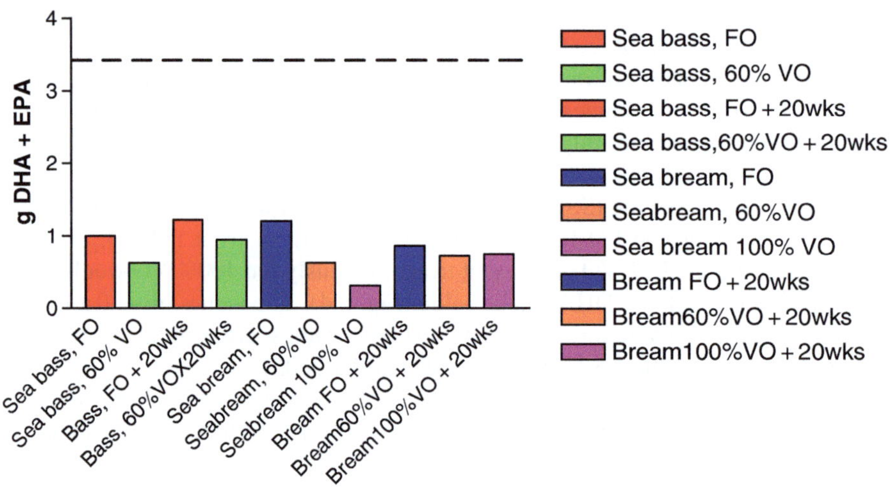

Fig. 6.5 Flesh EPA + DHA in 200 g of sea bass or sea bream fed 100% fish oil (FO), 60% vegetable oil (VO) or 100% VO for the whole production (64 weeks, bass; 40 weeks, bream) cycle and following 20 weeks on a fish oil finishing diet. The dotted line shows the ISSFAL recommended intake of EPA + DHA of 3.5 g/week

A very similar situation was seen in sea bass where 200 g of flesh, from fish grown on FO diets, can provide 29% of the ISSFAL weekly recommended intake (3.5 g) of EPA + DHA and around 690 g/week would be needed to fully meet the ISSFAL recommendation. As with bream, just over 1 kg of sea bass grown on 60% VO would meet the ISSFAL weekly intake value for EPA + DHA (Fig. 6.5). Following a FO finishing diet period 570 g of sea bass fed FO would supply the weekly intake value or 740 g of sea bass fed 60% VO would meet the recommended EPA + DHA intake value (Fig. 6.5). In contrast to bream, but similar to salmon and trout, the concentrations of EPA + DHA increase in the post-finishing diet period. This is due to a lower lipid deposition, seen in larger fish, in bass compared to bream (Mourente et al. 2005; Izquierdo et al. 2005).

While these results clearly demonstrate that fish with oily flesh, such as salmon and trout, can provide a higher dietary intake of EPA and DHA, compared with leaner fish, the contribution of beneficial fatty acids from the leaner species is still significant. Even in very lean white fish species, such as cod and plaice, where flesh lipid values may only be 1% of wet weight, the EPA + DHA content of ~0.25 g/100 g can still make an important contribution to the human diet (SACN/COT 2004; EFSA 2005). This is especially important as the HUFA in lean fish are concentrated in the phospholipid fraction which is thought to be more easily digested and deposited in cell membranes than HUFA supplied as triglyceride oils. For this reason, the advice of the UK Food Standards Agency that we should consume two portions of fish per week, of which one should be lean and the other oily, is sound nutritional advice (www.food.gov.uk).

6.7 Health Benefits of Consuming Fish Protein

In addition to the widely documented benefits of the n-3 HUFA, found in fish oils, there is also considerable evidence to support the role of fish protein, peptides and amino acids as valuable nutrients for animal and human health, especially cardiovascular health. Feeding rats cod protein reduced hepatic triacylglycerol (TAG) concentrations and reduced TAG secretion rates compared to rats fed casein (Demonty et al. 2003) while a mixture of cod protein and menhaden oil resulted in 50% lower plasma TAG compared to rats fed casein and beef tallow (Demonty et al. 2003). In rats fed fish, soybean or casein as protein source, the plasma cholesterol concentrations were reduced for fish and soya compared to casein (Iritani et al. 1985) while the former two also increased excretion of bile salts and cholesterol in faeces. In another study, feeding fish protein to rats, rabbits and humans resulted in reduced plasma cholesterol concentrations in all cases (Jacques 1990). In addition, feeding cod and soy proteins resulted in lower fasting plasma glucose and insulin levels, in rats, compared to those fed casein (Lavigne et al. 2000). Fish protein has also been shown to affect blood coagulation by increasing fibrinolysis in rats (Murata et al. 2004).

Fish protein also contains bioactive peptides that can be released by fermentation. Feeding a fermented fish protein concentrate (FPC) to mice for 7 days resulted in enhanced phagocytic activity of peritoneal macrophages (Duarte et al. 2006). In addition, intestinal IgA + cells were increased along with concentrations of IL-4, IL-6 and IL-10 in mice fed FPC.

Clearly consumption of fish protein and hydrolysates can have a wide range of benefits for cardiovascular function as well as enhancing non-specific immune defence and glucose metabolism. There is also more recent evidence that supplementing fish protein hydrolysates, from blue whiting, cod, plaice and salmon, can have anti-proliferative activity against human breast cancer cell lines grown in vitro (Picot et al. 2006).

6.8 Micronutrients in Feed and Fish

The intuitive opinion, that fish and seafood are healthy and nutritious relies not only on the content of n-3 HUFA, as discussed above, but also on the balanced content of micronutrients. Selected vitamins and minerals, present in high concentrations in seafood, are of special interest for their roles in prevention of life style disorders, such as CHD (vitamin B12, McCully 1991; selenium, Suadicani et al. 1992), osteoporosis (calcium, vitamin D), cancer (selenium, Albanes et al 1999), impaired vision (vitamin A), neurological and neuropsychiatric disorders (B vitamins), developmental disorders (vitamin A) and iodine deficiency disorders (iodine). For these reasons, national food administrations routinely advise people to increase their seafood consumption (SACN/COT 2004; Norwegian Scientific Committee for Food Safety 2006).

Aquaculture research has shown that fish flesh reflects the feed content of selected micronutrients and this allows the opportunity to tailor products both according to market preferences and as healthy added value food (Baker 2001; Lie 2001). Conversely, changes in feed micronutrient compositions and their bioavailabilities, arising from the use of FM and FO substitutes, could constitute an element of risk both for suboptimal micronutrient supply for the fish, as well as introduce undesirable changes in the composition of the fish for the consumer.

6.8.1 Essential Minerals and Trace Elements

A vegetable FM substitute like soybean meal often contains lower concentrations of minerals and in less bioavailable forms for the fish, for example, phosphorous in the form of phytate (Table 6.3). Phytate is also regarded an ANF, since it seems to impact on the digestibility of nutrients by chelating essential minerals, especially zinc, magnesium and phosphorous (Richardson et al. 1985; Storebakken et al. 1998; Francis et al. 2001; Denstadli et al. 2006), and are thereby potentially harmful by affecting growth and health of the fish. This effect could be overcome, for example, by enzymatic treatment (phytase) of the soy product before inclusion in fish feed (Storebakken et al. 1998; Vielma et al. 1998). As reviewed by Francis et al. (2001) several other ANFs from plant derived FM substitutes impact on element bioavailability. Strategies to overcome this problem include actions to eliminate these ANFs or compensate for their reduced element bioavailability. Despite the fact that elevated feed levels of the essential trace elements like copper, zinc, manganese and iron are reflected in respective target organs in the fish, they are not easily enriched in the fillet (Lall 2002; Lorentzen 1998; Maage 1994).

Table 6.3 Vitamin and mineral concentration ranges in South American and herring fish meal compared to soya meal and known requirements for cold water fish including the possibility to tailor fish fillets by dietary supplementation. (Data from Hertrampf and Piedad-Pascual, 2000; INRA, 1986; Lall, 2002; Halver, 2002)

Micronutrient	Fish meal (mg/kg)	Soya meal (mg/kg)	Requirement (mg/kg)	Possibility to tailor fillet?
Vitamins				
Thiamine (B1)	0.7–1.9	10	1–15	
Riboflavin (B2)	6.6–7.3	2.6	7–30	
Niacin	95–126	23	14–200	
Pantothenic acid	9–31	16	25–50	
Pyridoxine	3.5–3.7	10	3–20	
Folic acid	0.16–0.50	3.5	6–10	Yes
Cobalamin (B12)	0.18–0.25	–	0.02	
Biotin	0.26–0.42	0.3	1.0–1.5	
Choline	4400	2000	1500–4000	
Inositol	700–800 [a]	–	200–900	
Ascorbic acid (C)	0	–	30–150	Yes
Vitamin A	3.9–8.9	–	0.30–0.75	(Yes) [d]
Vitamin D	0.01–0.18 [b]	–	0.013–0.06	Yes
Vitamin E	3–4	55	30–100	Yes
Vitamin K	–	–	10	(Yes)
Minerals				
Calcium (%)	20–40	2.5	2.7–3.4	
Phosphorous (%)	19–26	5.7	3–9	
Magnesium (%)	–	2.9	0.4–0.6	
Potassium	7–12	–	8	
Iron	150–246	90	30–170	
Zinc	111–120	40	15–67	
Copper	5–11	15	3–5	
Manganese	2–10	25	2.4–13	
Selenium	1.4–2.2	0.5	0.15–0.25	Yes
Iodine	5–90 [c]	0.05	0.6–1.1	Yes

[a] Waagbø et al. (1998); [b] Horvli and Lie, (1994); [c] Lall, (2002); [d] including through retention of provitamin A carotenoids, like astaxanthin and canthaxanthin in salmonids muscle

Besides being rich in calcium, fish products contain considerable amounts of iodine (I) and selenium (Se) that can contribute a significant part of the human recommended daily intakes. It is possible to tailor fillet Se concentration by dietary supplementation in salmonids, however both the retained form and efficacy of retention depends on the dietary chemical form of the element (Bell and Cowey 1989; Lorentzen et al. 1994). It also appears that the bioavailability of Se, as observed in rats (Ørnsrud and Lorentzen 2002), and the benefits of Se in human nutrition and health depend upon the chemical form supplied (Drake 2006). For example, the oxidative nature of selenite seems to exert higher cancer chemopreventive effects than the amino acid forms, selenomethionine and Se-methyl-selenocysteine, which lack oxidation capability (Drake 2006). FM contains considerably more Se compared to soybean meal (Table 6.3). Recently, Polatajko et al. (2006) reviewed the complexity of chemical Se species in biological samples. In this regard, speciation of Se in feed and food is necessary to evaluate risks and benefits of using FM substitutes, in addition to considering the total Se content (Table 6.3).

Iodine deficiency disorders, such as goitre, hypothyroidism, cretinism and related mental effects occur frequently in humans, especially in developing countries, with estimates of 1 billion at risk (Hetzel and Clugston 1999). Fish and seaweed are among the food items with the highest naturally occurring iodine contents, however, there is large variation among fish species and even between individuals. Despite relatively low levels in salmonids, it has recently been demonstrated that it is possible to increase fillet iodine levels three fold in adult Atlantic salmon farmed in seawater (from 0.25 to 0.9 mg I kg^{-1} wet weight) by feeding diets supplemented with high levels of an iodine salt (0–80 mg I kg^{-1}) (Julshamn et al. 2006). Freshwater char (*Salvelinus* sp.) seem to respond in a similar range when using a marine algae in the feed (fillet conc. 0.14 to 0.54 mg I kg^{-1} wet weight) (Schmid et al. 2003). Other species show higher levels of fillet iodine, for example, wild caught Atlantic cod (*Gadus morhua*) from the Barents Sea, range between 0.34 and 12.7 mg I kg^{-1} (Julshamn et al. 2001), and may be an even better species for tailoring muscle iodine content than Atlantic salmon. Replacing FM with plant meals may introduce variations in fillet iodine content, even though the minimum iodine requirement for growth of fish (1 mg/kg diet; National Research Council 1993) may easily be covered through uptake from seawater and diet (Lall 2002).

6.8.2 Fish Oil and Fish Meal Substitutes – Consequences for Lipid Soluble Vitamins A and D

Fish are among the few natural sources of vitamins A and D, originating from the lower trophic levels in the marine food web. Fish species with oily flesh, like the salmonids, contain considerable amounts of vitamin D in their fillets (Ostermeyer and Schmidt 2006) and less vitamin A, while lean fish species normally have higher concentrations of fat soluble vitamins in their liver stores (for example, cod liver oil used as human vitamin A and D supplements). FO-based aquafeeds normally supply sufficient amounts of the lipid soluble vitamins A and D to support fish growth and product quality, even at elevated levels that have been of concern for fish bone health (Graff et al. 2002; Ørnsrud et al. 2002). Both FM (Table 6.3) and FO contain considerable amounts of these vitamins, so there should be no risks for vitamin deficiencies by use of vegetable feed substitutes. However, the benefits of vitamin rich seafood would be reduced, since vitamin supplementations to gain similar feed levels would not currently be supported by EC legislation (EC directive 1970).

6.8.3 Concerns on B-vitamins in Feed and Seafood

For water soluble B-vitamins, storage capacities in farmed fish are normally limited and the muscle tissue will easily reach saturation level at moderate feed intake levels. The possibility to manipulate the product through dietary means is therefore limited. However, several fold variations in B vitamins in fish fillets may be seen between species and relative to fillet muscle type (red or white muscle) and lipid content, as well as relative to environmental factors, sexual maturation and annual cycle (Brækkan 1959; Sandnes et al. 1998; Waagbø unpublished data). As a traditional major protein raw material, FM supplies many of these vitamins in adequate amounts and in readily available forms providing essential requirements for growth and muscle saturation, including biotin (Mæland et al. 1998), vitamin B12 (Mæland A, Sandnes, K and Waagbø R, unpublished data), panthotenic acid (Sandnes et al. 1998) and riboflavin (Brønstad et al. 2002). Table 6.3 illustrates differences in gross vitamin content in FM and soybean meal, the latter representing an important candidate among FM substitutes. Besides observed differences in content, vitamins from plant raw materials may occur in other chemical forms (pyridoxine, riboflavin, niacin, folic acid, vitamin B12) or together with anti-nutrients that results in lower bioavailabilities than vitamins from animal derived raw materials (Machlin 1991). Even though this information is derived from feeding studies or *in vitro* experiments in humans and terrestrial animals, this may also be true for carnivorous fish species. Thus, care should be taken to fulfill the optimal supply of these vitamins in aquafeeds containing FM substitutes through micronutrient supplementation or by using selected vitamin-rich raw materials.

6.8.4 Antioxidant Vitamins and Pigments

The success of micronutrient tailoring of farmed fish fillet depends on the ability of the fish species to handle the dietary intakes, through absorption, retention, metabolism and excretion. The concentrations of the antioxidant vitamins E and C in the fish fillet are important for ensuring the oxidative storage stability of the highly susceptible HUFAs as well as vitamins available to fish consumers (Hamre et al. 1998; Ng et al. 2004b; Waagbø et al. 1993; Yildiz et al. 2006). In a multivariate

23-week feeding study on the impact of dietary pro- and antioxidants on product quality and health of adult Atlantic salmon, feed vitamins E (α-tocopherol acetate) and C (ascorbate polyphosphate) supplementations of 69 and 430 mg/kg, and 52 and 1940 mg/kg, respectively, increased fillet α-tocopherol three fold (from 12–31 μg/g) and vitamin C two fold (15–31 μg/g), respectively (Waagbø unpublished data). In the same study, flesh astaxanthin varied two fold (1.3–2.5 μg/g), when salmon were fed diets containing 11 or 48 mg astaxanthin/kg (Hamre et al. 2004). Even though the flesh concentrations of antioxidant vitamins reflected the respective dietary levels, fillet α-tocopherol alone was the major determinant of oxidative stability after an *in vitro* oxidative challenge of muscle tissue (Hamre et al. 2004). The antioxidant nutrients occur normally at low concentrations in feed ingredients (Table 6.3), and even more may be lost through heat treatment and refining procedures. Therefore, these are routinely supplied in fish feed production through stabilized additives. There are no indications of increased requirement for these antioxidants when using FM and FO substitutes. Indeed, some vegetable oils contain plant derived antioxidants (vitamin E, carotenoids and xanthophylls), as well as n-3 PUFA less susceptible to oxidation which can reduce the oxidative challenge in feed and tissues (Hertrampf and Piedad-Pascual 2000; Ng et al. 2004b).

6.9 The Impact of Vegetable Oil Inclusion on Organic Contaminant Concentrations in Salmon Flesh

Polychlorinated dibenzodioxins and polychlorinated dibenzofurans, collectively known as dioxins, can arise from natural processes such as forest fires and incomplete combustion of organic matter, as well as from industrial processes. The dioxin-like polychlorinated biphenyls (DL-PCBs) are synthetic products used in electrical transformers, heat exchange fluids, hydraulic oils and plastic manufacturing. Although production of PCBs is now banned, they have been deposited in the oceanic benthos, due to industrial activity over the last century, and they are widely distributed across the marine biota (North Sea Task Force 1993). Dioxins and DL-PCBs are highly lipophilic with biological half-lives of several decades, which means they can accumulate in predators at the top of the food chain (Froescheis et al. 2000). However, levels of both dioxins and PCBs in the environment have been declining since the 1950s, although, due to their persistent nature, they will remain in the biota for a considerable period (Brevik et al. 1990; Bignert et al. 1998).

There are around 210 known dioxin and furan congeners and, of these, 17 have been shown to be toxic although individual congeners have different levels of toxicity. For this reason the World Health Organisation (WHO) have established toxic equivalency factors (TEFs), according to their relative toxicity, enabling the calculation of toxic equivalents (TEQs; Van den Berg et al. 1998). Similarly, of the 209 PCB congeners 12 have known dioxin-like toxicity and have been assigned WHO-TEQs. In 2001 the EU introduced new limits on dioxins and furans in fish feeds and fish for human consumption (SCAN 2000; SCF 2001). These values are

2.25 ng dioxin toxic equivalents (TEQ)/kg feed and 4.0 ng TEQ/kg fish. The EU has recently revised dioxin limits and assigned new limits for the 12 dioxin-like (DL) PCBs, such that combined values of 8 ng TEQ/kg for fish products and 7 ng/TEQ for fish feeds have now been introduced (EC 2006a,b).

The polybrominated diphenyl ethers (PBDEs) have been used as flame retardants in a wide variety of household furniture and electrical equipment since their introduction in the 1980s and global annual production currently exceeds 70 kt, with around 75% produced in North America (Hites et al. 2004b). Numerous PBDE congeners exist with different levels of bromination ranging from the tetra through to the deca-brominated products. The less brominated congeners tend to be more persistent in the environment and in 2004 the EU banned production of the tetra to nona-brominated products in favour of the less persistent deca-brominated congeners (Covaci et al. 2003; SACN/COT 2004). However, due to their lipophilic and persistent characteristics the PBDEs are currently increasing in the environment including fish products (FSA 2004; EFSA 2005).

As described above replacement of significant amounts of either FO or FM can be achieved without loss of growth performance or affects on fish health (Kaushik et al. 2004; Torstensen et al. 2005; Mourente et al. 2005). However, significant inclusion of dietary VO results in reduction of the flesh content of n-3 HUFA content, although this could potentially be offset by reduced levels of organic contaminants, as vegetable oils generally contain lower levels of these pollutants than most marine fish oils (SACN/COT 2004).

In the RAFOA II trial in salmon fed 100% FO, the flesh dioxin/furan content was 0.58 and the DL-PCBs 1.18 ng TEQ/kg making a total of 1.76 ng TEQ/kg that was well within existing and proposed EU limit values for dioxins and DL-PCBs (EC, 2006a). By contrast, salmon produced on diets containing 75% VO had reduced values of 0.21 and 0.42 ng TEQ/kg, for dioxins/furans and DL-PCBs, respectively, compared to fish fed FO. The total of 0.63 TEQ/kg flesh for dioxin + DL-PCBs represents a 64% reduction compared to fish fed FO (Fig. 6.6). The flesh dioxin + DL-PCB concentration in fish fed 100% FO of 1.76 ng TEQ/kg is lower than those reported by Lundebye et al. (2004) of 2.5 and by Bell et al. (2005) of 2.01 ng TEQ/kg. Indeed, all of these values are lower than those reported for Scottish farmed salmon by Hites et al. (2004a) of ~3 ng TEQ/kg. The value of 0.63 ng TEQ/kg for dioxin + DL-PCBs in salmon fed 75% VO for the whole production cycle is slightly lower than that reported by Bell et al. (2005) for salmon grown for the whole production cycle on diets containing 100% VO (0.68 ng TEQ/kg) but higher than the value found by Berntssen et al. (2005) in fish from the RAFOA II study conducted in Norway and fed 100% VO (0.30 ng TEQ/kg). The differences found between these studies are due to different dioxin + DL-PCB concentrations in the oils used in the dietary formulations. In the two RAFOA studies capelin oil was used whereas the study of Bell et al. (2005) used a mixture of capelin and herring oils as the FO component. The concentration of dioxins and DL-PCBs in capelin are generally lower than in herring, although there is considerable geographical and seasonal variation for both species (Lundebye-Haldorsen and Lie 1999; SCAN 2000). The dioxin + DL-PCB concentrations found in salmon flesh in

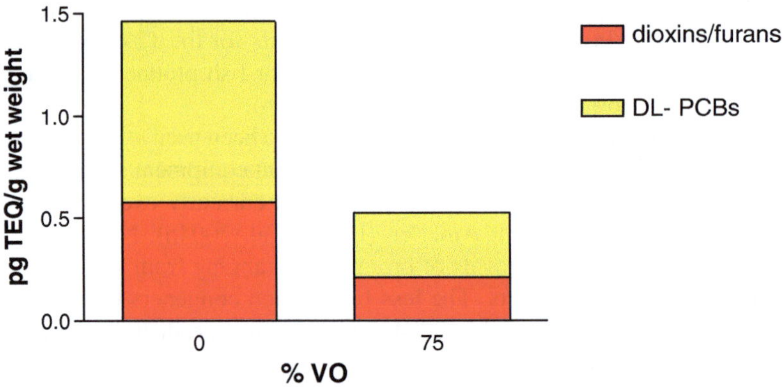

Fig. 6.6 Concentrations of flesh dioxins/furans and dioxin-like PCBs (pg TEQ/g wet weight) in salmon fed diets containing 100% fish oil (FO) or 75% vegetable oil blend (VO) for 24 months post-first feeding

the two RAFOA studies and in the study by Bell et al. (2005), where 75–100% of FO was replaced by VO, are similar or less than those found by Hites et al. (2004a) in wild Pacific salmon. In conclusion, these values confirm that replacing marine fish oils with VO in aqua feed formulations can significantly reduce dioxin and DL-PCB concentrations in farmed salmon flesh. While the reduction in flesh concentrations of dioxins and PCBs arising from the use of VO in aqua feeds is to be welcomed, the reduction in n-3 HUFA that accompanies the use of VO is potentially detrimental. However, careful use and choice of FO towards the end of the production cycle, including the use of FO "finishing diets" can largely restore n-3 HUFA levels in fish pre-market (Bell et al. 2004a; Torstensen et al. 2005; Mourente et al. 2005; Montero et al. 2005). In addition, there is the possibility that, by using more FO at the end of the production cycle to increase n-3 HUFA, organic contaminant concentrations might be elevated. In the study by Bell et al. (2005), a 16–24 week finishing diet period successfully restored flesh DHA and EPA concentrations to > 80% of the value seen in fish fed FO throughout the production cycle. However, the flesh dioxin and DL-PCB concentrations in the fish previously cultured on 100% VO diets, following the 24 week finishing diet period, were still 60 and 47% lower, respectively, than the values seen in fish cultured on FO throughout (Bell et al. 2005). This suggests that finishing diets can still be used to successfully restore n-3 HUFA levels while producing fish with significantly reduced contaminant concentrations.

While over 40 PBDE congeners have been identified (Covaci et al. 2003), the contribution to overall tissue concentrations in fish is largely due to six congeners namely PBDEs 28, 47, 99, 100, 153 and 154 (Jacobs et al. 2002b; Hites et al. 2004b). In salmon flesh from the RAFOA II study in Scotland, the concentrations of the 6 major PBDE congeners were 3819 ng/kg in fish fed FO compared to 1083 ng/kg in fish fed 75% VO (Fig. 6.7). The flesh PBDE concentration found in the RAFOA fish fed FO was similar to that observed by Hites et al. (2004b) for

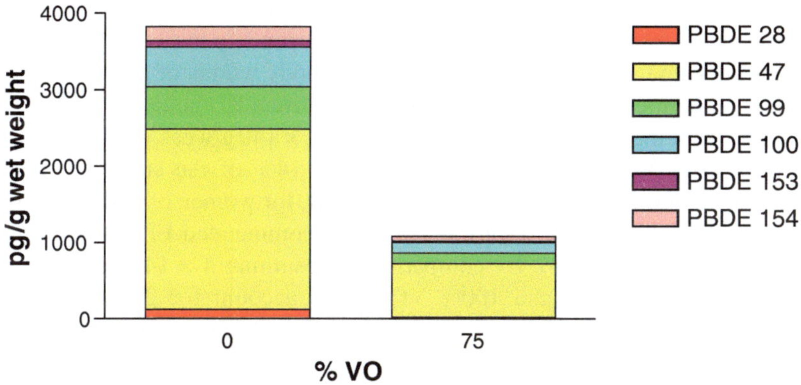

Fig. 6.7 Concentrations of the six principal PBDE congeners (pg/g wet weight) in flesh of salmon fed diets containing 100% fish oil (FO) or 75% vegetable oil blend (VO) for 24 months post-first feeding

farmed salmon sourced in Europe in 2002 and also similar to wild chinook salmon from British Columbia but was higher than farmed Atlantic salmon from Chile or North America. However, the flesh concentration in RAFOA fish fed FO was lower than the concentration found in salmon obtained in Scotland in 1999 by Jacobs et al. (2002b). The flesh concentration in salmon fed 75% VO was only 28% of that found in fish fed FO and the levels in the former were comparable to salmon sourced in Chile but still significantly higher than wild chum, pink, coho and sockeye sourced from British Columbia and Alaska (Hites et al. 2004b). In comparison to dioxins and DL-PCBs there are no TEQ values assigned to any PBDE congeners or any tolerable daily intake (TDI) values calculated at the present time. However, there is some evidence from studies with mice and rats, fed high doses (30 mg/kg/day) of PBDE during gestation, that neurodevelopmental and thyroid hormone defects could result (Fowles et al. 1994; Zhou et al. 2002). Similar effects on thyroid hormones, as well as increased oxidative stress and altered retinol concentrations, were seen in American kestrels when eggs were dosed and nestlings were fed doses of PBDE congeners between 100 and 1500 ng/g with these concentrations being similar to values found in Great Lakes trout and Great Lakes gull eggs respectively (Fernie et al. 2005). However, the values used in this dosing study are 33–500 times greater than those found in salmon fed diets containing FO and the recent SACN/COT (2004) report stated that levels found in fish in the UK were unlikely to present any risk to human health. In addition, the ban on production of penta- to nono-BDEs, introduced in EU countries in 2004, should prevent any future increases in PBDEs in the European food chain.

Following the recent SACN/COT (2004) report the UK Food Standards Agency (FSA, www.food.gov.uk) revised their advice to consumers regarding consumption of fish. The new recommendation suggests a weekly intake of up to four 140 g portions of oily fish per week for men, boys and women over reproductive age, up to a maximum of 8 pg TEQ/kg body wt/day of dioxins and PCBs. However, girls and women of reproductive age are advised to consume up to two 140 g portions of oily fish/week up to a maximum of 2 pg TEQ/kg body wt/day of dioxins and PCBs. The latter recommendation is in agreement with current EU guidelines. Thus, based on the data from the RAFOA studies, consuming 2 × 140 g/week portions of salmon fed FO or 75% VO would account for 39 and 14% of the suggested maximal weekly intake of dioxins + DL-PCBs, respectively, for women of child bearing age, while providing 128 and 60% of the ISSFAL recommended EPA + DHA intake, respectively (Fig. 6.8a & b). By comparison, consuming 4 × 140 g portions/week of salmon fed FO or 75 and 100% VO would account for 20 and 7% of the suggested weekly maximal intake of dioxins + DL-PCBs, respectively, for men, boys and women over child bearing age while providing 256 and 128% of ISSFAL recommended EPA + DHA intake, respectively (Fig. 6.8a & b).

The cautionary caveat, that women of childbearing age should consume less oily fish, is unfortunate but understandable given that unborn children and young infants may be particularly sensitive to environmental pollutants. However, they are also the most likely to benefit from the positive effects of increased intake of n-3 HUFA. We feel that the data presented here confirms that farmed fish can be regarded as a safe and healthy food option for human consumers. However, future strategies for aquaculture production must aim to reduce current dependence on fish oil and fish meal while taking all possible steps to reduce contaminant levels in fish and, at the same time ensure that we preserve current levels of n-3 HUFA in farmed fish. This can be achieved by judicious use of terrestrial plant products to replace marine raw materials during the main grow out phase of production but should also include the investigation of cleaned or decontaminated FO as a means of restoring n-3 HUFA without increasing the contaminant burden. Such decontamination processes are

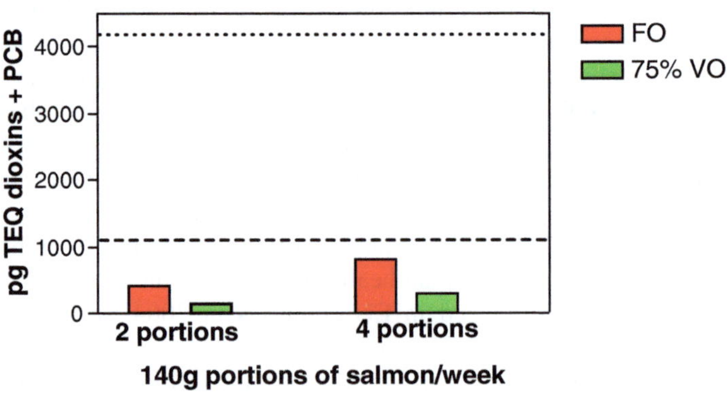

Fig. 6.8a Amount of dioxin + DL-PCBs present in 2 or 4 × 140 g portions of salmon, produced using 100% fish oil (FO) or 75% vegetable oil blend (VO) diets for the full production cycle

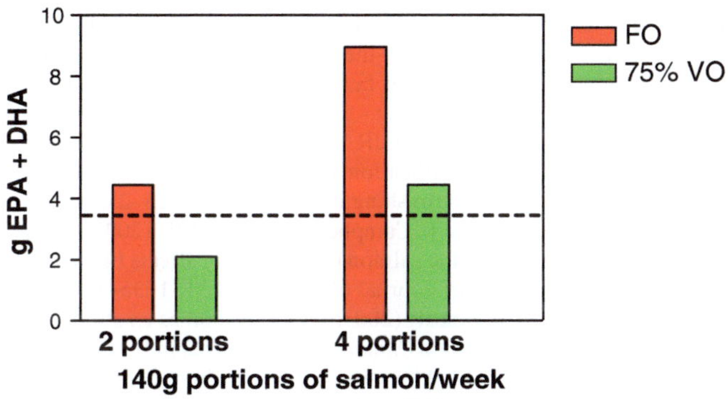

Fig. 6.8b Amount of EPA + DHA (g) provided by 2 or 4 × 140 g portions of salmon, produced using 100% fish oil (FO) or 75% vegetable oil blend (VO) diets for the full production cycle

currently being developed (Breivik and Thorstad 2004; Maes et al. 2005) and, with economy of scale, the cost implications for the industry are minor. The future of aquaculture production, as well as the improved health and well being of human consumers, depends on the investigation and implementation of new sustainable aquafeeds that are safe and nutritionally optimal for both fish and consumers.

References

Albanes D, Hartman TJ (1999) Antioxidants and cancer: evidence from human observational studies and intervention trials. In: Papas, A.M., ed.), Antioxidant status, diet, nutrition, and health. CRC, Boca Raton, FL, pp. 497–544.

Arslan G, Brunborg LA, Frøyland L, Brun JG, Valen M, Berstad A (2002) Effects of duodenal seal oil administration in patients with inflammatory bowel disease. Lipids 37: 935–940

Arzel J, Cardinal M, Cornet J, Metailler R, Guillaume JC (1993) Nutrition of brown trout (*Salmo trutta*) reared in seawater, effect of dietary lipid on growth performances, body composition and fillet quality. Special publication European Aquaculture Society 19: 309

Arzel J, Martinez Lopez FX, Metailler R, Stephan G, Viau M, Gandemer G, Guillaume J (1994) Effect of dietary lipid on growth performance and body composition of brown trout (*Salmo trutta*) reared in seawater. Aquaculture 123: 361–375

Baker RTM (2001) The effect of certain micronutrients on fish flesh quality. In: Kestin SC, Warriss PD (eds.), Farmed fish quality. Blackwell, Oxford, UK, pp. 180–191

Bang HO, Dyerberg J (1972) Plasma lipids and lipoproteins in Greenlandic west coast Eskimos. Acta Medical Scandinavia 192: 85–94

Bang HO, Dyerberg J, Sinclair HM (1980) The composition of the Eskimo food in north western Greenland. The American Journal of Clinical Nutrition 33: 2657–2661

Barlow S, Pike IH (2001) Aquaculture feed ingredients in year 2010: fish meal and fish oil. In: Nash, CE, Julien, V (eds.), Aquavision'98. 2nd Nutreco Aquaculture Business Conference. Stavanger, Norway, 13–15th May 1998 Nutreco Aquaculture, Stavanger, Norway, pp. 71–74

Bell JG, Cowey CB (1989) Digestibility and bioavailability of dietary selenium from fishmeal, selenite, selenomethionine and selenocystine, in Atlantic salmon (*Salmo salar*). Aquaculture 81: 61–68

Bell JG, Henderson RJ, Tocher DR, McGhee F, Dick JR, Porter A, Smullen RP, Sargent JR (2002) Substituting fish oil with crude palm oil in the diet of Atlantic salmon (*Salmo salar*) affects muscle fatty acid composition and hepatic fatty acid metabolism. Journal of Nutrition 132: 222–230

Bell JG, Henderson RJ, Tocher DR, Sargent JR (2004a) Replacement of dietary fish oil with increasing levels of linseed oil: Modification of flesh fatty acid compositions in Atlantic salmon (*Salmo salar*) using a fish oil finishing diet. Lipids 39: 223–232

Bell JG, McEvoy J, Tocher DR, McGhee F, Campbell PJ, Sargent JR (2001) Replacement of fish oil with rapeseed oil in diets of Atlantic salmon (*Salmo salar*) affects tissue lipid compositions and hepatocyte fatty acid metabolism. Journal of Nutrition 131: 1535–1543

Bell JG, McGhee F, Campbell PJ, Sargent JR (2003b) Rapeseed oil as an alternative to marine fish oil in diets of post-smolt Atlantic salmon (*Salmo salar*): changes in flesh fatty acid composition and effectiveness of subsequent fish oil "wash out". Aquaculture 218: 515–528

Bell JG, McGhee F, Dick JR, Tocher DR (2005) Dioxin and dioxin-like polychlorinated biphenyls (PCBs) in Scottish farmed salmon (*Salmo salar*): effects of replacement of dietary marine fish oil with vegetable oils. Aquaculture 243: 305–314

Bell JG, MacKinlay EE, Dick JR, MacDonald DJ, Boyle RM, Glen ACA (2004b) Essential fatty acids and phospholipase A_2 in autistic spectrum disorders. Prostaglandins, Leukotrienes and Essent. Fatty Acids 71: 201–204

Bell JG, McVicar, AH, Park, MT, Sargent, JR (1991) High dietary linoleic acid affects the fatty acid compositions of individual phospholipids from tissues of Atlantic salmon (*Salmo salar*): association with stress susceptibility and cardiac lesion. Journal of Nutrition 121: 1163–1172

Bell JG, Tocher DR, Farndale BM, Cox DI, McKinney RW, Sargent JR (1997) The effect of dietary lipid on polyunsaturated fatty acid metabolism in Atlantic salmon (*Salmo salar*) undergoing parr-smolt transformation. Lipids 32: 515–525

Bell JG, Tocher DR, Henderson RJ, Dick JR, Crampton VO (2003a) Altered fatty acid compositions in Atlantic salmon (*Salmo salar*) fed diets containing linseed and rapeseed oils can be partially restored by a subsequent fish oil finishing diet. Journal of Nutrition 133: 2793–2801

Bell JG, Youngson A, Mitchell AI, Cowey CB (1989) The effect of enhanced intake of linoleic acid on the fatty acid composition of tissue polar lipids of post smolt Atlantic salmon (*Salmo salar*). Lipids 24: 240–242

Bell MV, Dick JR (1991) Molecular species composition of the major diacyl glycerophospholipids from muscle, liver retina, and brain of cod (*Gadus morhua*). Lipids 26: 565–573

Bell MV, Dick JR, Porter AEA (2001b) Biosynthesis and tissue deposition of docosahexaenoic acid (22:6n-3) in rainbow trout (*Oncorhynchus mykiss*). Lipids 36: 1153–1159

Bell MV, Dick JR, Porter AEA (2003c) Pyloric ceca are a major site of 22:6n-3 synthesis in rainbow trout (*Oncorhynchus mykiss*). Lipids 39: 39–44

Belluzzi A, Miglio F (1998) n-3 fatty acids in the treatment of ulcerative colitis. In: Kremer J (ed.), Medicinal fatty acids in inflammation. Birkhauser Verlag, Basel, pp.103–109

Bendiksen EÅ, Berg OK, Jobling M, Arnesen AM, Måsøval K (2003) Digestibility, growth and nutrient utilisation of Atlantic salmon (*Salmo salar* L.) in relation to temperature, feed fat content and oil source. Aquaculture, 224: 283–299

Berntssen MHG, Lundebye A-K, Torstensen BE (2005) Reducing the levels of dioxins and dioxin-like PCBs in farmed Atlantic salmon by substitution of fish oil with vegetable oil in the feed. Aquaculture Nutrition 11: 219–231

Bignert A, Olsson M, Persson W, Jensen S, Zakrisson S, Litzen K, Eriksson U, Haggberg L, Alsberg T (1998) Temporal trends of organochlorines in Northern Europe, 1967–1995. Relation to global fractionation, leakage from sediments and international measures. Environmental Pollution 99: 177–198

Billman GE, Kang JX, Leaf A (1999) Prevention of sudden cardiac death by dietary pure omega-3 polyunsaturated fatty acids in dogs. Circulation 99: 2452–2457

Brækkan OR (1959) A comparative study of vitamins in the trunk muscles of fishes. Reports on technological research concerning Norwegian fish industry. Vol. III (8): 1–42

Brevik EM, Biseth A, Oehme M (1990) Levels of polychlorinated dibenzofurans and dibenzo-*p*-dioxins in crude and processed fish oils in relation to origin and cleaning method. Organohalogen Compound. 1: 467–470

Brevik H, Thorstad O (2004) Removal of organic environmental pollutants from fish oil by short path distillation. The effect of a working fluid. Eurofed Lipid Conference, 5–8th September, Edinburgh, UK

Brønstad I, Bjerkås I, Waagbø R (2002) The need for riboflavin supplementation in high and low energy diets for Atlantic salmon *Salmo salar* L. parr. Aquaculture Nutrition. 8: 209–220

Broughton KS, Johnson CS, Pace BK, Liebman M, Kleppinger KM (1997) Reduced asthma symptoms with n-3 fatty acid ingestion are related to 5-series prostaglandin production. The American Journal of Clinical Nutrition 65: 1011–1017

Burel C, Boujard T, Kaushik SJ, Boeuf G, Van Der Geyten S, Mol KA, Kuhn ER, Quinsac A, Krouti M, Ribaliller D (2000a) Potential of plant protein sources as fish meal substitutes in diets for turbot (*Psetta maxima):* growth, nutrient utilisation and thyroid status. Aquaculture 188: 363–382

Burel C, Boujard T, Tulli F, Kaushik SJ (2000b) Digestibility of extruded peas, extruded lupin, and rapeseed meal in rainbow trout (*Oncorhynchus mykiss*) and turbot (*Psetta maxima*). Aquaculture 188: 285–298

Caballero MJ, Obach A, Rosenlund G, Montero D, Gisvold M, Izquierdo MS (2002) Impact of different dietary lipid sources on growth, lipid digestibility, tissue fatty acid composition and histology of rainbow trout, *Oncorhynchus mykiss*. Aquaculture 214: 253–271

Calder PC (2001) Polyunsaturated fatty acids, inflammation and immunity. Lipids 36: 1007–1024

Calder PC, Zuricr RD (2001) Polyunsaturated Fatty Acids and Rheumatoid Arthritis. Current opinion in clinical nutrition and metabolic care 4: 115–121

Connor WE (2000) Importance of n-3 fatty acids in health and disease. The American Journal of Clinical Nutrition 71: 171S-175S

Coorigan FM, Horrobin DF, Skinner ER, Besson JA, Cooper MB (1998) Abnormal content of n-6 and n-3 long-chain unsaturated fatty acids in the phosphoglycerides and cholesterol esters of parahippocampus cortex from Alzheimer's disease patients and its relationship to acetyl CoA content. The International Journal of Biochemistry & Cell Biology 30: 197–207

Covaci A, Voorspoels S, de Boer J (2003) Determination of brominated flame retardants, with emphasis on polybrominated diphenyl ethers (PBDEs) in environmental and human samples – a review. Environment International 29: 735–756

de Francesco M, Parisi G, Medale F, Lupi P, Kaushik SJ, Poli BM (2004) Effect of long-term feeding with a plant protein mixture based diet on growth and body/fillet quality traits of large rainbow trout (*Oncorhynchus mykiss*). Aquaculture 236: 413–429

DeLany JP, Windhauser, MM, Champagne, CM, Bray, GA (2000) Differential oxidation of individual dietary fatty acids in humans. The American Journal of Clinical Nutrition 79: 905–911

Demonty I, Deshaies Y, Lamrache B, Jacques H (2003) Cod protein lowers the hepatic triglyceride secretion rate in rats. Journal of Nutrition 133: 1398–1402

Denstadli V, Skrede A, Krogdahl Å, Sahlstrøm S, Storebakken T (2006) Feed intake, growth, feed conversion, digestibility, enzyme activities and intestinal structure in Atlantic salmon (*Salmo salar* L.) fed graded levels of phytic acid. Aquaculture 256: 365–376

Dewailly EE, Blanchet C, Gingras S, Lemieux S, Sauve L, Bergeron J, Holub BJ (2001a) Relations between n-3 fatty acid status and cardiovascular disease risk factors among Quebecers. The American Journal of Clinical Nutrition 74: 603–611

Dewailly EE, Blanchet C, Lemieux S, Sauve L, Gingras S, Ayotte P, Holub BJ (2001b) n-3 fatty acids and cardiovascular disease risk factors among the Inuit of Nunavik. The American Journal of Clinical Nutrition 74: 464–473

DH (Department of Health) (1994) Nutritional aspects of cardiovascular disease. Report of the cardiovascular review group committee on Medical aspects of food policy. Reports on health and social subjects No. 46. HMSO: London

Dias J (1999) Lipid deposition in rainbow trout (*Oncorhynchus mykiss*) and European seabass (*Dicentrarchus labrax* L.): nutritional regulation of hepatic lipogenesis. Dr. thesis, Univ. Porto (Portugal) and Univ. Bordeaux I (France), 190 pp

Dosanjh BS, Higgs DA, Plotnikoff MD, Markert JR, Buckley JT (1988) Preliminary evaluation of canola oil, pork lard and marine lipid singly and in combination as supplemental dietary lipid sources for juvenile fall Chinook salmon (*Oncorhychus tshawytscha*). Aquaculture 68: 325–343

Drake EN (2006) Cancer chemoprevention: selenium as a prooxidant not an antioxidant Medical Hypotheses 67: 318–322

Duarte J, Vinderola G, Ritz B, Perdidon G, Matar C (2006) Immunomodulating capacity of commercial fish protein hydrolysate for diet supplementation. Immunobiology 211: 341–350

Dyerberg J, Bang HO, Hjorne N (1975) Fatty acid composition of the plasma lipids in Greenland eskimos. The American Journal of Clinical Nutrition 28: 958–966

Easton MDL, Luszniak D, Von der Geest E (2002) Preliminary examination of contaminant loadings in farmed salmon, wild salmon and commercial salmon feed. Chemosphere 46: 1053–1074

EFSA (European Food Safety Authority) (2005) The EFSA Journal 236, 123 pages http://www.efsa.eu.int/science/contam/contam_opinions/1007_en.html (Consulted 12th July 2005)

European Commission Directive 70/524/EEC of 23 November 1970 concerning additives in feeding-stuffs

European Commission Regulation 199/2006a amending Regulation (EC) 466/2001 setting maximum levels for certain contaminants in foodstuffs as regards dioxins and dioxin-like PCBs

European Commission Directive (2006b)/13/EC amending Annexes I and II to Directive 2002/32/EC of the European Parliament and of the Council on undesirable substances in animal feed as regards dioxins and dioxin-like PCBs

FAO (2005) FAO Fisheries Department, Fishery Information, Data and Statistics Unit. Fishstat Plus; Universal software for fishery statistical time series, Aquaculture production; quantities 1950–2003, Aquaculture production values 1984–2003; Capture production; 1950–2003; Commodities production and trade; 1950–2003; Total production 1970–2003, Vers. 2.30

Ferreti A, Flanagan VP (1996) Antithromboxane activity of dietary α-linolenic acid: A pilot study. Prostaglandins, Leukotrienes and Essential Fatty Acids 54*:* 451–455

Fernandes G, Chandrasekar B, Luan X, Troyer DA (1996) Modulation of antioxidant enzymes and programmed cell death by n-3 fatty acids. Lipids 31: S91-S96

Fernie KJ, Shutt JL, Mayne G, Hoffman D, Letcher RJ, Drouillard KG, Ritchie IJ (2005) Exposure to polybrominated diphenyl ethers (PBDEs): Changes in thyroid, vitamin A, glutathione homeostasis and oxidative stress in American kestrels (*Falco sparverius*). Toxicological Sciences 88: 375–383

Figueiredo-Silva A, Rocha E, Dias J, Silva P, Rema P, Gomes E, Valente LMP (2005) Partial replacement of fish oil by soybean oil on lipid distribution and liver histology in European sea bass (*Dicentrarchus labrax*) and rainbow trout (*Oncorhynchus mykiss*) juveniles. Aquaculture Nutrition 11: 147–155

FIN (2004) Contaminants in fish and fishmeal – a fish meal information network summary

Finnen MJ, Lovell CR (1991) Purification and characterisation of phospholipase A2 from human epidermis. Biochemical Society Transactions 19: 91S

Fonseca-Madrigal J, Karalazos V, Campbell PJ, Bell JG, Tocher DR (2005) Influence of dietary palm oil on growth, tissue fatty acid compositions, and fatty acid metabolism in liver and intestine in rainbow trout (*Oncorhynchus mykiss*). Aquaculture Nutrition 11: 241–250

Food Standards Agency (2004) Brominated flame retardants in trout and eels from the Skerne-Tees river system and total diet study samples. Food Surveillance Information Sheet No. 52/04, April 2004, FSA, London, available at http://www.food.gov.uk/science/surveillance/fsis2004branch/fsis5204 (consulted 11th July 2005)

Foran JA, Good DH, Carpenter DO, Hamilton MC, Knuth BA, Schwager SJ (2005) Quantitative analysis of the benefits and risks of consuming farmed and wild salmon. Journal of Nutrition 135: 2639–2643

Fournier V, Huelvan C, Debruyeres E (2004) Incorporation of a mixture of plant feedstuffs as a substitute for fish meal in diets of juvenile turbot (*Psetta maxima*). Aquaculture 236: 451–465

Fowles JR, Fairbrother A, Baecher-Steppan L, Kerkvliet NI (1994) Immunologic and endocrine effects of flame retardant pentabrominated diphenyl ether (DE-71) in C57BL/6J mice. Toxicology 86: 49–61

Fox M, Benjamin J, Nemets B (2004) A placebo-controlled cross-over trial of adjunctive EPA in OCD. Journal of Psychiatric Research 38: 323–325

Francis G, Makkar HPS, Becker K (2001) Antinutritional factors present in plant-derived alternate fish feed ingredients and their effects in fish. Aquaculture 199: 197–227

Frasure-Smith N, Lesperance F, Julien P (2004) Major depression is associated with lower omega-3 fatty acid levels in patients with recent acute coronary symptoms. Biological Psychiatry 55: 891–896

Froescheis O, Looser R, Cailliet GM, Jarman WM, Ballschmiter K (2000) The deep-sea as a final global sink of semivolatile persistent organic pollutants? Part 1: PCBs in surface and deep-sea dwelling fish of North and South Atlantic and the Monterey Bay Canyon (California). Chemosphere 40: 651–660

Frøyland L, Lie Ø, Berge RK (2000) Mitochondrial and peroxisomal beta-oxidation capacities in various tissues from Atlantic salmon, *Salmo salar*. Aquaculture Nutrition 6: 85–89

Givens DI, Gibbs RA (2006) Very long chain n-3 polyunsaturated fatty acids in the food chain in the UK and the potential of animal-derived foods to increase intake. Nutrition Bulletin 31: 104–110

Glencross BD, Carter CG, Duijster N, Evans DR, Dods K, McCafferty P, Hawkins WE, Maas R, Sipsas S (2004) A comparison of the digestibility of a range of lupin and soybean protein products when fed to either Atlantic salmon (*Salmo salar*) or rainbow trout (*Oncorhynchus mykiss*). Aquaculture 237: 333–346

Glencross BD, Hawkins WE, Curnow JG (2003) Restoration of the fatty acid composition of red seabream (*Pagrus auratus*) using a fish oil finishing diet after grow-out on plant oil based diets. Aquaculture Nutrition 9: 409–418

Goldburg R, Naylor R (2005) Future escapes, fishing, and fish farming. Frontiers in Ecology and the Environment 3: 21–28

Gomes EF, Rema P, Kaushik SJ (1995) Replacement of fish meal by plant proteins in the diet of rainbow trout (*Oncorhynchus mykiss*): digestibility and growth performance. Aquaculture 130: 177–186

Gomez-Requeni P, Mingarro M, Kirchner S, Calduch-Giner JA, Medale F, Martin SAM, Houlihan DF, Kaushik SJ, Perez-Sanchez J (2004) Protein growth performance, amino acid utilisation and somatotrophic axis responsiveness to fish meal replacement by plant protein sources in gilthead sea bream (*Sparus aurata*). Aquaculture 232: 493–510

Gomez-Requeni P, Mingarro M, Kirchner S, Calduch-Giner JA, Medale F, Corraze G, Panserat S, Martin SAM, Houlihan DF, Kaushik SJ, Perez-Sanchez J (2003) Effects of dietary amino acid profile on growth performance, key metabolic enzymes and somatotrophic axis responsiveness of gilthead sea bream (*Sparus aurata*). Aquaculture 220: 749–767

Graff IE, Waagbø R, Fivelstad S Vermeer C, Lie Ø Lundebye A-K (2002) A multivariate study on the effects of dietary vitamin K, vitamin D_3 and calcium, and dissolved carbon dioxide on growth, bone minerals, vitamin status and health performance in smolting Atlantic salmon (*Salmo salar* L.). Journal of Fish Diseases 25: 1–16

Greene DHS, Selivonchick DP (1990) Effects of dietary vegetable, animal and marine lipids on muscle lipid and haematology of rainbow trout (*Oncorhynchus mykiss*). Aquaculture 89: 165–182

Guillou A, Soucy P, Khalil M, Adambounou L (1995) Effects of dietary vegetable and marine lipid on growth, muscle fatty acid composition and organoleptic quality of flesh of brook charr (*Salvelinius fontalis*). Aquaculture 136: 351–329

Haby MM, Peat JK, Marks GB, Woolcock AJ, Leeder SR (2001) Asthma in preschool children: prevalence and risk factors. Thorax 56: 589–595

Hamre K, Berge RK, Lie O (1998) Oxidative stability of Atlantic salmon (*Salmo salar*, L.) fillet enriched in alpha-, gamma-, and delta-tocopherol through dietary supplementation. Food Chemistry 62: 173–178

Hamre K, Christiansen R, Waagbø R, Maage A, Torstensen B, Lygren B, Lie Ø, Wathne E, Albrektsen S (2004) Antioxidant vitamins, minerals and lipid levels in diets for Atlantic salmon (*Salmo salar*, L.): effects on growth performance and fillet quality. Aquaculture Nutrition 10: 113–123

Hardman WE (2002) Omega-3 fatty acids to augment cancer therapy. Journal of Nutrition 132: 3508S-3512S

Hardman WE, Moyer MP, Cameron IL (2002) Consumption of an omega-3 fatty acid product, INCELL AAFA, reduced side-effects of CPY-11 (irinotecan) in mice. The British Journal of Cancer 86: 983–988

Hardy RW, Scott TM, Harrell LW (1987) Replacement of herring oil with menhaden oil, soybean oil, or tallow in the diets of Atlantic salmon raised in marine net-pens. Aquaculture 65: 267–277

Harris WS, von Schacky MD (2004) The Omega-3 index: a new risk factor for death from coronary heart disease? Preventive Medicine 39: 212–220

Heimli H, Giske C, Naderi S, Drevon CA, Hollung K (2002) Eicosapentaenoic acid promotes apoptosis in Ramos cells via activation of caspase-3 and -9. Lipids 37: 797–802

Hemre GI, Sandnes K (1999) Effect of dietary lipid level on muscle composition in Atlantic salmon *Salmo salar*. Aquaculture Nutrition 5: 9–16

Henderson RJ (1996) Fatty acid metabolism in freshwater fish with particular reference to polyunsaturated fatty acids. Archives of Animal Nutrition 49: 5–22

Hertrampf JW, Piedad-Pascual F (eds.) (2000) Handbook on ingredients for aquaculture feeds. Kluwer, Dordrecht, The Netherlands, 573 pp

Hetzel BS, Clugston GA (1999) Iodine. In: Shils, Olaon, Shine and Ross (eds.), Modern nutrition in health and disease. Lippincott Williams & Wilkins, Philadelphia, pp. 253–264

Hibbeln JR (2002) Seafood consumption, the DHA content of mother's milk and prevalence rates of postpartum depression: a cross-national, ecological analysis. Journal of Affective Disorders 69: 15–29

Hites RA, Foran JA, Carpenter DO, Hamilton MC, Knuth BA, Schwager SJ (2004a) Global assessment of organic contaminants in farmed salmon. Science 303: 226–229

Hodge L, Salome CM, Hughes JM, Liu-Brennan D, Rimmer R, Allman M, Pang D, Armour C, Woolcock AJ (1998) Effect of dietary intake of omega-3 and omega-6 fatty acids on severity of asthma in children. The European Respiratory Journal 11: 361–365

Hodge L, Salome C, Peat JK, Haby MM, Xuan W, Woolcock AJ (1996) Consumption of oily fish and childhood asthma risk. Medical Journal of Australia 164: 137–140

Horrocks LA, Yeo YK (1999) Health benefits of docosahexaenoic acid (DHA). Pharmacological Research 40: 211–225

Huntington T, Frid C, Banks R, Scott C, Paramor O (2004) Assessment of the sustainability of industrial fisheries producing fishmeal and fish oil. Report to the Royal Society for the Protection of Birds (RSPB), Poseidon Aquatic Resource Management Ltd., Lymington, Hampshire, UK, 62 pp, Appendices

Iribarren C, Markovitz JH, Jacobs DR, Schreiner PJ, Daviglus M, Hibbeln JR (2004) Dietary intake of n-3, n-6 fatty acids and fish: relationship with hostility in young adults – the CARDIA study. European Journal of Clinical Nutrition 58: 24–31

Iritani N, Narita R, Fujita T, Tanaka T (1985) Effects of dietary fish protein, soybean protein and casein on cholesterol turnover in rats. Journal of Nutritional Science and Vitaminology 31: 385–392

Izquierdo MS, Montero D, Robaina L, Caballero MJ, Rosenlund G, Gines R (2005) Alterations in fillet fatty acid profile and flesh quality in gilthead sea bream (*Sparus aurata*) fed vegetable oils for a long term period. Recovery of fatty acid profiles by fish oil feeding. Aquaculture 250: 431–444

Izquierdo MS, Obach A, Arantzamendi L, Montero D, Robaina L, Rosenlund G (2003) Dietary lipid sources for sea bream and sea bass: growth performance, tissue composition and flesh quality. Aquaculture Nutrition 9: 397–407

Jacobs MN, Covaci A, Schepens P (2002b) Investigation of selected organic pollutants in farmed Atlantic salmon (*Salmo salar*), salmon aquaculture feed, and fish oil components of the feed. Environmental Science & Technology 36: 2797–2805

Jacobs MN, Ferrario J, Byrne C (2002a) Investigation of polychlorinated dibenzo-*p*-dioxins dibenzo-*p*-furans and selected coplanar biphenyls in Scottish farmed Atlantic salmon (*Salmo salar*). Chemosphere 47: 183–191

Jacques H (1990) Effects of dietary proteins on plasma cholesterol and lipoproteins in animal models and in humans. In: Sugano and Beynen (eds.), Monographs on atherosclerosis. Dietary Proteins, Cholesterol Metabolism and Atherosclerosis VII. Karger AG, Basel, Switzerland, New York, pp. 59–70

Jobling M (2003) Do changes in Atlantic salmon, *Salmo salar* L., fillet fatty acids following a dietary switch represent wash out or dilution? Test of a dilution model and its application. Aquaculture Research 34: 1215–1221

Julshamn K, Dahl L, Eckhoff K (2001) Determination of iodine in seafood by inductively coupled plasma/mass spectrometry. Journal of AOAC International 84: 1976–1983

Julshamn K, Maage A, Waagbø R, Lundebye A-K (2006) A preliminary study on tailoring of fillet iodine concentrations in adult Atlantic salmon (*Salmo salar* L.) through dietary supplementation. Aquaculture Nutrition 12: 45–51

James MJ, Cleland LG (1997) Dietary n-3 fatty acids and therapy for rheumatoid arthritis. Seminars in Arthritis and Rheumatism 27: 85–97

Kaushik S, Corraze G (2004) Substitution des huiles de poisson dans les aliments pour poissons. AquaFilia 2: 5–9

Kaushik SJ, Coves D, Dutto G, Blanc D (2004) Almost total replacement of fish meal by plant protein sources in the diet of a marine teleost, the European sea bass, *Dicentrarchus labrax*. Aquaculture 230: 391–404

Kaushik SJ, Cravedi JP, Lalles JP, Sumpter J, Fauconneau B, Laroche M (1995) Partial or total replacement of fishmeal by soybean protein on growth, protein utilisation, potential estrogenic or antigentic effects, cholesterolemia and flesh quality in rainbow trout, *Oncorhynchus mykiss*. Aquaculture 133: 257–274

Kiessling K-H, Keissling A (1993) Selective utilisation of fatty acids in rainbow trout (*Oncorhynchus mykiss,* Walbaum) red muscle mitochondria. Canadian Journal of Zoology 71: 248–251

Kelley VE, Ferretti A, Izui S, Strom TB (1985) A fish oil diet rich in eicosapentaenoic acid reduces cyclooxygenase metabolites and suppresses lupus in MRL-lpr mice. Journal of Immunology 134: 1914–1919

Keys A, Anderson JT, Grande F (1957) Serum cholesterol response to dietary fat. Lancet 1: 787

Kris-Etherton PM, Harris WS, Appel LJ (2002) Fish consumption, fish oil, omega-3 fatty acids, and cardiovascular disease. Circulation 106: 2747–2757

Lall SP (2002) The minerals. In: Halver, JE, Hardy, RW (eds.), Fish nutrition, 3rd edn. Academic Press, San Diego, CA, pp. 259–308

Lands WEM (1992) Biochemistry and physiology of n-3 fatty acids. FASEB J. 6: 2530–2536

Lands WEM (2003) Diets could prevent many diseases. Lipids. 38: 317–321

Lands WEM, Libelt B, Morris A, Kramer NC, Prewitt TE, Bowen P, Schmeisser D, Davidson MH, Burns JH (1992) Maintenance of lower proportions of n-6 eicosanoid precursors in phospholipids of human plasma in response to added dietary n-3 fatty acids. Biochimica et Biophysica Acta 1180: 147–162

Lavigne C, Marette A, Jacques H (2000) Cod and soy proteins compared with casein improve glucose tolerance and insulin sensitivity in rats. American Journal of Physiology, Endocrinology and Metabolism 278: E491-E500

Leaf A, Weber PC (1987) A new era for science in nutrition. The American Journal of Clinical Nutrition 45: 1048–1053

Lie Ø (2001) Flesh quality – the role of nutrition. Aquaculture Research 32: 341–348

Lombardo YB, Chicco AG (2006) Effects of dietary polyunsaturated n-3 fatty acods on dyslipidemia and insulin resistance in rodents and humans. A review. The Journal of Nutritional Biochemistry 17: 1–13

Lorentzen M (1998) Trace elements in practical diets to Atlantic salmon (*Salmo salar* L.). Dr Scient Dissertation, University of Bergen, Bergen, Norway

Lorentzen M, Maage A, Julshamn K (1994) Effects of dietary selenite or selenomethionine on tissue selenium levels of Atlantic salmon (*Salmo salar*). Aquaculture, 121: 359–367

Lundebye AK, Berntssen MHG, Lie Ø, Richie G, Isosaari P, Kiviranta H, Vartiainen T (2004) Dietary uptake of dioxins (PCDD/PCDFs) and dioxin-like PCBs in Atlantic salmon (*Salmo salar*). Aquaculture Nutrition 10: 199–207

Lundebye-Haldorsen A-K, Lie Ø (1999) An overview of dioxins in Norwegian fish and marine products. Norwegian Agricultural Inspection Service, Oslo, Norway

Maage A (1994) Trace elements in Atlantic salmon *Salmo salar* nutrition. Dr Scient Dissertation, University of Bergen, Bergen, Norway

Machlin LJ (ed. 1991) Handbook of vitamins, 2nd edn. Marcel Dekker, New York, 595 pp

Mæland A, Sandnes K, Hjeltnes B, Waagbø R (1998) Biotin in practical fish meal based diet for Atlantic salmon, Salmo salar L., fry. Aquaculture Nutrition 4: 241–247

Maes J, De Meulenaer B, Van Heerswynghels P, Dr Greyt W, Eppe G, De Pauw E, Huyghebaert A (2005) Removal of dioxins and PCB from fish oil by activated carbon and its influence on the nutritional quality. Journal of the American Oil Chemists Society 82: 593–597

Maillard V, Bougnoux P, Ferrari P, Jourdain ML, Pinault M, Lavillonniere F, Body G, Le Folch O, Chajes V (2002) N-3 and n-6 fatty acids in breast cancer in case-control study in Tours, France. International Journal of Cancer 98: 78–93

Marchioli R, Barzi F, Bomba E, Chieffo C, Di Gregorio DDMR, Franzosi MG (2002) Early protection against sudden death by n-3 polyunsaturated fatty acids after myocardial infarction: time course analysis of the results of the Gruppo Italiano per lo Studio della Sopravvivenza nell'Infarto Miocardico (GSSI)-Prevenzione. Circulation 105: 1897–1903

McCully KS (1991) Micronutrients, homocysteine metabolism, and atherosclerosis. In: Bendich A, Butterworth CE (eds.), Micronutrients in health and disease prevention. Marcel Dekker, New York, pp.69–93

Medale F, Bouchard T, Vallee F, Blanc D, Mambrini M, Roem A, Kaushik SJ (1998) Voluntary feed intake, nitrogen and phosphorous losses in rainbow trout, *Oncorhynchus mykiss*. Aquaculture 124: 117–126

Menoyo D, Izquierdo MS, Robaina L, Gines R, Lopez-Bote CJ, Bautista JM (2004) Adaptation of lipid metabolism, tissue composition and flesh quality in gilthead sea bream (*Sparus aurata*) to the replacement of dietary fish oil by linseed and soybean oils. The British Journal of Nutrition 92: 41–52

Ministry of Agriculture, Fisheries and Food (1999) Dioxins and polychlorinated biphenyls in marine fish, fish and fish products. Food Safety Information Sheet 184, MAFF, London, UK

Montero D, Robaina L, Caballero R, Gines R, Izquierdo MS (2005) Growth, feed utilization, and flesh quality of European sea bass (*Dicentrarchus labrax*) fed diets containing vegetable oils: A time course study on the effect of a re-feeding period with a 100% fish oil diet. Aquaculture 248: 121–134

Morris MC, Evans DA, Bienias JL, Tangney CC, Bennett DA, Wilson RS, Aggarwal N, Schneider J (2003) Consumption of fish and n-3 fatty acids and risk of incident Alzheimer's disease. Archives of Neurology 60: 940–946

Morris PC, Gallimore P, Handley J, Hide G, Haughton P, Black A (2005) Full-fat soya for rainbow trout (*Oncorhynchus mykiss*) in freshwater: Effects on performance, composition and flesh fatty acid profile in absence of hind gut enteritis. Aquaculture 248: 147–161

Mourente G, Good JE, Bell JG (2005) Partial substitution of fish oil with rapeseed, linseed and olive oils in diets for European sea bass (*Dicentrarchus labrax* L.) effects on flesh fatty acid composition, plasma prostaglandins E_2 and $F_{2\alpha}$, immune function and effectiveness of a fish oil finishing diet. Aquaculture Nutrition 11: 25–40

Mundheim H, Aksnes A, Hope B (2004) Growth, feed efficiency and digestibility in salmon (*Salmo salar* L.) fed different dietary proportions of vegetable protein sources in combination with two fish meal qualities. Aquaculture 237: 315–331

Murata M, Sano Y, Bannai S, Ishihara K, Matsushima R, Uchida M (2004) Fish protein stimulated the fibrinolysis in rats. Annals of Nutrition & Metabolism 48: 348–356

National Research Council (1993) Nutrient requirements of fish. National Academy Press, Washington DC, USA

Newcomer LM, King IB, Wicklund KG, Stanford JL (2001) The association of fatty acids with prostate cancer. The Prostate 47: 262–268

Ng W-K, Sigholt T, Bell JG (2004a) The influence of environmental temperature on the apparent nutrient and fatty acid digestibility in Atlantic salmon (*Salmo salar* L.) fed finishing diets containing different blends of fish oil, rapeseed oil and palm oil. Aquaculture Research 35: 1228–1237

Ng W-K, Wang Y, Ketchimenin P, Yuen K-H (2004b) Replacement of dietary fish oil with palm fatty acid distillate elevates tocopherol and tocotrienol concentrations and increases oxidative stability in the muscle of African catfish (*Clarias gariepinus*). Aquaculture 233: 423–437

Noaghiul S, Hibbeln JR (2003) Cross national relationship of seafood consumption and rates of bipolar disorders. American Journal of Psychiatry 160: 2222–2227

North Sea Task Force (1993) North Sea Quality Status Report. Oslo and Paris Commissions/ International Council for Exploration of the Sea, London

Norwegian Scientific Committee for Food Safety (2006) Et helhetssyn på fisk og annen sjømat i norsk kosthold. Report (in Norwegian), 171 pp

Ørnsrud R, Graff LE, Høie S, Totland GK, Hemre GI (2002) Hypervitaminosis A in first-feeding fry of the Atlantic salmon (*Salmo salar* L.). Aquaculture Nutrition 8: 7–13

Ørnsrud R, Lorentzen M (2002) Bioavailability of selenium from raw or cured selenomethionine-enriched fillets of Atlantic salmon (*Salmo salar*) assessed in selenium-deficient rats. The British Journal of Nutrition 87: 13–20

Ostermeyer U, Schmidt T (2006) Vitamin D and provitamin D in fish. European Food Research and Technology, 222: 403–413

Peet M, Brind J, Ramchand CN, Shah S, Vankar GK (2001) Two double blind placebo-controlled pilot studies of eicosapentaenoic acid in the treatment of schizophrenia. Schizophrenia Research 49: 243–251

Peet M, Murphy B, Shay J, Horrobin DF (1998) Depletion of omega-3 fatty acid levels in red blood cell membranes of depressive patients. Biological Psychiatry 43: 315–319

Pereira TG, Oliva-Teles A (2002) Preliminary evaluation of pea seed meal in diets for gilthead sea bream (*Sparus aurata*) juveniles. Aquaculture Research 33: 1183–1189

Picot L, Bordenave S, Didelot S, Fruitier-Arnaudin I, Sannier F, Thorkelsson G, Berge JP, Guerard F, Chabeaud A, Piot JM (2006) Anti-proliferative activity of fish protein hydrolysates on human breast cancer cell lines. Process of Biochemistry. 41: 1217–1222

Pike IH (2005) Eco-efficiency in aquaculture: global catch of wild fish used in aquaculture. International Aquafeed 8 (1): 38–40

Polatajko A, Jakubowski N, Szpunar J (2006) State of the art report of selenium speciation in biological samples. Journal of Analytical Atomic Spectrometry 21: 639–654

Pusztai A, Bardocz S (2006) GMO in animal nutrition: potential benefits and risks. In: Mosenthin R, Zentek J, Zebrowska T (eds.), Biology of nutrition in growing animals. Elsevier, Edinburgh, UK, pp. 513–540

Refstie S, Korsoen OJ, Storebakken T, Baeverfjord G, Lein I, Roem AJ (2000) Differing nutritional responses to dietary soybean meal in rainbow trout (*Oncorhynchus mykiss*) and Atlantic salmon (*Salmo salar*). Aquaculture 190: 49–63

Richard, N., Kaushik, S., Larroquet, L., Panserat, S. and Corraze, G. (2006) Replacing dietary fish oil by vegetable oils had little effects on lipogenesis, lipid transport and tissue lipid uptake in rainbow trout (*Oncorhynchus mykiss*). Brit. J. Nut. 96, 299–309.

Richardson AJ (2004) Clinical trials of fatty acid treatment in ADHD, dyslexia, dyspraxia and the autistic spectrum. Prostaglandins, Leukotrienes, and Essential Fatty Acids 2004, 70: 383–390

Richardson AJ, Cyhlarova E, Ross MA (2003) Omega-3 and omega-6 fatty acid concentrations in red blood cell membranes relate to schizotypal traits in healthy adults. Prostaglandins, Leukotrienes, and Essential Fatty Acids 69: 461–466

Richardson NL, Higgs DA, Beames RM, McBride JR (1985) Influence of dietary calcium, phosphorous, zinc and sodium phytate level on cataract incidence, growth and histopathology in juvenile chinook salmon (*Oncorhynchus tshawytscha*). Journal of Nutrition 115: 553–567

Rogers JB (1998) The role of n-3 fatty acids in psoriasis, In: Kremer J (ed.), Medicinal fatty acids in inflammation. Birkhauser Verlag, Basel, pp. 45–53

Rosenlund G, Obach A, Sandberg MG, Standal H, Tveit K (2001) Effect of alternative lipid sources on long-term growth performance and quality of Atlantic salmon (*Salmo salar* L.). Aquaculture Research 32: 323–328

Roulet M, Frascarolo P, Rappaz I, Pilet M (1997) Essential fatty acid deficiency in well nourished young cystic fibrosis patients. European Journal of Pediatrics 156: 952–956

Ruyter B, Røsjø C, Grisdale-Helland B, Rosenlund G, Obach A, Thomassen MS (2003) Influence of temperature and high dietary linoleic acid content on esterification, elongation, and desaturation of PUFA in Atlantic salmon hepatocytes. Lipids 38: 833–840

SACN/COT (Scientific Advisory Committee on Nutrition and Committee on Toxicity) (2004) Advice on fish consumption: benefits and risks. The Stationary Office: Norwich

Sanden M, Bruce IJ, Rahman MA, Hemre G-I (2004) The fate of transgenic sequences present in genetically modified plant products in fish feed, investigating the survival of GM soybean DNA fragments during feeding trials in Atlantic salmon, *Salmo salar* L. Aquaculture 237: 391–405

Sanderson P, Finnegan YE, Williams CM, Calder PC, Burdge GC, Wootton SA, Griffin BA, Millward DJ, Pegge NC, Bemelmans WJE (2002) UK Food standards agency alpha-linolenic acid workshop report. The British Journal of Nutrition 88: 573–579

Sandnes K, Rosenlund G, Mangor-Jensen A, Lie Ø (1998) Contents and organ distribution of pantothenic acid in maturing turbot (*Psetta maxima*). Aquaculture Nutrition 4: 285–286

Sargent JR, Tocher DR, Bell JG (2002) The Lipids. In: Halver, JE, and Hardy, RW (eds.), Fish nutrition, 3rd edn. Elsevier Science, New York, pp. 181–257

Sastry PS (1985) Lipids of nervous tissue: composition and metabolism. Progress in Lipid Research 24: 69–176

SCAN (2000) Dioxin contamination of feeding stuffs and their contribution to the contamination of food of animal origin. Opinion of the Scientific Committee on Animal Nutrition, adopted on November 6th 2000. 105 pages. European Commission for Health and Consumer Protection Directorate-General, Brussels, Belgium. http://europa.eu.int/comm/food/fs/sc/scf/index_en.html (consulted 15th May 2004)

SCF (2001) Update of the Risk assessment of dioxins and dioxin-like PCBs in food based on new scientific information available since adoption of the SCF opinion of 22nd November 2000. Opinion of the Scientific Committee on food, adopted on May 30th 2001. European Commission for Health and Consumer Protection Directorate-General, Brussels, Belgium http://europa.eu.int/comm/food/fs/sc/scf/index_en.html (consulted 15th May 2004)

Schmid S, Ranz D, He ML, Burkard S, Lukowicz MV, Reiter R, Arnold R, Le Deit H, David M, Rambeck WA (2003) Marine algae as natural source of iodine in the feeding of freshwater fish – a new possibility to improve iodine supply in man. Revue de Medecine Veterinaire 154: 645–648

Simopoulos AP (1991) Omega-3 fatty acids in health and disease and in growth and development. The American Journal of Clinical Nutrition 54: 438–463

Simopoulos AP (1999) Essential fatty acids in health and chronic disease. The American Journal of Clinical Nutrition 70: 560S-569S

Simopoulos AP, Leaf A, Salem N (1999) Essentiality of and recommended dietary intakes for omega-6 and omega-3 fatty acids. Annals of Nutrition & Metabolism 43: 127–130

Simopoulos AP, Leaf A, Salem N (2000) Workshop statement on the essentiality of and recommended intakes for omega 6 and omega 3 fatty acids. Prostaglandins, Leukotrienes and Essential Fatty Acids 63: 119–121

Singh RB, Niaz MA, Sharma JP, Kumar R, Rastogi V, Moshiri M (1997) Randomized double-blind, placebo-controlled trial of fish oil and mustard oil in patients with suspected acute myocardial infarction: The Indian experiment of infarct survival-4. Cardiovascular drugs and therapy 11: 485–491

Sitja-Bobadilla A, Pena-Llopis S, Gomez-Requeni P, Medale F, Kaushik S, Perez-Sanchez J (2005) Effect of fishmeal replacement by plant protein sources on non-specific defence mechanisms and oxidative stress in gilthead seabream (*Sparus aurata*). Aquaculture 249, 387–400.

Storebakken T, Shearer KD, Roem AJ (1998) Availability of protein, phosphorous and other elements in fish meal, soy-protein concentrate and phytase-treated soy-protein-concentrate-based diets to Atlantic salmon, *Salmo salar*. Aquaculture 161: 365–379

Storlien LH, Hulbert AJ, Else PL (1998) Polyunsaturated fatty acids, membrane function and metabolic diseases such as diabetes and obesity. Current Opinion in Clinical Nutrition and Metabolic Care 1: 559–563

Stevens LJ, Zentall SS, Deck JL, Abate ML, Watkins BA, Lipp SR, Burgess JR (1995) Essential fatty-acid metabolism in boys with an attention-deficit hyperactive disorder. The American Journal of Clinical Nutrition 62: 761–768

Suadicani P, Hein HO, Gyntelberg F (1992) Serum selenium concentration and risk of ischaemic heart disease in a prospective cohort study of 3000 males. Atherosclerosis 96: 33–42

Tacon AGJ (1997) Feeding tomorrow's fish: keys for sustainability. In: Tacon AGJ, Basurco B (eds.), Feeding tomorrow's fish. Cahiers options Mediterraneennes, Vol. 22. Zaragoza, Spain, pp.11–33

Tacon AGJ (2003) Global trends in aquaculture and compound aquafeed production. International Aquafeed Directory 2003: 8–23

Tacon AGJ (2004) Use of fish meal and fish oil in aquaculture: a global perspective. Aquatic Resources, Culture & Development 1 (1): 3–14

Tacon AGJ (2005) Salmon aquaculture dialogue: Status of information on salmon aquaculture feed and the environment. International Aquafeed 8: 22–37

Taylor KE, Higgins CJ, Calvin CM, Hall JA, Easton T, McDaid AM, Richardson AJ (2000) Dyslexia in adults is associated with clinical signs of fatty acid deficiency. Prostaglandins, Leukotrienes and Essential Fatty Acids 63: 75–78

Teskeredzic Z, Higgs DA, Dosanjh BS, McBride JR, Hardy RW, Beames RM, Jones JD, Simell M, Vaara T, Bridges RB (1995) Assessment of undephytinised and dephytinised rapeseed protein concentrate as sources of dietary protein for juvenile rainbow trout, *Oncorhynchus mykiss*. Aquaculture 131: 261–277

Tibaldi E, Tulli F, Amerio M (1999) Feed intake and growth responses of sea bass (*D. labrax*) fed different plant protein sources are not affected by supplementation with a feeding stimulant. In: Piva G, Bertoni G, Satoh S, Bani P, Calamari L (eds.), Recent Progress in Animal Production Science: I. Proceedings. A.S.P.A. XIII Congress, Piacenza, Italy, 21–24 June 1999. Assn. Sci. Anim. Production, Italy, pp. 752–754

Tidwell JH, Allan GL (2002) Fish as food: Aquaculture's contribution. World Aquaculture 33, 44–48

Tocher DR, Bell JG, Dick JR, Henderson RJ, McGhee F, Mitchell DF, Morris PC (2000) Polyunsaturated fatty acid metabolism in Atlantic salmon (*Salmo salar*) undergoing parr-smolt transformation and the effects of dietary linseed and rapeseed oils. Fish Physiology and Biochemistry 23: 59–73

Torstensen BE, Bell JG, Sargent JR, Rosenlund G, Henderson RJ, Graff IE, Lie Ø Tocher, DR (2005) Tailoring of Atlantic salmon (*Salmo salar* L.) flesh lipid composition and sensory quality by replacing fish oil with a vegetable oil blend. Journal of Agricultural and Food Chemistry 53: 10166–10178

Torstensen BE, Frøyland L, Lie Ø (2004a) Replacing dietary fish oil with increasing levels of rapeseed oil and olive oil – effects on Atlantic salmon (*Salmo salar*) tissue and lipoprotein composition and lipogenic enzyme activities. Aquaculture Nutrition 10: 175–192

Torstensen BE, Frøyland L, Ørnsrud R, Lie Ø (2004b) Tailoring of a cardioprotective fillet fatty acid composition of Atlantic salmon (*Salmo salar*) fed vegetable oils. Food Chemistry 87: 567–580

Torstensen BE, Lie O, Froyland L (2000) Lipid metabolism and tissue composition in Atlantic salmon (*Salmo salar* L.) – effects of capelin oil, palm oil and oleic acid – enriched sunflower oil as dietary lipid sources. Lipids 35: 653–664

Torstensen BE, Lie Ø, Hamre K (2001) A factorial experimental design for investigation of effects of dietary lipid content and pro-and antioxidants on lipid composition in Atlantic salmon (*Salmo salar*) tissues and lipoproteins. Aquaculture Nutrition 7: 265–276

Tulli F, Tibaldi E, Comin A (1999) Dietary protein sources differently affect plasma lipid levels and body fat deposition in juvenile sea bass. In: Piva G, Bertoni G, Satoh S, Bani P, Calamari L (eds.), Recent Progress in Animal Production Science: I. Proceedings. A.S.P.A. XIII Congress, Piacenza, Italy, 21024 June 1999. Assn. Sci. Anim. Production, Italy, pp. 782–784

Van den Berg M, Birnbaum L, Bosveld ATC, Brunstrom B, Cook P, Feeley M, Glesy JP, Hanberg A, Hasegawa R, Kennedy SW, Kubiak T, Larsen JC, van Leewen FXR, Liem AKD, Nolte C, Peterson RE, Peollinger L, Safe S, Schrenk D, Tillit D, Tysklind M, Younes M, Waern F, Zacharewski T (1998) Toxic equivalency factors (TEFs) for PCBs, PCDDs, PCDFs for humans and wildlife. Environmental Health Perspectives 106: 775–792

Verlengia R, Gorjao R, Kanunfre CC, Bordin S, Martins de Lima T, Fernandes-Martins E, Newsholme P, Curi R (2004) Effects of EPA and DHA on proliferation, cytokine production and gene expression in Raj cells. Lipids 39: 857–864

Vielma J, Lall SP, Koskela J, Schöner F-J, Mattila P (1998) Effects of dietary phytase and cholecalciferol on phosphorous bioavailability in rainbow trout (*Oncorhynchus mykiss*). Aquaculture 163: 309–323

Visentainer JV, de Souza NE, Makoto M, Hayashi C, Franco MRB (2005) Influence of diets enriched with flaxseed oil on the linolenic eicosapentaenoic and docosahexaenoic fatty acid in Nile tilapia (*Oreochromis niloticus*). Food Chemistry 90: 557–560

Waagbø R, Sandnes K, Lie Ø, Roem A (1998) Effects of inositol supplementation on growth, chemical composition and blood chemistry in Atlantic salmon, *Salmo salar* L., fry. Aquaculture Nutrition 4: 53–59

Waagbø R, Sandnes K, Sandvin A, Lie Ø (1991) Feeding three levels of n-3 polyunsaturated fatty acids at two levels of vitamin E to Atlantic salmon (*Salmo salar*). Growth and chemical composition. Fisk. Dir. Skr., Ser. Ernaering. 4: 51–63

Waagbø R, Sandnes K, Torrisen OJ, Sandvin A, Lie Ø (1993) Chemical and sensory evaluation of fillets from Atlantic salmon (*Salmo salar*) fed three levels of n-3 polyunsaturated fatty acids at two levels of vitamin E. Food Chemistry 46: 361–366

Wang L, Folsom AR, Eckfeldt JH (2003) Plasma fatty acid composition and incidence of coronary heart disease in middle aged adults: the Atherosclerosis Risk in Communities (ARIC) Study. Nutrition, Metabolism, and Cardiovascular Diseases 13: 256–266

Watanabe T (1977) Sparing action of lipid on dietary protein in fish. Low protein diet with high calorie content. Technocrat 10: 34–39

Watanabe T, Verakunpiriya V, Watanabe K, Viswanath K, Satoh S (1998) Feeding raonbow trout with non-fishmeal diets. Fisheries Science 63: 258–266

Yaqoob P (2004) Fatty acids and the immune system: from basis science to clinical applications. Proceedings of the Nutrition Society 63: 89–104

Yildiz M, Sener E, Gun H (2006) Effect of refrigerated storage on fillet lipid quality of rainbow trout (*Oncorhynchus mykiss* W.) fed a diet containing different levels of DL alpha – tocopherol acetate. Turkish Journal of Veterinary & Animal Sciences 30: 143–150

Young G, Conquer J (2005) Omega-3 fatty acids and neuropsychiatric disorders. Reproduction Nutrition Development 45: 1–28

Zheng ZJ, Croft JB, Giles WH, Mensah GA (2001) Sudden cardiac death in the United States, 1989 to 1998. Circulation 104: 2158–2163

Zhou T, Taylor MM, DeVito MJ, Crofton KM (2002) Developmental exposure to brominated diphenyl ethers results in thyroid hormone disruption. Toxicological Sciences 105–116

Ziboh VA (1998) The role of n-3 fatty acids in psoriasis. In: Kremer J (ed.), Medicinal fatty acids in inflammation. Birkhauser Verlag, Basel, pp. 45–53

List of abbreviations Bell

ADHD – Attention Deficit/Hyperactivity Disorder
ALA – alpha-linolenic acid
ANF – anti-nutritional factors

ARA – arachidonic acid
CHD – coronary heart disease
CVD – cardio-vascular disease
DHA – docosahexaenoic acid
DL – dioxin-like
DL-PCBs – dioxin-like polychlorinated biphenyls
EFA – essential fatty acid
EPA – eicosapentaenoic acid
FCR – Feed conversion ratio
FM – fish meal
FO – fish oil
FPC – fish protein concentrate
HUFA – highly unsaturated fatty acids
IL-10 – interleukin 10
INF-γ – interferon-gamma
LA – linoleic acid
LO – linseed oil
OO – olive oil
PBDE – polybrominated dephenylether
PC – phosphatidylcholine
PCBs – polychlorinated biphenyls
PE – phosphatidylethanolamine
PFF – post-first feeding
PLA_2 – phospholipase A_2
PO – palm oil
POPs – persistent organic pollutants
PUFA – polyunsaturated fatty acid
RO – rapeseed oil
SO – soybean oil
SGR – specific growth rate
TAG – triacylglycerol
TDI – tolerable daily intake
TEQ – toxic equivalents
TEFs – toxic equivalency factors
TGC – thermal growth coefficient
TNF-α – tumour necrosis factor-alpha
VO – vegetable oils

Chapter 7
NGO Approaches to Minimizing the Impacts of Aquaculture: A Review

Katherine Bostick

Abstract The rapid growth of the aquaculture industry and its associated environmental and social impacts have brought aquaculture to the forefront of debates about the long term viability of global food production systems. Environmental non-governmental organizations (NGOs) are involved in aquaculture from the point of view of environmental sustainability at the local farm level, the larger ecosystem level, and the level of international trade. NGOs are driven by their missions, which may be focused on conserving nature and biodiversity, protecting the marine environment, improving the quality of food, or creating a sustainable society. NGO concerns with aquaculture are related to their missions and are often derived from related conservation themes such as agriculture, fisheries, or other marine or freshwater issues or develop out of specific local concerns of their communities. NGO activities range from research, information distribution and training to local organization, lobbying for legislative change, and partnering with industry to minimize impacts. Increasingly, the aquaculture industry is recognizing the diversity of NGOs and that each organization addresses aquaculture's impacts through a unique combination of approaches. This chapter discusses a variety of specific approaches used by NGOs to influence the aquaculture industry including encouraging the use of better management practices, educating consumers, and developing standards, with WWF-US as a specific example.

Keywords NGO, environment, aquaculture certification, aquaculture standards, BMP

7.1 Introduction

Aquaculture is the fastest growing food production system on the planet. From 1970 to 2005, aquaculture's share of global fisheries landings increased from 5% to approximately 33% of total product and 43% of food fish (Fig. 7.1, FAO 2007a).

Aquaculture Program Officer, World Wildlife Fund-US, 1250 24th St, NW, Washington DC, 20037. Tel: (202) 293-4800; Fax: (202) 293-9211; E-mail: katherine.bostick@wwfus.org

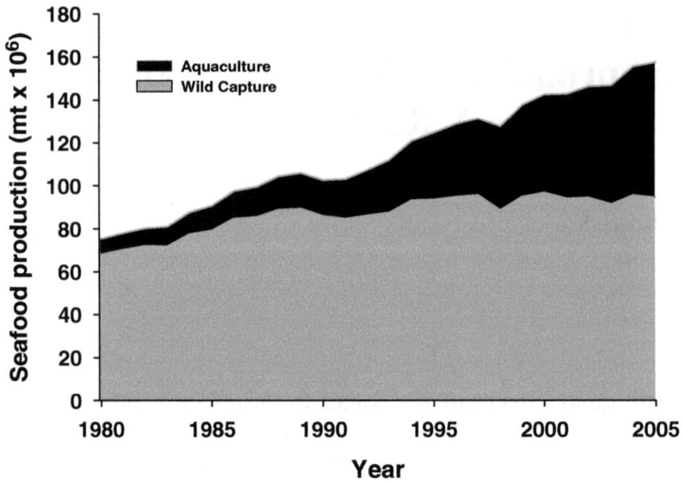

Fig. 7.1 Comparison of aquaculture production and capture fisheries production. Production values are in millions of metric tons (mt) (FAO 2007b)

Aquaculture has grown at an average rate of 8.9% per year since 1970 and has doubled since the mid-1990s (FAO 2004; Johnson 2005). Growth in the aquaculture sector is projected to continue well into the future. By comparison, harvests from wild fisheries have remained relatively flat since the early 1990s. The U.N. Food and Agriculture Organization (FAO 2004) estimates that about three-quarters of wild stocks are fully fished, overfished, or depleted. As a result, any future increases in seafood production are expected to come from aquaculture.

Aquaculture plays an increasingly important role in the global seafood market and in seafood trade not only because it accounts for a significant and increasing percentage of seafood production, but also because it consumes the bulk of global pelagic fish catch. Approximately one-third of wild fisheries catch is reduced to fishmeal and fish oil. Aquaculture consumes more fishmeal and fish oil than any other industry, using half of total global fishmeal and more than 80% of total fish oil (Tacon 2005). Thus, the combined production and consumption of aquaculture accounts for more than half of all fisheries' production in the world.

Recently, high-value aquaculture species including shrimp and salmon have received significant press related to the social and environmental impacts of production as well as food safety and public health issues. These species are primarily imported by the United States, Europe, and Japan for consumers who are relatively wealthy. However, by volume, the production of these species is quite low in comparison to many of the other species groups that are cultivated (Fig. 7.2). For example, aquatic plants account for almost 25% of total aquaculture production by weight. Production of carp is thirteen times that of salmon, and almost 10 times that of shrimp (FAO 2007b).

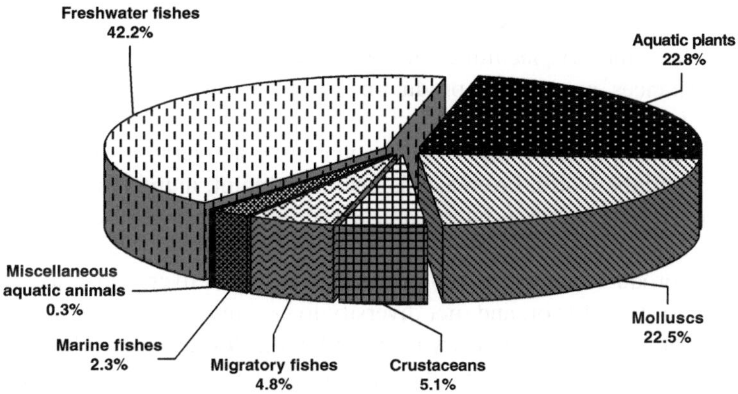

Fig. 7.2 Comparison of the main species groups produced by aquaculture on a wet-weight basis (FAO 2006)

Aquaculture production has become of active interest to a number of non-governmental organizations (NGOs) around the world. Their interest is to a large extent aimed at addressing the on-the-ground environmental or social impacts that threaten or undermine their ability to deliver on their overall missions of conservation or social welfare. In this regard, as the aquaculture industry continues to grow, a number of tangible, negative impacts are increasing. The problems that are most often cited with regard to aquaculture are food safety and the environmental and social impacts of the different facets of the industry.

An NGO is defined by the World Resources Institute (1992) as "a non-profit group or association organized outside of institutionalized political structures to realize particular social objectives (such as environmental protection) or serve particular constituencies (such as indigenous peoples). NGO activities range from research, information distribution, training, local organization, and community service to legal advocacy, lobbying for legislative change, and civil disobedience. NGOs range in size from small groups within a particular community to huge membership groups with a national or international scope."

NGOs often serve as self-appointed watchdogs for society and at the heart of their interests is the sustainability of human activities, including food production. Sustainability incorporates economic, social and environmental components, and for conservation-focused NGOs, environmental sustainability serves as a base on which economic and social sustainability can be built. For aquaculture, environmental sustainability is necessary at the local farm level, the larger ecosystem level, and the level of international trade.

Some environmental NGOs are interested in aquaculture not only because they feel it currently has unacceptable negative impacts, but also because of its potential to reduce pressure on the world's oceans, and in at least some cases, replace over-fished species in the marketplace. While this is certainly not always the case, it is or can be the case for species such as tilapia, catfish, oysters, mussels, clams, and scallops in the future. Aquaculture can be a very efficient food production system. In certain instances, aquaculture production can have fewer environmental impacts than wild catch of the same species.

The aquaculture industry also has the potential to deliver many benefits to communities around the world. From a consumer standpoint, individuals in both developed and developing countries have increased access to seafood because aquaculture production has increased supplies and caused an overall reduction in prices. Aquaculture products can be of consistent quality and deliver year-round, healthy, high quality protein, flavor, and diet diversity to consumers.

Much of the aquaculture industry is located in the developing world and in addition to serving as an important food source, it can be a source of employment for skilled and non-skilled workers. It can replace more destructive, resource-dependent industries and can serve as an alternative income source for fishermen in regions where wild fish stocks have become depleted. Aquaculture can also be a driver of regional economic growth and trade. At present, some species cultured in developing countries are wholly or partially exported to high value markets in the U.S., Europe, and Japan. According to the FAO, by 2020, nearly 80% of all seafood will come from developing countries and aquaculture will account for more than 40% of all fisheries production. Thus, aquaculture production is extremely important for developing countries where, for example, by 2001 revenues from fishery and aquaculture exports were greater than those of coffee, bananas, rubber, tea, rice and meat combined (Ruckes 2003).

Carp and other aquaculture products of low market value are most often consumed in the country where they are produced, providing much needed protein to communities in developing countries. Many development groups have promoted and provided technical assistance to start small-scale aquaculture operations, such as tilapia grown in ponds, in areas in Africa and Latin America as a means of providing poor rural families with both a source of food and a source of income.

Both technologically advanced and traditional aquaculture operations have a range of costs and benefits. There are now increased incentives to improve overall industry performance as global trade expands, international food safety and environmental regulations are harmonized, and as retailers increasingly require traceability for all the food they sell. These conditions create new opportunities to halt the destruction of biologically diverse coastal, marine, and freshwater habitats, while raising consumer awareness of the impact that their food purchases have on their own health and the health of the environment.

While aquaculture has caused and could continue to lead to significant environmental damage and social conflict, it can also produce food in a more responsible manner. Thus it remains an issue where NGOs strive to eliminate worse practices, reduce negative impacts, and increase the benefits of aquaculture. This chapter outlines the impacts of aquaculture which most concern NGOs, describes a variety of the approaches taken by these groups, and details the work of one international environmental NGO working to minimize the environmental and social impacts of aquaculture.

7.2 Key Concerns of NGOs Related to Aquaculture

There are hundreds of small, negative impacts that can be associated with aquaculture and they vary by species produced and production system. However, the major impacts of aquaculture have been clearly documented and remain a relatively short list, which can be broken down into environmental, social, and food safety impacts. There is an extensive body of literature as well as significant ongoing research on these impacts, which are briefly described here.

It is important to note that although a number of major impacts of aquaculture have been clearly identified, there is much debate about the extent and frequency of these impacts. There remain a number of issues where scientific consensus has not yet been reached and where there is poor and often contradictory data.

7.2.1 Environment Impacts

Impacts of aquaculture on wild fauna and flora are the cause of many of the concerns about the industry. It is important to note that the bulk of the concerns relate to the production of fish and shrimp, especially carnivorous species, rather than to filter-feeding shellfish and seaweeds. Though not without potential negative impact, production of filter-feeding bivalve molluscs and seaweeds result in a different suite of positive and negative impacts than the production of other species because they remove nutrients from the water column and do not require feed (Boyd et al. 2005; Shumway et al. 2003). These species typically rank well in NGO evaluations of culture at a broad species level, and are not an area of focus for NGOs concerned with minimizing impacts of aquaculture.

Effluents from fish and shrimp farms alter the natural nutrient levels of local waters and can lead to eutrophication (Chapter 1). Chemicals used in aquaculture production, including antibiotics, are released into local waters along with effluents. Effluents and chemicals can be released from both open systems in rivers, lakes, or coastal sites as well as from closed ponds or raceways that exchange water or drain at harvest. Sensitive benthic ecosystems below marine culture sites are impacted by these chemicals, organic matter, and accompanying sedimentation. Impacts range from increased benthic productivity in oligotrophic waters to changes in the benthic communities and local losses of biodiversity to anoxic sediments and even anoxia in the lower levels of the water column (Black et al. 2002).

Wild fish populations can be impacted by aquaculture production as a result of escapes and disease introduction or transmission. Escaped farmed species can interbreed with wild populations, altering the genetic pool and lowering the fitness of these populations, or they can become established and compete with wild species for food and habitat (Chapter 4, Black et al. 2002; Boyd et al. 2005). In 2005, it was estimated that one in four salmon found in Norwegian seas was of farmed origin (Esmark et al. 2005). In the same year, an average of 13% of salmon found in rivers during spawning season were escaped fish, which is as low as it has been since the

1980s (Hansen et al. 2006). Escaped farmed fish are of particular concern in areas where there are sensitive wild endangered fish populations.

Diseases can be both introduced and transmitted through aquaculture. Without proper controls and quarantines, it is possible for diseases or parasites to be introduced to a region through the importation of juveniles. In cases of disease outbreak on a farm, the disease can be transmitted to the wild if it is an open production system or if contaminated water from a closed system is released into the environment. Just as pathogens and parasites can be transferred from farms into the wild, disease free farmed species can be infected from the wild and in open production systems there is flow in both directions. Diseases can also be transmitted from farm to farm, and the shrimp industry in certain parts of the world has collapsed due to severe disease outbreaks. Disease introduction and transfer can also be a concern in shellfish and seaweed culture systems (Boyd et al. 2005).

Other key farm-level impacts that concern many environmental NGOs are land conversion and predator control. The construction of shrimp ponds has been linked with the conversion of hundreds of thousands of hectares of coastal mangrove forests (Naylor et al. 2000). Though it appears that this trend has decreased significantly, NGOs remain concerned with the loss of these forests because they provide a wide range of important ecosystem services as well as economic benefits to coastal communities. In inland, coastal, and ocean production systems, predator control is another issue of concern to some NGOs, especially in cases where the wild predating species is endangered. Birds, seals, and sharks are among the species that are drawn to aquaculture ponds or pens in search food. Lethal control of these species is sometimes used by producers to protect their farmed stocks. As with many of these farm-level impacts, the impact of one farm on the environment may be relatively small. Yet when there are a number of farms in a region, impacts accumulate to a level where they are of considerable concern.

The impact of fish farms extends beyond a local or regional level. NGOs commonly hold that aquaculture should produce more protein than it consumes. On a global scale, aquaculture relies on high-protein, fishmeal and fish oil-based feed for carnivorous species that often requires multiple kilograms (kg) of wild fish to produce one kg of edible aquaculture product (Tacon 2005). Although the production of omnivorous or vegetarian fish such as tilapia and carp can result in a net increase in fish protein, they are also often fed a small percentage of fishmeal and fish oil in their diets. This can be cumulatively significant for species that are produced in large volumes. Given the tenuous state of global pelagic fisheries, environmental NGOs are concerned that increased pressure on and demand for these fisheries will lead to their collapse. The overfishing of small pelagic fish affects species up the marine food web because they are critical components of the diets of marine mammals, seabirds, and large carnivorous fish species (Naylor et al. 2000). By way of example, sandeel stocks in the North Sea are depleted which has been demonstrated to have detrimental effects on seabird populations such as the sandwich tern (ICES 2006). Most of the species reduced into fishmeal and fish oil for aquaculture feeds are also consumed directly by humans. Some NGOs are also concerned that the use of these pelagic fish for aquaculture, and any consequent collapse in fisheries, could have significant impacts on poor communities that rely on this protein source.

Wild species are also impacted when producers rely on wild stock for juveniles rather than using hatchery juveniles. For example, juvenile wild bluefin tuna in the Mediterranean are being caught, "fattened" in net-pen cages, and then typically sold at a great profit to the high end Japanese market. The growth of the tuna fattening industry has placed additional pressure on Eastern Atlantic bluefin tuna stocks, which have been declining for years and are now believed to be in danger of collapse (Tudela and García 2004). Similarly, the European eel population has seen a marked decrease over the last few decades, a trend that is also driven in part by the catch of juvenile eel for farming (Dekker 2007).

In theory, aquaculture has the potential to take pressure off wild fisheries. This is a source of hope but also frustration for many in the NGO community. A number of conditions would need to be met in order for this to happen, several of which are mentioned here. First, to reduce pressure on wild fisheries, aquaculture products would need to replace wild fish in the market. Unfortunately, to date, it appears that much aquaculture is supplementing rather than replacing wild catch in the market, which has been demonstrated to be the case for salmon (Naylor and Burke 2005). Secondly, society would need to produce and consume molluscs and omnivorous or herbivorous fish preferentially over species that use more fishmeal and fish oil in their production. Nevertheless, the farming of carnivorous marine species is one of the fastest growing segments of the aquaculture industry. The industry is making strides towards reducing the percentage of fishmeal and fish oil in feeds for carnivorous species and this would need to continue as well. Thirdly, aquaculture would need to continue to move away from the use of wild juveniles in cases where wild populations are impacted by this use.

7.2.2 Social Impacts

Over the last 15 years, the aquaculture industry has been the focus of negative press about its social impacts, most of which focused on shrimp farming. While aquaculture can and does in many cases provide needed income and food security, there exist a number of consistent and significant areas of social impact that result from aquaculture production. Labor issues and concerns related to workers rights and welfare both on farms and in processing plants have been a prime concern of some NGOs, such as Fundación Terram in Chile, which has reported on the salmon industry in southern Chile.

The use and conversion of habitat and resources for farms, as well as reduced access of communities to remaining resources was one primary area of concern relating to shrimp farming (World Bank et al. 2002). It is not uncommon for the privatization of public commons such as mangrove areas or inland coastal waters to lead to social conflict. The degree of conflict varies greatly, ranging from mild animosity between farmers and coastal land-owners wishing to protect pristine ocean views and high property values, to trespassing, theft, and even murder in areas where the survival of poor communities is threatened by farms that restrict access to critical natural resources.

In many regions, fishermen will oppose aquaculture concessions in or near their fishing grounds for a variety of reasons. Fishermen want to ensure continued access to fishing grounds and are also concerned with the possibility of declines in wild fish populations due to direct environmental impacts. Aquaculture production has also been linked to an overall decline in market prices, and in some regions fishermen blame aquaculture for this decrease in the value of their catch.

The growth of any new industry, especially in rural areas, can lead to inflation in the cost of key local goods such as food, labor, and land. This inflation can disproportionately affect poor people who are not involved with the industry. Additionally, an increase in job opportunities can draw individuals away from traditional manual labor such as fishing or farming. While individuals then earn a full time income, they also have less time for subsistence activities and increase dependence on purchased food items. In many regions the aquaculture industry has been shown to pay better wages than other comparable employers (World Bank et al. 2002), yet in some areas, low wages, worker rights, and regional socio-economic benefits have been called into question (Pinto and Kremerman 2005).

7.2.3 Food Safety Concerns

Food safety concerns are increasing in the aquaculture industry as new research and some NGO activism focuses on this issue. One of these concerns is the accumulation of heavy metals and persistent organic pollutants (POPs) in the edible flesh of farmed seafood, especially in carnivorous species. Chemical residues and the presence of hormones or antibiotics in aquaculture products is another concern, especially in cases where industry is suspected of using antibiotics or other chemicals prophylactically. For antibiotics, there is concern related not only to residues in product, but also to potential development of antibiotic resistance. The use and residues of chemicals that were recently banned in a number of countries, such as malachite green, are also a concern.

Today's market is increasingly vigilant, and producers are being asked to comply with food safety standards of importing countries, chain of custody requirements to ensure the traceability of their product, and additional product quality or safety standards for certain retailers. NGOs concerned with human health implications of aquaculture production have a receptive audience of retailers and consumers concerned with food safety.

7.3 NGO Approaches to Aquaculture

All human activities have impacts and many NGOs including conservation, human rights, or animal welfare organizations have become involved in evaluating these impacts. Generally this entails identifying and vocalizing concerns with those impacts that they feel are "unacceptable" and working to either minimize these impacts to fall within an acceptable range, or, in some cases, campaigning to close down an industry or operation.

Many NGO concerns with aquaculture began through related conservation themes such as agriculture, fisheries, or other marine or freshwater issues. Alternatively, they can develop out of specific local concerns of their communities. The goal or mission of the aquaculture industry is to run a profitable business producing fish. NGOs, in contrast, are driven by their missions, which may be focused on conserving nature and biodiversity, protecting the marine environment, improving the quality of food, or creating a sustainable society. This "agenda" drives NGO decision-making, and every NGO will base their actions on their specific mission, thus leading each NGO to approach aquaculture from a different viewpoint. Increasingly, the aquaculture industry and others are recognizing the diversity of NGOs and that each organization addresses aquaculture's impacts through a unique combination of approaches.

A small NGO which focuses on preserving a single bay, fjord, or river may call for aquaculture production to cease in that specific area if it is seen as an immediate and significant threat to the local environment or livelihoods. A large NGO that works on aquaculture from a global perspective may be more likely to call for improvements in the industry because they are working in a larger landscape than a local NGO. These approaches can be complementary.

The majority of the impacts of aquaculture production are local, as are many of the economic benefits. However, the markets for the products are regional, national, and in many cases, international (Chapter 9). The market affects the way some NGOs approach the issues. In addition to working to strengthen regulations related to aquaculture, many use market-based approaches to improve the sustainability of the aquaculture industry. Many NGOs that work on aquaculture and on sustainable seafood as a whole are based in North America or Europe, and target producers and buyers through market pressures that allow them to influence aquaculture around the globe, not only in their base countries.

7.3.1 Identify Tools to Minimize Impacts

It is clear that aquaculture can have a number of impacts, both negative and positive, related to the environment, communities, and human health. It is also clear that aquaculture is here to stay, in one form or another. Many NGOs focus their efforts on improving the environmental and social performance of the aquaculture industry, though there is not always agreement among NGOs on how to achieve this goal.

Better management practices (BMPs) are one tool for reducing negative impacts and existing research has documented their efficacy in many cases. The implementation of BMPs can significantly decrease the negative impacts of production on many farms, though there must be monitoring to ensure a BMP is having the desired effect. Monitoring is especially important because BMP efficacy can vary significantly across farms. One concern with BMP schemes is that they are typically industry developed and implemented, and thus can be seen as the industry policing itself, especially where there are no clear monitoring measures in place.

Some BMPs are simple to implement, and they begin at the planning stages for a farm. Proper siting is a key factor in preventing and mitigating many environmental and social impacts across aquaculture. It has been suggested that for shrimp aquaculture, 90% of all impacts result from initial siting decisions (World Bank et al. 2002). As important as identifying appropriate production sites is recognizing where not to site farms. NGOs typically agree that highly sensitive ecological areas exist where aquaculture, and other activities, should not be allowed.

Many better management practices that minimize impacts are related to the efficiency of production, and therefore are economically beneficial to producers. A prime example of this is feeding practices. Feed conversion ratios for salmon farming have improved significantly over the last 20 years (Tacon 2005) and the same pattern holds for many species as producers experiment and the industry matures. On farm feed management, improved feeding techniques, and changes in feed formulations, allow producers to use less feed than they previously required to produce species such as salmon. This corresponds with less feed going uneaten and passing through the net-pens where it would impact the environment.

Social BMPs can have great impact and can also prove beneficial to companies financially. Costs of security guards for a shrimp farm can be high, and in Honduras over 25% of the costs of one farm were related to security (World Bank et al. 2002). Improved relations with local community would reduce this cost by reducing the need to protect against trespassing, vandalism, and theft.

BMPs are developed as producers experiment and identify better, more efficient ways of farming. This means that BMPs are most effective when farming established species, and that as producers move into different species they will need to develop different BMPs. For example, salmon net-pen technology is being used to farm cod, which have different behavioral patterns than salmon. Better management practices for preventing and mitigating escapes, as well as for feeding, will be need to be adjusted to take into account the physiological and behavioral differences between cod and salmon. Research has demonstrated that a cod is significantly more likely to escape from a net-pen than a salmon because they stay close to the nets and have been found to chew through the nets (Esmark et al. 2005). Yet NGOs such as WWF-Norway are concerned that adapting the technology will take too long, especially given evidence that escaped cod from farms in Norway could significantly impact already declining wild coastal cod stocks (Esmark et al. 2005). As another example, vaccines for new species may not be developed until the production of that species has proven viable and the industry is already well underway. For some species, producers are likely to use high levels of antibiotics until vaccines and better management practices to minimize stress on the fish are developed. Identifying these types of differences and developing solutions takes time, and NGOs are concerned that unacceptably high levels of impact will result from production of new species before better practices and technologies are developed.

7.3.2 Target Consumers and Communities

An educated consumer can alter their eating habits and perhaps those of a few of their friends or family. A number of NGOs work to shift the demand of seafood consumers in the U.S. and Europe. This is done through general education on the impacts of fisheries and aquaculture, seafood guides which highlight better and worse seafood choices, and campaigns. NGOs such as the Monterey Bay Aquarium in the U.S., the Marine Conservation Society in the UK, and the North Sea Foundation in the Netherlands produce pocket or wallet-sized seafood guides that consumers can use to make choices on seafood purchases, both wild and farmed. This provides consumers with information that is relatively simple and in a useful form.

Campaigns and guides that generalize the aquaculture industry play an important role in overall education. However, such work requires NGOs to aggregate information on producers and to base their recommendations on a broad brush stroke of the industry. This works better for fisheries, where a guide can highlight a specific population of a given species, and in specific cases for aquaculture, such as tuna fattening, where it is generally agreed that current production practices are unsustainable. The impacts of aquaculture production vary in part by species and production system, but depend heavily on farm management. The sustainability of a farm depends on management practices as well as the local or regional ecosystem. These generalized seafood guides play an important role in raising consumer awareness, but they are of limited utility as a guide for selecting farmed products since the practices used by the aquaculture industry vary greatly from farm to farm and region to region.

The percentage of consumers that can be reached through NGO activities varies from country to country, and European consumers tend to be most receptive to changing their purchasing habits based on such education. However, this is a niche market and the percentage of consumers who change their habits will always remain relatively low. Consumers can be overwhelmed and confused by the information they receive about seafood and aquaculture. Frequently, they know that there are sustainability issues related to seafood, but are not clear on specific facts and therefore do not know how to act on their concerns. This holds true for many seafood buyers, wholesalers, and chefs (Bridgespan 2005).

Many NGOs of varying size develop and distribute electronic or hard copy newsletters with the aim of educating the public about the environmental and social impacts of aquaculture. Aquaculture related information is often included in broader organizational newsletters that cover the entire range of interest of an NGO. Smaller local NGOs based in aquaculture producing regions often organize local communities through campaigns that identify impacts of the industry on them both as consumers of the product and residents of the production zone. They call on residents to take actions such as changing their purchasing habits and writing letters to their local government representatives. In regional and local efforts, tourism and fishing interests frequently align themselves with NGOs against aquaculture due to concerns that aquaculture may impact their livelihoods.

7.3.3 *Influence Retailers*

Increasingly, NGOs are working directly with retailers and other buyers to influence purchasing policies. The global food retail sector has, as part of a global industrial trend, consolidated over the last 20–30 years. For example, in the U.S. the top five grocery retailers controlled 19% of the market in 1992 and by 2005 their market share was almost 50% (Konefal 2006). An educated retailer has the capacity to make sourcing and marketing decisions that influence hundreds of thousands or even millions of consumers. Additionally, retailers can send strong signals to seafood producers, including the aquaculture industry without necessarily educating consumers or burdening them with more information and choices.

The role of retailers in the sustainable seafood market has grown considerably. Retailers can influence production wherever they source their products. With consolidation, their role in the supply chain has shifted and they are now able to set prices, production methods, and quality standards. They are increasingly developing private standards, labels, and product lines (Konefal 2006). In some European countries, retailers even set requirements on the percentages of fishmeal and fish oil in the feed used in the salmon farms from which they source (Tacon 2005). NGOs have recognized this new role of retailers and increasingly are working with them to influence the production practices of the aquaculture industry.

Some NGOs have conducted general assessments of supermarket seafood policies and sales. For example, Greenpeace and the Marine Conservation Society in the UK have both developed rankings of supermarkets, which can pressure retailers to examine their policies as well as help consumers make decisions about where to shop. Other NGOs have vocally campaigned to try to influence specific retailers to change their purchasing policies. For example, the member NGOs of the Canadian Alliance for Aquaculture Reform (CAAR) targeted Safeway Canada for their sales of farmed salmon through billboards, advertisements, and demonstrations.

A number of NGOs work directly with retailers. One-on-one relationships between NGOs and retailers allow NGOs to advise seafood purchasers not only on the larger picture of which species tend to be more sustainable, but also to provide advice on the development of purchasing policies. For example, in early 2006 Wegmans Food Markets, a regional U.S. retailer, announced new environmental and health purchasing standards for farmed salmon that were developed in collaboration with Environmental Defense, an NGO based in New York. A former Wegman's supplier, Marine Harvest Canada, also participated in the project. Wegmans and Environmental Defense are now working with Wegmans' current farmed salmon suppliers to meet these standards. Similarly, NGOs collaborate with processors and distributors to strengthen their sourcing policies. For example, Oxfam and IUCN in the Netherlands are working with Heiploeg, a European shrimp processor, to improve the social and environmental components of their internal policy for sourcing shrimp products.

Such NGO activities can complement general consumer education by going one step further and singling out environmentally and socially friendly producers of a wide range of species. These producers in return benefit from enhanced relationships with their customers, potential access to new markets and, in some cases, prices premiums. This approach, however, has some drawbacks. Retailer purchasing policies developed in coordination with NGO and industry partners are most often developed in private. The groups developing the policies may or may not consult with a range of interested parties including scientists, communities, and other NGOs. The focus and strength of the standards vary, as do the methods for auditing and enforcing the implementation of the standards. Moreover, it remains unclear whether these relationships have the ability to push the aquaculture industry as a whole to more sustainable production. To be transformative, a handful of small retailers purchasing from producers that are already environmentally and socially responsible will not be sufficient. Leading retailers and buyers will need to implement strong and enforceable purchasing policies, encourage more retailers to take action, and partner with NGOs, producers, extension agencies, and others to measurably improve the performance of the industry.

7.3.4 Influence Governments

Most NGOs use a range of approaches in their aquaculture work, and a number of NGOs work to influence government policies and regulations. Strong and targeted regulation, when effectively enforced, can ensure that aquaculture producers meet a minimum environmental and social performance level. Government officials in the U.S., the U.K., Canada, and other countries maintain regular communication with environmental groups about aquaculture related issues, but the formality of the role of NGOs in governments varies widely among NGOs and countries. In the U.S., a representative from Environmental Defense served on the U.S. Department of Agriculture National Organic Program's Aquaculture Working Group, which drafted standards for farmed aquatic foods in 2007. Representatives from Canada's Department of Fisheries and Oceans are regular participants in the WWF-initiated Salmon Aquaculture Dialogue. The Marine Conservation Society consults formally and informally on government policies in the U.K. In Norway, WWF-Norway is appointed to sit on the Governmental Commission to reduce escapes. Additionally, there are a host of NGOs that work with country governments to develop protected or aquaculture free areas both on land and in the oceans. IUCN and WWF participate on the UN Food and Agriculture Organization's Committee on Fisheries (COFI) which is composed of representatives from the member country governments. Both WWF and IUCN work to influence the COFI and subsequent changes in member country policies. Although regular contact and communication exists among governments and NGOs, the degree of influence NGOs have on government decision-making varies widely.

7.3.5 Conduct Research

Many NGOs conduct or commission aquaculture related research in order to clarify impacts, explore alternative technologies, identify better producers, and inform programmatic strategies. For example, Pendleton et al. (2005) conducted an analysis of the economic viability of closed containment production systems for salmon culture for CAAR. CAAR is one of a number of NGOs that promote closed containment technology as a means of eliminating or minimizing the key impacts of salmon aquaculture. An understanding of the economic costs and benefits of such systems better enables NGOs to debate their use. As another example, WWF-US and partners are conducting farm-by-farm surveys of catfish producers in Alabama and shrimp producers in Madagascar and Belize to collect baseline data on production practices and impacts in order to then develop standards for production.

NGOs will partner with other organizations to identify research priorities, and to commission research. One example of this is the World Bank/FAO/NACA/WWF-US Consortium on Shrimp Farming and the Environment. The organizations collaborated in order to identify gaps in information related to the environmental and social impacts of shrimp culture. The Consortium then funded research and case studies to fill these gaps, resulting in over 35 original reports.

NGOs also support, communicate, and partner with independent scientists conducting research that is related to the impacts of aquaculture. Scientists from universities around the world are conducting research from which NGOs will draw information. Stakeholders in WWF-initiated Aquaculture Dialogues have jointly commissioned independent reviews of impacts and the drafting of standards for specific species. In another example, marine biologists from the Fundación Huinay research station are documenting the oceanographic conditions and benthic communities in several fjords in southern Chile. NGOs, governments, and industry can then use this baseline data to help identify types and extent of impacts of aquaculture.

7.3.6 Negotiate and Partner with Industry

NGOs and industry representatives can also work together to reach mutually beneficial arrangements. NGOs try to enlist support from the aquaculture industry for conservation efforts such as marine protected areas, coastal zone management, and improved management of reduction fisheries. In 2003 in Norway, a major aquaculture producer issued a public corporate statement supporting the creation of marine protected areas. The statement helped WWF-Norway influence parliament to pass related legislation and later that year the Norwegian government prohibited salmon aquaculture in 13 rivers in order to protect wild Atlantic salmon populations (Esmark personal Communication; Porter 2003). Both parties have something to gain from such initiatives. In the case of reduction fisheries management, NGOs are aiming to prevent overfishing and influence governmental decisions while the industry is seeking to ensure sustainable sources of fishmeal and fish oil well into the future.

NGOs and industry representatives can sign memorandums of understanding or other agreements that benefit both organizations. For example, CAAR recently signed a Framework for Dialogue with Marine Harvest Canada in which they agreed to jointly conduct further research on both closed containment production and sea lice impacts. Marine Harvest also agreed to remove salmon from a farm located on a migratory wild salmon route and where impacts of the farm on the migrating salmon have been contentiously debated. Consequently, CAAR agreed not to specifically target Marine Harvest in their campaigns. Agreements such as this have the potential to result in mutually beneficial long-term solutions.

Collaboration between industry and NGOs can include more than one company and one NGO. As detailed below, WWF-US is working together with industry and other stakeholders to agree on key impacts and ways to reduce their impacts to acceptable levels. Participants in the WWF initiated-aquaculture dialogues have agreed to mutually beneficial goals and objectives as a basis for working together.

7.4 WWF-US and Aquaculture

WWF-US (hereafter WWF) first began to focus on aquaculture in the 1990s. In 1994, WWF supported a research project to compare the environmental impacts of shrimp aquaculture and shrimp trawling. Through this work, it was decided that while neither was sustainable, shrimp farming had the potential to become more sustainable through technological innovation and improved management strategies. The main recommendation from the study was that WWF identify strategies to both reduce the major impacts of the shrimp aquaculture industry and engage shrimp producers and governments alike in a productive dialogue. To carry out this work, WWF joined with the World Bank, the Food and Agriculture Organization of the United Nations, and the Network of Aquaculture Centres in Asia Pacific to form the Consortium on Shrimp Farming and the Environment. The work of the Consortium, from 1999–2002, generated analyses of the negative impacts of shrimp aquaculture, identified better practices around the world to reduce such impacts while ensuring financial viability, and demonstrated the financial business case for the use of better practices. Through consultation with a broad range of stakeholders the Consortium created a consensus on the key impacts of shrimp culture and their analyses became regarded as the most credible data on these impacts to date. The information was used to develop the International Principles for Responsible Shrimp Farming and is currently being used as the basis for the development of performance based standards for responsible shrimp farming.

7.4.1 Collaboration

Learning from their early experiences with the Consortium, WWF continues to focus on collaborating with producers and other stakeholders to develop voluntary measures that are appropriate globally and which, when implemented, move the aquaculture industry toward sustainability. WWF's mission is the conservation of nature around the globe, using the best available scientific knowledge to preserve the diversity and abundance of life on Earth and the health of ecological systems. As a large international conservation organization, WWF works on a broad range of environmental issues and can do so from a broad perspective. Helping to ensure that human needs are met in harmony with nature, WWF work's includes a focus on minimizing the negative environmental impacts of food production systems.

Aquaculture is here to stay, and some forms of it are already broadly recognized as being responsible. WWF focuses on more sustainable species such as molluscs and seaweeds as well as more controversial species like shrimp and salmon. The approach attempts to encourage both the production of more sustainable species and the use of better practices for all species and production systems.

7.4.2 Complementing Other Initiatives and Regulations

WWF aims to shift the entire aquaculture industry to more sustainable production. For all production systems and species produced, there exists a range of environmental performance from producers. There will always be better and worse producers of any crop from an environmental perspective. In Fig. 7.3, a theoretical bell curve of producers demonstrates that a small percentage of producers tend to have the worst environmental performance levels, a small percentage have high environmental performance, and the large majority of producers fall somewhere in the middle.

Governments aim to prevent the worst production practices and most severe impacts through regulation. Government policies, regulations, and enforcement play a critical role in protecting the environment through maintaining a minimum standard for production.

More traditional conservation approaches such as the designation of protected areas also play a critical role in ensuring the sustainability of aquaculture and other human activities. Land-based or marine protected areas frequently come about from

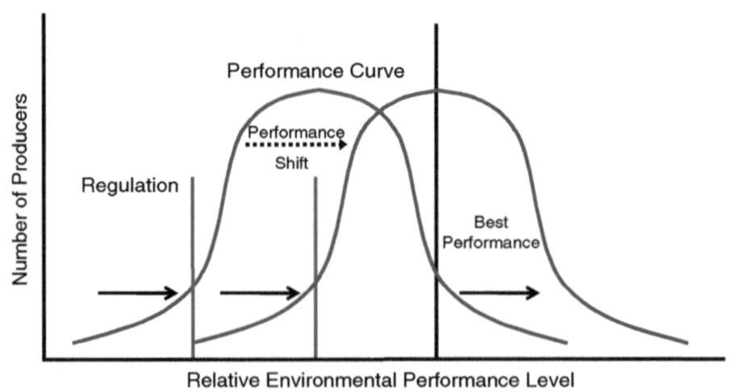

Fig. 7.3 Accelerating adoption of better practices

the collaboration of NGOs and other stakeholders with governments. Marine protected areas are an important tool for coastal zone management and the designation of areas where aquaculture is prohibited can ensure the protection of sensitive ecological areas such as breeding grounds and migration routes for wild fish or sea mammals.

Thoughtful coastal zone planning, undertaken by governments in collaboration with stakeholders, can help to ensure not only that critical habitats are protected, but that fish farms are sited in the most appropriate areas: the existence of "no-go" areas for aquaculture implies that there are also "go" areas. Producers should have a strong incentive to ensure that farms are appropriately sited. WWF's work with agricultural producers has shown that farming marginal areas is not only bad for the environment, but a financial loss for producers. There is extensive anecdotal and some empirical evidence that shows that there is a business case for taking these marginal areas out of production (Clay 2004; Fransen 2005). It is believed the same will hold true for aquaculture.

Voluntary measures can complement regulation by encouraging innovation and the best production practices. WWF believes that through collaboration and the implementation of voluntary changes in the industry, the entire range of producers can be shifted forward to a better performance level. The better performance of today will be the average performance level in the future. When the performance curve is shifted forward, regulation can then be updated, strengthened, and shifted as well.

7.4.3 Standard Development

WWF is convening multi-stakeholder dialogues to develop standards for the responsible production of over a dozen aquaculture species. These standards can then be used in a variety of ways, including producer or buyer screens, as part of government permitting processes, and as the basis for a certification program or ecolabel.

Numerous standards and codes of conduct for aquaculture production have been developed in recent years. These include industry developed codes and standards, retailer purchasing standards, chain of custody standards and organic production standards. This proliferation of standards as well as their complexity has become problematic in today's market. It can be burdensome on producers and confusing to consumers.

By way of example, WWF found that some banana producers in Central America are certified and audited by up to 8 different certification programs in order to sell into different markets. Each of the 8 signifies paperwork, audits, and in most cases a financial cost to the producer. Half of the items audited are required by multiple certification programs but still must be undertaken each time and the harmonization of programs could reduce costs significantly.

New standards continue to be developed, in part because retailers or producers are looking for a way to distinguish their products, and in part because those developing the standards are not satisfied with the existing standards. This holds true for

WWF, which has a short list of criteria on the characteristics of a credible standard. WWF believes that standards should be measurable rather than prescriptive. Measurable standards not only ensure that performance improvements can be demonstrated against a baseline, but also allow producers to meet the standards using a wide variety of practices, encouraging innovation. The flexibility in deciding how best to meet a quantitative standard is especially important for small farmers who may not have access to certain technologies. No eco-label or organic standard exists today for aquaculture that focuses on measurable rather than prescriptive standards for key impacts.

WWF believes that standards should be targeted to focus on key environmental and social impacts while providing the traceability necessary to ensure food safety, food quality, and chain of custody. Though there are numerous potential impacts of aquaculture production, a very small number are likely to be responsible for the majority of the environmental and social concerns. The Consortium on Shrimp Farming and the Environment found that 6 – 10 key impacts accounted for most of the concerns related to the global shrimp farming industry. Any given farm might have 3–4 activities that account for the majority of the impact of that operation (World Bank et al. 2002). The Consortium also found that for over two-thirds of these impacts, implementing better management practices and mitigating the impacts could pay for themselves within three years. Identifying these key impacts, determining ways producers have found to reduce them, and focusing standards on the handful of key impacts of production can in most cases significantly improve the sustainability of a farm and reduce impacts to acceptable levels.

Finally, a credible standard must have broad stakeholder support, must have been developed in a transparent manner. WWF is developing standards in collaboration with a large number of partners through international, multi-stakeholder Aquaculture Dialogues. Each Aquaculture Dialogue is focused on a species group and works to agree upon key impacts of production for that species group, identify better management practices and key gaps in the research, and develop metric-based standards for the production of the species-group that are acceptable to stakeholders. A broad range of stakeholders, including producers, NGOs, researchers, governments, retailers and other members of the market chain are involved in each of the Dialogues.

It is relatively easy for stakeholders to come to agreement on standards for cultured species that do not depend on food supplements, use little by way of fishmeal or fish oil, or are already highly regulated with regard to effluent pollution. It is much harder to reach consensus for carnivorous species cultured in open systems and species which rely on wild stocks for juveniles. For such species there is more contention on where to draw the line on impacts.

Generally speaking, there exists tremendous debate around agricultural, fisheries, and aquaculture related eco-labels regarding the strength of the standards that form the basis of these labels. Groups that support the strictest of standards are concerned that anything less would allow unsustainable production to receive an eco-label. Groups that support less stringent standards, with a requirement for continuous improvement, are concerned that if standards are too strict, producers will not be motivated to improve in order to meet the standards. Organics and eco-labels have traditionally been niche markets, but a growing number of NGOs are looking to these tools to influence the industry as a whole, not simply the top 1% of producers.

7.4.4 Certification

WWF wants to see the transformation of the aquaculture industry. The organization believes that independent, third-party certification can play an important role in this transformation. This is a strategy that has been implemented for a range of industries including forestry, fisheries, and agriculture. WWF was a key player in the creation and implementation of the Forest Stewardship Council, the Marine Stewardship Council, Protected Harvest, and the Marine Aquarium Council.

Product certification provides a label that informs the buyer of product qualities and complements other information on a label regarding price, quantity, identity, government standards, health and nutritional information, and producer name. An eco-label typically certifies qualities that may not be readily apparent to the buyer, such as product safety, the process of production, the place of origin, and the social and environmental qualities of the product and production process.

Through purchasing labeled product, buyers, be they distributors or end-consumers, express preferences such as their beliefs and personal ethics on health and safety risks, treatment of workers, and environmental stewardship. An independent and recognized eco-label will provide consumers with a product that they trust to be sustainable without requiring them to remember details about better and worse choices. In exercising choice of certified products over non-certified products, the buyer rewards certain producers and creates a market demand that provides incentives for an industry to adopt preferred production practices, materials, and processes.

Certification complements other strategies to improve the environmental, social, and food safety performance of the aquaculture industry. Social action and activism, mandatory or command and control regulatory systems, or first- or second-party certification can each play a role in improving overall industry performance. To the extent that independent, performance-based certification programs are successful they can provide a more effective option than any of the above because they provide a credible, concrete alternative by identifying the best of the industry and using that benchmark as the basis for change. Aquaculture producers throughout the world producing a wide range of different products have publicly stated that they want certification programs that will help them gain market advantage.

WWF believes that certification is a unique and complementary strategy to drive economic, social, and environmental improvements in the aquaculture industry. Just as the standard development process needs to be credible, an aquaculture certification body must be independent and transparent. Aquaculture production should not be certified by WWF nor by industry interests, but by independent, third-party certifiers. However, given the numerous certification programs currently in existence, such an entity would require a critical mass of supporters across a wide variety of stakeholders to be financially viable over the long term.

7.5 Trends, Obstacles and Uncertainties for NGO Activities

A number of trends, obstacles, and uncertainties influence NGOs working on aquaculture. The aquaculture industry is dynamic, yet forming working relationships between NGOs and industry and then addressing key issues are not short term projects. The consolidations, mergers, and bankruptcies of producers, feed manufacturers, or retailers can make it difficult for NGOs to maintain long-term relationships and agreements with companies.

The majority of the NGOs that work on aquaculture are small, local organizations. There are relatively few international groups working globally. And with the exception of seafood guides which identify ecologically friendly choices, NGO work focuses on the segments of the aquaculture industry where there are the greatest perceived impacts. This is reflected above in discussions of NGO activities related to aquaculture, which most frequently focus on shrimp and salmon. When WWF began their Mollusc Aquaculture Dialogue in 2004, it rapidly became clear that only a few NGOs had thought about the sustainability of these systems or had the time and funding to do so. In order to develop independent, industry and NGO supported standards for mollusc production, WWF will need to work with the NGO community to identify any locally specific negative impacts of these production systems.

NGO ability to address issues is dependent on funding, which can be unpredictable. Funding from foundations is primarily focused on shorter term projects lasting no more than two or three years. Yet the most meaningful and credible relationships with industry, retailers, or other partners often take several years to develop. Industry partners are another potential funding source, and companies often provide some support to NGOs with whom they partner. However, receiving major financial support from companies can compromise credibility of a project or NGO, especially in developing countries.

A lack of data can be an obstacle for both NGOs and industry, and interpretations of data can be problematic when these and other stakeholders work together. Generally speaking, there exists a large body of scientific literature related to the environment and aquaculture. For some geographic regions, such as China, or for some issues, such as many social impacts, there are gaps or contradictions in the information, leading to uncertainty regarding trends and the extent of impacts.

In cases where there is abundant data, different parties interpret data in different ways. Additionally, the science can be quite controversial. On some topics there is a perception or reality that the literature is biased either towards industry interests or NGO beliefs. In the WWF-initiated Salmon Aquaculture Dialogue, stakeholders commissioned state of information reports completed by teams of scientists that they have agreed are independent and unbiased. These reports then serve as a shared basis of knowledge from which they can move forward.

Finally, there are two large uncertainties related to aquaculture. With over 60% of world production, China is the leading aquaculture producing country in the world, and production is increasing (Johnson, 2005). The direction China takes in terms of both species produced and production practices will have extraordinary impacts on fishmeal and fish oil demand, seafood markets, and the environment. Further analyses and projections of the future of aquaculture in China would give some insight into the global implications of China's actions. A second large uncertainty which will influence the future of aquaculture and its impacts is climate change. Climate change is predicted to significantly alter sea levels, water and air temperatures, ocean currents and salinity, and increase the frequency and severity of extreme weather events. The implications of these changes for coastal and marine aquaculture are unclear, but these factors are critical in determining where species will be produced and will also play an important role in disease and parasite control and transmission.

7.6 Conclusions and Looking Forward

Many NGOs are looking for the aquaculture industry to adopt a truly environmentally and socially centered approach to addressing the impacts of aquaculture production. Ecological systems are complex and the extent of certain impacts may remain unclear until the effects are extensive and irreversible. Both industry and NGOs can examine the past for lessons learned. By way of example, exotic species have been purposefully or accidentally introduced around the world, sometimes with dramatic ecological and economic impact (Chapter 5). When farmed species escape, there is a small chance for a disastrous outcome. If the escapees are exotic and become established, their presence can dramatically alter the ecosystem balance. If the escapees are native, they can interbreed, compete with, and decimate a small wild population (Chapter 4). In aquaculture, the role of the NGO has frequently been to identify this type of worst case scenario and work to reduce the chances of worst happening. In addition to maintaining this watchdog role, NGOs should continue to become more proactive in identifying solutions and opportunities for more sustainable aquaculture.

Despite increasing instances of collaboration, there exists significant mistrust between NGOs and industry. Though this is natural to some extent, high levels of mistrust and animosity will impede the road to a more responsibly managed industry. Collaborative work on shrimp through the Consortium on Shrimp Farming on the Environment proved that industry, NGOs, and other stakeholders were for the most part able to work past the mistrust and contention in order to achieve a common goal. The WWF-initiated Aquaculture Dialogues have also demonstrated that it is possible for a wide range of industry, NGO, government, and market chain representatives to sit together and candidly discuss issues. Not all NGOs and industry players will choose to work together, but many do, even though an acceptable resolution cannot be guaranteed.

A more sustainable aquaculture industry could exist in the view of the NGO community at large. In addition to taking steps to minimize a range of impact areas including escapes, disease, pollution, and chemicals, most NGOs feel it is important for the industry to move away from the production of carnivores or at least the dependence on fishmeal and fish oil and to halt the fattening of species from depleted wild stocks such as tuna and eel. The industry would also need to keep a watchful eye on human health and social impacts, which would require innovation and the adoption of new technologies. As the industry adopts change, NGOs and industry will need to ensure that we are not creating a new problem each time we solve an old one. Producers would need to both obey the law and go beyond it in some countries using voluntary codes practices. Multinational companies would be asked to meet the same high standards no matter where they were producing. The industry would need to demonstrably benefit communities. Examples of these production systems exist around the world, and together governments, industry, and NGOs can shift the industry so that these better systems are the norm and not the exception.

References

Black KD, Cook EJ, Jones KJ, Kelly MS, Leakey RJ, Nickell TD, Sayer MDJ, Tett P, Willis K (2002) Review and synthesis of the environmental impacts of aquaculture. Report for the Scottish Executive Central Research Unit

Boyd CE, McNevin AA, Clay J Johnson HM (2005) Certification issues for some common aquaculture species. Reviews in Fisheries Science 13:231–279

Bridgespan Group (2005) Seafood choices initiative evaluation. Prepared for the David & Lucile Packard Foundation

Clay J (2004) World Agriculture and the Environment: a commodity-by-commodity guide to impacts and practices. World Wildlife Fund and Island Press, Washington DC

Dekker W (2007) Eel stocks dangerously close to collapse. ICES Article. Downloaded from http://www.ices.dk/marineworld/eel.asp May 28, 2007

Esmark M, Stensland S, Synnove Lilleeng M (2005) On the run—escaped farmed fish in Norwegian waters. WWF-Norway, Oslo

Fransen T (2005) The economic and environmental case against farming marginal land in commercial agriculture operations. In: Clay J, Rajbhandari B (eds), Agriculture and the Environment Volume II, Thematic Issues: A WWF Handbook on Agriculture Policies and Issues. World Wildlife Fund, Washington DC

FAO (Food and Agriculture Organization of the United Nations) (2004) The State of World Fisheries and Aquaculture 2004. FAO, Rome

FAO (2007a) The State of World Fisheries and Aquacutlure 2006. FAO, Rome

FAO (2007b) FISHSTAT Plus. UN Food and Agriculture Organization, Fisheries Aquaculture Information and Statistics Service

Hansen LP, Fiske P, Holm M, Jensen AJ, og Sægrov H (2006) Bestandsstatus for laks. Rapport fra arbeidsgruppe. Utredning for DN 2006-3:48 sider

ICES (2006) Report of the ICES Advisory Committee on Fishery Management., Advisory Committee on the Marine Environment and Advisory Committee on Ecosystems, 2006. ICES Advice Books 1–10:6, 310 pp

Johnson HM (2005) Annual Report on the United States Seafood Industry, Thirteenth Edition. H.M. Johnson & Associates, USA

Konefal J (2006) Protecting the Oceans and Marine Life via the Market and Consumption: The Sustainable Seafood Movement. Submitted for the Sustainable Consumption and Society Conference, Univ. Wisconsin, 24 pp

Naylor R, Goldburg R, Primavera J, Kautsky N, Beveridge M, Clay J, Folke C, Lubchenco J, Mooney H, Troell M (2000) Effect of aquaculture on world fish supplies. Nature 405:1017–1024

Naylor R, Burke M (2005) Aquaculture and Ocean Resources: Raising Tigers of the Sea. Annual Review of Environment and Resources 30:185–218

Pendleton, L, Whitney J, Bonine K, Schmelz K, Ritchlin J, Rothenbush T, Wahlberg L (2005) Closing in on Environmentally Sound Salmon Aquaculture: a fresh look at the economics of closed tank systems. Produced for the Coastal Alliance of Aquaculture Reform

Pinto F, Kremerman M (2005) Cultivando Pobreza: Condiciones laborales en la salmonicultura. Fundación Terram, Santiago, Registro de Problemas Públicos 18

Porter G (2003) Protecting Wild Atlantic Salmon from Impacts of Salmon Aquaculture:A Country-by-country Progress Report. World Wildlife Fund, Washington DC and Atlantic Salmon Federation, St. Andrews, NB Canada

Ruckes E (2003) World fish trade demand forecasts and regulatory framework. In: FAO (2004) Report of the Expert Consultation on International Fish Trade, Rio de Janeiro, Brazil, 3–5 December 2003. *FAO Fisheries Report*. No. 744. Rome, FAO, pp 27–36

Shumway S, Davis C, Downey R, Karney R, Kraeuter J, Parsons J, Rheault R, Wikfors G (2003) Shellfish aquaculture—in praise of sustainable economies and environments. World Aquaculture 34(4):15–18

Tacon GJ (2005) State of information on salmon aquaculture feed and the environment. WWF report, New York, pp 80

Tudela S, García R (2004) Tuna farming in the Mediterranean: the bluefin tuna stock at stake. World Wide Fund for Nature, Spain

World Bank, NACA, WWF, FAO (2002) Shrimp Farming and the Environment, A World Bank, NACA, WWF and FAO Consortium Program: To analyze and share experiences on the better management of shrimp aquaculture in coastal areas. Synthesis report. Consortium

World Resources Institute, IUCN-The World Conservation Union, United Nations Environment Programme (UNEP) in consultation with the Food and Agriculture Organization (FAO) and the United Nations Education, Scientific and Cultural Organization (UNESCO) (1992) Global Biodiversity Strategy: Guidelines for action to save, study and use Earth's biotic wealth sustainably and equitably. WRI, IUCN, and UNEP, 260 pp

Chapter 8
Aquaculture in the Coastal Zone: Pressures, Interactions and Externalities

David Whitmarsh[1] and Maria Giovanna Palmieri[2]

Abstract This chapter focusses on the way aquaculture interacts with other users of the coastal zone, and specifically the side effects (externalities) of these interactions. The DPSIR (Driving forces – Pressure – State – Impact – Response) paradigm is used here to explore the externalities problem and to suggest policy solutions. Two aspects of the problem are considered, on the one hand those externalities generated by coastal zone activities that affect aquaculture, and on the other those originating from aquaculture itself. Monetary valuation is one way to assess externalities, and an example is provided for shrimp farming in mangroves (see appendix). The benefits and risks of different strategies for policy solutions based on assessments of externalities are discussed.

Keywords Socio-economic indicators, externalities, market failure, non-marketed goods, property rights

8.1 Introduction

Aquaculture worldwide has grown rapidly, and while this has undoubtedly brought benefits in the form of increased food supplies and employment creation it has also been matched by concern over its environmental impact and sustainability. Expansion of marine aquaculture is seen as especially problematic, not least because it has to compete for resources and space with other coastal activities, and the scope for conflict generated by this growing pressure is thus considerable (Bailly and Paquotte 1996). The focus of this chapter is on the way aquaculture interacts with other users of the coastal zone, and specifically with the question of

[1]CEMARE, University of Portsmouth, Burnaby Terrace, 1-8 Burnaby Road, Portsmouth, PO1 3AE, UK. Tel: +44 (0)2392 844084; E-mail: David.Whitmarsh@port.ac.uk

[2]CEMARE, University of Portsmouth, Burnaby Terrace, 1-8 Burnaby Road, Portsmouth, PO1 3AE, UK. Tel: +44 (0)2392 844036; E-mail: Gianna.Palmieri@port.ac.uk

M. Holmer et al. (eds.), *Aquaculture in the Ecosystem.*

externalities – the side effect of actions by individuals and firms that impact on the well-being of others. The DPSIR (Driving forces – Pressure – State – Impact – Response) paradigm, which has been applied to a range of issues concerning the sustainability of coastal resources (Ledoux and Turner 2002), is used here to explore the externalities problem and to suggest policy solutions. In formulating policy it needs to be understood that marine aquaculture not only creates externalities but is also the 'victim' of external costs generated elsewhere by other activities in the coastal zone. Indeed, the fact that externalities associated with the exploitation of marine and coastal resources typically arise from multiple activities, amongst which aquaculture is but one, provides a strong rationale for integrated approaches to coastal zone management. We therefore start by outlining the concept of externalities, specifically the use of monetary valuation, and this is followed by a review of the empirical evidence as it relates to aquaculture.

8.2 Externalities Caused by Marine Environmental Disturbance: An Overview

The marine environment provides goods and services, which support economic activities and the welfare of individuals directly. These resources include commodities such as fish and raw materials, and services ranging from nutrient cycling, disturbance regulation, and biological control, to recreational and cultural services. Many of these resources are unmarketed, which means that no property rights are assigned over their use and that there are no markets reflecting their scarcity. This implies that users can make resources scarce for others but this is not reflected in any change in their cost of access to the resources. The price mechanism has therefore failed in one of its basic functions, which is to signal to society the real value of resources and the services they supply. Externalities are the symptom of this market failure. In economic theory a negative externality is said to occur when the production or consumption decisions of an economic agent have an unintended adverse impact on the utility or profit of a third party, and the generator of the impact offers no compensation to the affected party (Perman et al. 2003). What happens is that some of the costs of private production or consumption decisions are 'external' to the economic agents making those decisions and, consequently, are not taken into account in their decision process. Externalities may affect a production activity by modifying the efficiency of the production process and consequently its profitability, or affect the satisfaction (utility) of a consumer. In the case of the marine environment, for example, an oil spill may reduce the yield for fishermen and fish farmers and the enjoyment of the marine landscape by visitors. Negative externalities imply that the social cost of an economic activity will be greater than the private cost, a fact which provides the rationale for environmental control measures such as taxes and charges that attempt to 'internalise' such externalities.

One way to account for the impact on human welfare of changes in the quantity or quality of environmental assets is to assign a monetary value to them. Many criticisms have arisen among those who consider such an approach unethical, but this is to miss the point. Monetary valuation is essentially a means of measuring the social importance of the marine environment in a way that enables choices to be made. To take an aquaculture example, deciding whether or not to impose (say) an effluent charge on salmon growers as a way of 'internalising' the externalities of nutrient release logically requires the decision-maker to know what economic costs such emissions impose on society. If it were shown that nitrate and phosphate release from fish farms represented a significant external cost to society, the legitimacy of imposing such a measure on the polluters would be upheld. Conversely, if the external cost were estimated to be trivial, the legitimacy of imposing an effluent charge would be called into question. Monetary valuation is thus a way of bringing the environment into the reckoning of cost–benefit analysis, which may then be used by decision-makers to evaluate policy options in a rational and consistent manner. This raises the question of how easy it is in practice to estimate such values, and what progress has been made in valuing the environment. To date, empirical studies related to the valuation of marine resources have focused primarily on wetlands or on particular resources (such as mangrove and coral) adjacent to the coast, and only rarely on ocean resources. A study by Costanza et al. (1997) on the value of the world's ecosystem services and natural capital estimates the value of marine systems to be US$20.9 trillion annually (63% of total ecosystem services and natural capital) within which coastal systems contribute some 50%. Although these are quite crude estimates and the authors themselves point out the limitations of their study, what is clear is that marine resources have an economic value, which can be affected by environmental disturbance of anthropogenic origin.

8.3 Interaction between Aquaculture and the Marine Environment

Aquaculture is affected by the externalities created by other activities, and is itself a contributor to such externalities. In this section we look more closely at these interactions and their socio-economic significance. The key linkages are illustrated in Fig. 8.1, which is based on the DPSIR framework. We may start by considering the operating performance of commercial aquaculture in terms of a number of standard indicators (e.g., productivity, costs, prices and profits), which for a given production system are determined by the prevailing technology and market conditions. If these conditions alter, which they will if any of the economic drivers (e.g., subsidies to aquaculture) change, production will adjust as firms respond to new opportunities and pressures. Figure 8.1 also suggests that operating performance of fish farms may be impacted by various types of perturbation, which can be thought of as a particular type of external driving force, and given what we know of their

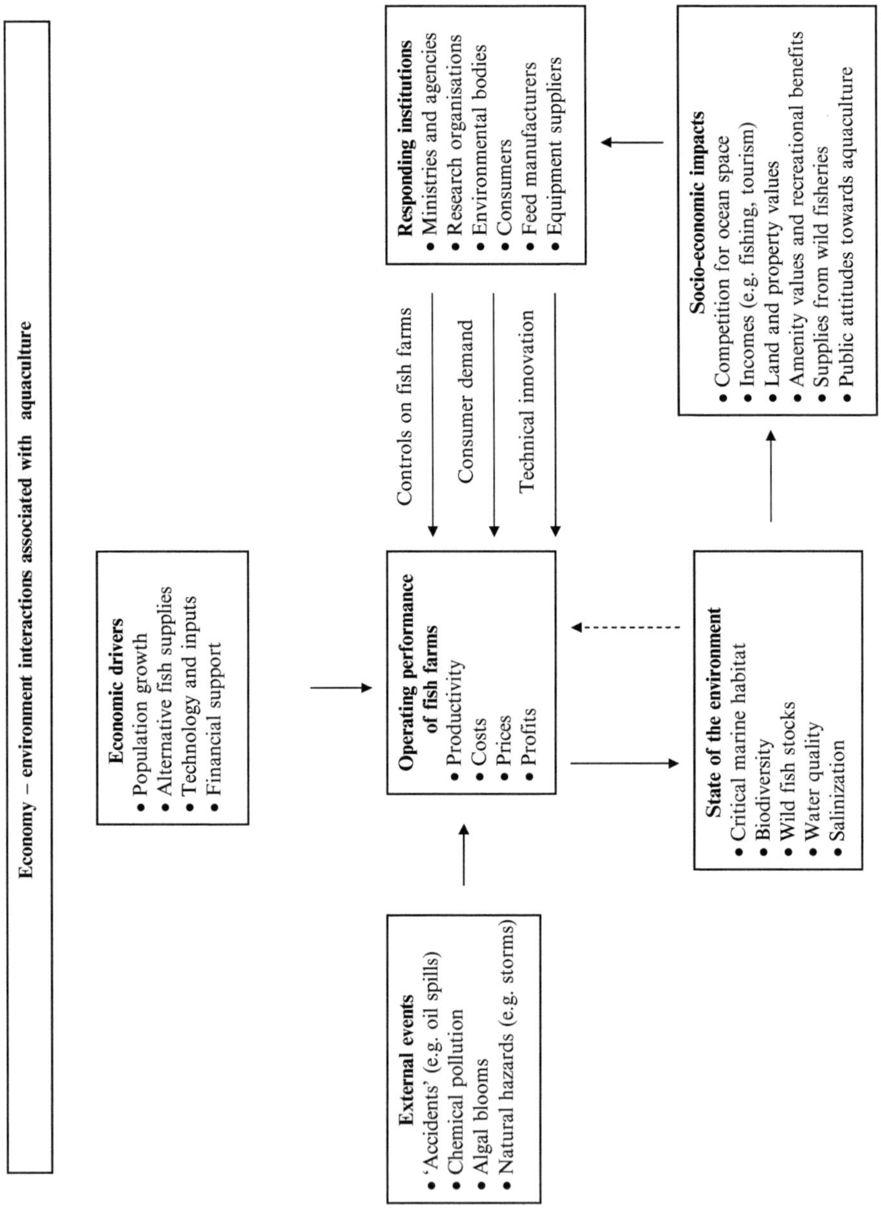

Fig. 8.1 Economy – environment interactions associated with aquaculture

economic significance for aquaculture it seems appropriate to start our discussion with these.

8.3.1 Impacts on Aquaculture of External Events

Some external events can be classed as either totally or partially anthropogenic (e.g., acute or chronic pollution, 'red tides' linked with eutrophication), while others are caused by natural hazards (e.g., storms). Our concern is mainly with the former, since these correspond most closely with the definition of externalities given earlier; that is, unintended side effects of human activity which have an economic impact on third parties. However, experience suggests that natural hazards may be just as financially ruinous to individual fish farmers as those that are anthropogenic, and within the DPSIR framework it seems sensible to consider both types of event. The focus of discussion in this section is on specific examples of environmental disturbance where a direct connection can be made with the operating performance of fish farms. Changes which might impact indirectly on aquaculture, such as global warming, are not considered though their potential importance is recognised.

Pollution incidents have impacted on aquaculture in a number of documented cases. The *Amoco Cadiz* oil spill tanker disaster in 1978 caused serious losses to the Brittany oyster farming industry, estimated at 107 million FF (= US $26 million) at 1978 equivalent values (Grigalunas et al. 1986), while the more recent *Prestige* incident in 2002/3 is reckoned to have led to a drop in the annual production value of Galician aquaculture (mussels and turbot) of approximately 9 million euros (Garza-Gil et al. 2006). As well as acute pollution incidents such as oil spills, chronic pollution has also been shown to be damaging to aquaculture. This is well illustrated by the effects of organotins on shellfish, a particularly noteworthy case being the culture of oyster (*Crassostrea gigas*) in Arcachon Bay (France) where reproductive failure and shell deformation caused by the presence of TBT resulted in very heavy financial losses by the shellfish industry in the early 1980s (Santillo et al. 2001). Environmental disturbances not directly linked with pollution have also been shown to have serious consequences for aquaculture, in some cases with wider implications for society at large. In the US, harmful algal blooms (HABs) are estimated to have caused losses to commercial fisheries (capture and culture) of between $14m. and $26m. annually for the period 1987–1992 (Anderson et al. 2000), with some notably severe incidents affecting farmed salmon in the state of Washington. In one year alone, HABs led to massive kills of net pen salmon in northern Puget Sound and caused losses in production value of approximately $11m. In the UK, the infectious salmon anaemia (ISA) virus was estimated to have caused a drop in turnover for Scottish salmon growers in the late 1990s of some 14%, due to a combination of enforced slaughter of stock and depressed prices (Scottish Parliament Information Centre 1999). Jobs in the industry were lost directly as a result, with knock-on effects for employment in related sectors upstream and downstream.

The financial burden of these external events arises in different ways. To start with, and perhaps most obviously, there may be biological injury (e.g., in the form of increased mortality or reduced growth) which in the extreme may cause a fish farm to cease production completely. A more common situation, however, is where a regulatory body responds to a pollution incident by closing a fishery and banning the sale of the product. In such cases there may be no biological injury *per se*, but the cost to the aquaculture producer in the form of lost revenue is nonetheless real. The financial impact of pollution incidents may also be felt where concern over the quality of the product translates into a fall in consumer demand, a situation which may itself be pre-empted by restrictions on production and sales in order to maintain public confidence. Such an event followed the *Braer* oil tanker disaster in 1993, which led to a ban on sales of farmed salmon from the affected area of the Shetlands as a move intended partly to protect the reputation of the industry. The impact of this action was also felt by smolt producers, who lost revenue as orders were cancelled by the salmon ongrowers. Water quality differences can impose indirect costs on producers, the best example being the need for shellfish growers in areas not meeting the Class A standard (<300 faecal coliforms or 230 *E. coli* per 100 g) to undertake compensatory investment in order to make their product saleable. In the UK this includes a requirement that shellfish undergo purification or relaying, which for Class C areas (<60,000 faecal coliforms or 46,000 *E. coli* per 100 g) must be a period of at least 2 months (Younger and Kershaw 2004). The final way in which external events may impact financially on aquaculture is through risk management costs, most obviously via insurance or else through physical precautions to reduce the probability of harmful incidents occurring (e.g., siting farms in low risk areas, vaccination of stock, etc.). Indeed, given that aquaculture is perceived by underwriters as a very high risk activity (Secretan and Nash 1989; van Anrooy et al. 2006), such precautions are generally a condition for obtaining insurance cover. In salmon farming during the mid-1990s, insurance represented some 3% of total costs for UK and 2% for Norwegian growers, though more recent data for Norway suggests that this proportion has now fallen to below 2% (PACEC and Stirling Aquaculture 1999; Bjorndal 2002; Anon 2004). However, these figures probably understate the true risk of salmon farming, if only because some producers may underinsure their operations whilst others may not buy cover at all.

8.3.2 External Effects of Aquaculture

Aquaculture may itself create externalities via its impact on the marine environment (Fig. 8.1), and though it is not always possible to quantify the importance of these effects in monetary terms there is little doubt that they can often be significant for human welfare. One of the most clearly demonstrated negative effects of aquaculture (i.e., external costs) is the degradation or loss of critical marine habitat, which may in turn lead to a reduction in biodiversity and in some cases the removal of an important natural resource providing a range of products and services.

The best documented example of this is the conversion of mangrove swamps to shrimp farming. Barbier and Strand (1998) estimate the effects of changes in mangrove area on the shrimp fisheries of Campeche State (Mexico) based on a production function methodology, and using data for the decade 1980–1990 find that a 1 km^2 decline in mangrove area leads to a loss of 14.4 t of commercial shrimp harvest worth some US $144,000. Though mangrove deforestation over that period was comparatively small, the authors point out that in the future this is likely to worsen due to urban expansion and mariculture development. In other words, shrimp aquaculture has the *potential* to impose significant external costs on society by indirectly causing a decline on the harvest of capture fisheries. Studies of coastal wetland systems in Thailand, where the area of mangrove has been severely reduced over the last 30 years, have focussed more specifically on the link with shrimp farming. Research based on a small local community in southern Thailand (Sathirathai 1998; Sathirathai and Barbier 2001) derived values for several of the functions supported by mangrove. Using field surveys the value of direct use of wood as well as other resources collected from the mangroves was estimated to be $88 per ha. Benefits in terms of off-shore fishery linkages were estimated using a production function approach and were in the range of $21–$69 per ha. The value of coastline protection was estimated using a replacement cost method (based on the cost of replacing mangroves with breakwaters to prevent erosion) and amounted to $3,679 per ha. Over a 20-year period, the total present value of the mangrove system to the local community was estimated to be as much as $27,264–$35,921. Of particular interest was the evidence showing the discrepancy between the private and social returns from converting mangrove wetland to commercial shrimp farms. Private returns were substantial, a result which clearly explains the incentive that has driven the expansion of the aquaculture industry hitherto. Once the external costs are accounted for, however, the returns to society from mangrove conversion are shown to be negative (Sathirathai 1998). A more recent study by Barbier and Cox (2004) looks more closely at the economic drivers behind mangrove deforestation in Thailand, and offers very clear evidence of the role played by shrimp farm profitability.

Habitat degradation caused by mariculture may also result from farming species other than shrimp, and we would point to the evidence which reveals the growing threat that cage aquaculture poses for seagrass meadows in the Mediterranean (Holmer et al. 2003). Sedimentation of waste products from fish farms has been shown to adversely impact the growth of seagrass (*Posidonia oceanica*) in the vicinity of sea cages, and given that seagrass is an important source of food and habitat for several varieties of marine organisms it seems likely that fish farms located inappropriately close to *P. oceanica* meadows may have indirect negative effects on the biodiversity and productive capacity of the marine environment. This is likely to translate into an external cost, most obviously a reduction in the harvest of capture fisheries, though the magnitude of this loss is as yet unknown. Evidence from the US, however, suggests that the economic value of seagrass habitat is substantial, and in the State of Florida it is recognised as being a key resource in supporting fisheries as well as having an important nutrient recycling function. The value of seagrass estimated by the Florida Department of Environmental Protection is $20,500 per acre per year (Anon 2003), a remarkably high figure which presumably owes much to the role of seagrass in maintaining water clarity (crucial for recreation). If the value of seagrass in the Mediterranean were shown to be of a similar order of magnitude it would clearly have important implications for policies aimed at protecting this habitat from anthropogenic threats, whether from aquaculture or some other source.

Land subsidence and saline intrusion have been shown to be linked to aquaculture, in both cases resulting in measurable economic costs. In Taiwan, growth of aquaculture has been accompanied by increasing demand for fresh water, which has led to overuse of underground supplies and consequent land subsidence (Huang 1990). This in turn has given rise to a wide range of negative externalities, notably: damage to property and agriculture due to sea water intrusion; salinization of fresh water supplies; production losses to fish farmers due to deteriorating drainage and sanitary conditions; and destruction of infrastructures (i.e., roads, ditches, sea dikes, etc.). Huang (1990) formulates an econometric model to estimate these effects in the case of one of the main species (grass shrimp), the results showing that the net social benefits of aquaculture vary widely across regions but are on average negative. These demonstrate the burden on society imposed by the growth of aquaculture, and 'strongly signal the necessity of reformulating the fishery structure and the natural resource policies in Taiwan.' In the Mekong Delta, the adoption of shrimp farming has seen the appearance of a number of environmental problems that have impacted negatively on rice growers as well as fish farmers themselves (Tran et al. 1999). Off-site impacts include the salinization of adjacent rice monoculture areas, while the main on-site impact has been siltation of ponds and fields due to turbid water inundation. To investigate the extent of these problems and to estimate their economic cost, Tran et al. (1999) conducted a survey of households engaged in shrimp production (either as monoculture or integrated with rice) and rice monoculture. The results revealed that salinization added to the costs per hectare in rice growing areas, and this was supported by other evidence showing that rice growers situated close to shrimp farms had reduced yields. Removal of sediment also imposed a significant financial burden on farmers, and while this might strictly be regarded as an 'internal' rather than an 'external' cost, the authors emphasise that the loss of land could have long term implications that farmers might disregard in their decision-making.

Aquaculture may interact with commercial fisheries to engender a range of externalities, though not all of these are negative. A number of studies have shown that the entry of aquaculture producers into a market may have a beneficial effect where, by diverting demand, it reduces harvesting pressure on over-exploited capture fisheries and allows natural stocks to recover. The result may thus be to increase overall supply and reduce the price of wild-caught fish, benefiting consumers (Anderson 1985). The magnitude of this effect will be stronger where the products from these two sources of supply, the capture fishery and the aquaculture sector, are regarded by the market as close substitutes (Ye and Beddington 1996). Interactions with the feed fishery add an additional layer of complication, and it is here that the negative effects of aquaculture become more conspicuous. It is argued that growth in demand for fishmeal used in intensive aquaculture increases pressure on stocks such as anchovy, sandeel, etc., with potentially adverse effects on the production of edible supplies from marine capture fisheries. The problem, however, has at least as much to do with the way capture fisheries are managed as with the growth of aquaculture *per se* (Hannesson 2003). Where fisheries are effectively open-access, the over-exploitation of stocks caused by increased demand for fishmeal is likely to be more severe. Nonetheless, the problem of higher fishmeal prices (and hence increased incentive to over-exploit) will be exacerbated if demand for fishmeal becomes more inelastic, a situation which seems likely as the global scale of the industry continues to enlarge and dependence on this source of raw material increases in the face of limited alternatives (Asche and Tveteras 2004). A quite separate mechanism by which aquaculture may impose external costs on capture fisheries is through the spread of disease and infestations, a particularly controversial example being the transmission of sea lice from salmon cages to the wild salmon stocks. While the evidence for this is still a matter of dispute, if the link were to be established it would be potentially serious in socio-economic as well as ecological terms.

Apart from its impact on fish stocks, cage aquaculture may affect the coastal environment in ways which reduce its amenity value, possibly with implications for tourism. One way is through reductions in water quality as a result of nutrient enrichment and eutrophication, another is through visual intrusion due to the inappropriate siting of sea cages. Despite the difficulties of assessing these types of environmental impact in monetary terms, it is clear that the public has a view about the social acceptability of aquaculture development in particular circumstances. Katrinidis et al. (2003) conducted a questionnaire survey of public attitudes towards coastal aquaculture development at two Greek islands and found that a negative attitude to development was more likely where respondents believed that pollution by fish farms would be high. Further evidence that people are not indifferent to the environmental effects of aquaculture comes from a survey of public attitudes towards salmon farming in Scotland and the perceived problem of organic pollution. (Whitmarsh and Wattage 2006). This was conducted through a postal questionnaire sent to a sample of households across the different Scottish regions. The results showed that respondents placed a relatively high priority on the desirability of minimising pollution from aquaculture, and this was matched by a Willingness to Pay (WTP) higher prices for salmon farmed in a more sustainable manner. Aside from this evidence, what we know about the externalities of pollution from marine aquaculture is very fragmentary. The external cost of eutrophication from coastal cage salmon farming in the early 1990s has been estimated using Swedish data to be between 50 and 100 SEK (= US $6.4 and US$12.8) per kg of nitrogen (Folke et al. 1994), figures which it is claimed would make salmon farming unsustainable if added to the cost of production (i.e., internalised). Though this particular study has been severely criticised for its methodology (Black et al. 1997), it is worth noting that an economic assessment of eutrophication in the Baltic based on a Willingness to Pay approach (Gren 2000) has estimated the external cost of nutrient release (N) from all sources to be 62 SEK per kg N and hence of a similar order of magnitude to that found by Folke et al. (1994). This finding might appear to support the claim that salmon farming is unsustainable, but it should be acknowledged that the external costs of nutrient release in a water body such as the Baltic are likely to be higher than in less eutrophic areas with greater assimilative capacity. Indeed, evidence showing that nutrients released by fish farms situated in oligotrophic waters may increase productivity in local capture fisheries (Machias et al. 2004) strongly implies that in such circumstances there will be positive externalities of cage aquaculture that will need to be set against any possible negative externalities associated with reduced amenity value.

8.4 Institutional Responses and Policy Implications

The DPSIR framework illustrated in Fig. 8.1 suggests that the response to the externalities of aquaculture will come from a variety of institutions, most obviously from government departments or agencies in the form of controls on fish farms intended to mitigate or prevent potential environmental damage. The external costs of shrimp farming are now widely recognised, and in several countries governments and NGOs are taking steps to redress the problem. In Thailand, shrimp farming in the non-coastal rice and fruit growing areas has been banned since 1998 (Barbier and Cox 2004), and a mangrove restoration program has recently been initiated. Environmental damage arising from the cultivation of species such as salmon is arguably of a lower order of magnitude compared to shrimp farming, but it is clear that here also governments are not indifferent to the possible risks. Marine aquaculture throughout Europe is subject to a wide range of regulatory controls and monitoring (Fernandes et al. 2000; Read and Fernandes 2003), which in some countries such as Denmark have imposed quite severe restraints on fish farm development. The institutional response may also derive from commercial firms linked with the supply side of aquaculture, for example by the provision of pollution-abatement technologies and inputs that are environmentally less damaging (e.g., altering the nutritional composition of pelleted feed to include less fishmeal). In the case of Norwegian salmon farming there is evidence that the adoption of such innovations has improved environmental performance (Asche et al. 1999), the rapid growth of the industry acting as an additional incentive for fish farmers to reduce pollution (Tveteras 2002). A further type of response may come from stakeholder groups not directly connected with aquaculture, and it is worth noting that consumers are increasingly being called upon to make ‘informed choices’ when they purchase seafood according to a set of environmental criteria. As well as established ecolabelling and certification schemes such as that operated by the Marine Stewardship Council, information is now available to consumers to enable them to evaluate fisheries products using an environmental ranking system. This approach has lately been pioneered by the Blue Ocean Institute and applied to farmed fish sold in the US, with species awarded a score based on their supposed environmental impact. An explanation of their scoring methodology can be found at the website http://www2.blueocean.org/.

Responses such as these may not, however, produce an optimal outcome as far as aquaculture development is concerned. Even assuming a correct assessment of the externalities, the most obvious difficulty is that actions designed to mitigate the environmental problems of aquaculture may simply be inadequate. This is the fate most likely to befall voluntary schemes based on best management practice, as shown by the general failure to adopt reduced water exchange amongst shrimp farmers in Asia and Latin America (Stanley 2000). In other circumstances, regulatory 'capture' by large producer interests may result in government agencies effectively acting on the side of the very groups they are supposed to be controlling. There is anecdotal evidence suggesting that this may have happened in some shrimp exporting countries where aquaculture producers are politically influential. However, quite the reverse kind of problem may occur where there is an *overreaction* to the perceived environmental damage caused by aquaculture. Within the DPSIR framework, the ways institutions respond is a function of the information that they receive, and if that information is inadequate or unbalanced the response will be misdirected. The messages emanating from some of the environmental groups concerning the supposed 'unsustainability' of aquaculture testifies to the fact that information in this area is far from objective, and if such messages are listened to and acted upon they carry the risk that decisions made by government agencies or consumers may be inappropriate. While almost all forms of aquaculture will impact on the environment, that in itself is not *ipso facto* evidence of social harm. This is the whole point about trying to measure externalities, because the monetary valuation of such effects provides a metric by which we can assess their severity and take appropriate action (Muir et al. 1999). A specific example of this rationale concerns the use of economic incentives such as taxes as charges in order to 'internalise' the externalities of aquaculture caused by pollution. The introduction of a tax or charge based on an estimate of the pollution damage would give an incentive to fish farmers to change their behaviour in a way which should lead to reduced pollution levels. For instance, the imposition of an effluent charge on production (e.g., at the rate of so many € per unit of nitrogen), would alter the cost structure of producers and confront them with the fact that pollution damage could no longer be disregarded. Their decisions and actions would then reflect not only their own private costs but all relevant costs including those attributable to nitrogen emissions. However, for this to happen, a reasonably reliable estimate of the pollution damage needs to be obtained.

These considerations underscore the important role of sustainability indicators, chosen and constructed so that unbiased and essential information can be communicated to policy makers. Several suggestions have been made on how such indicators might be applied to coastal resources, including fisheries and aquaculture (Bowen and Riley 2003; Caffey et al. 2001; Garcia et al. 2000; Grieve et al. 2003; Liu et al. 2005; Muir 2005), and it is generally agreed that socio-economic as well as environmental dimensions of sustainability need to be accounted for. It is our contention that, in the case of aquaculture, the set of socio-economic indicators should include some measure of the external costs where these can reliably be assessed. Where these cannot be quantified in monetary terms then some attempt should at least be made to derive a non-monetary metric of the social acceptability of particular types of fish farming practice. Without this information, decisions affecting the scale and development of the aquaculture sector run the risk of being incorrect.

Acknowledgements Research for this chapter was conducted as part of the European Commission project ECASA (Ecosystem Approach for Sustainable Aquaculture), Contract No. 006540.

References

Anderson JL (1985) Market interactions between aquaculture and the common-property commercial fishery. Marine Resource Economics 2:1–24

Anderson DM, Hoagland P, Kaoru Y, White AW (2000) Estimated annual economic impacts from harmful algal blooms (HABs) in the United States. Woods Hole Technical Report WHOI-2000-11

Anon (2003) Conserving Florida's seagrass resources: developing a coordinated statewide program. Florida Fish and Wildlife Conservation Commission, Florida Marine Research Institute, St Petersburg FL

Anon (2004) Lønsedsundersøkjing for Matfiskanlegg (Survey of Profitabililty of Fish Farms). Norwegian Directorate of Fisheries

Asche F, Tveteras S (2004) On the relationship between aquaculture and reduction fisheries. J Agricultural Economics 55:245–265

Asche F, Guttormsen AG, Tveteras R (1999) Environmental problems, productivity and innovations in Norwegian salmon aquaculture. Aquaculture Economics and Management 3:19–29

Bailly D, Paquotte P (1996) Aquaculture and environment interactions in the perspective of renewable resource management theory. Coastal Management 24:251–269

Barbier EB, Strand I (1998) Valuing mangrove-fishery linkages. A case study of Campeche, Mexico. Environmental and Resource Economics 12:151–166

Barbier EB, Cox M (2004) An economic analysis of shrimp farm expansion and mangrove conversion in Thailand. Land Economics 80:389–407

Bjorndal T (2002) The competitiveness of the Chilean salmon aquaculture industry. Aquaculture Economics and Management 6(1/2):97–116

Black E, Gowen R, Rosenthal H, Roth E, Stechy D, Taylor F (1997) The costs of eutrophication from salmon farming: implications for policy – a comment. Journal of Environmental Management 50:105–109

Bowen RE, Riley C (2003) Socio-economic indicators and integrated coastal management. Ocean and Coastal Management 46:299–213

Burbridge P, Hendrick V, Roth E, Rosenthal H (2001) Social and economic policy issues relevant to marine aquaculture. Journal of Applied Icthyology 17:194–206

Caffey RH, Kaxmierczak RF, Avault JW (2001) Developing consensus indicators of sustainability for southeastern United States aquaculture. LSU AgCenter No. 879, Baton Rouge, LA, USA

Costanza R, D'Arge R, de Groot R, Farber S, Grasso M, Hannon B, Limburg K, Naem S, O'Neill V, Paruelo J, Raskin RG, Sutton P, Van Belt M (1997) The value of the world's ecosystem services and natural capital. Nature 387:253–260

Fernandes TF, Miller KL, Read PA (2000) Monitoring and regulation of marine aquaculture in Europe. Journal of Applied Icthyology 16:138–143

Folke C, Kautsky N, Troell M (1994) The costs of eutrophication from salmon farming: implications for policy. Journal of Environmental Management 40:173–182

Garcia SM, Staples DJ, Chesson J (2000) The FAO guidelines and use of indicators for sustainable development of marine capture fisheries and an Australian example of their application. Ocean and Coastal Management 43:537–556

Garza-Gil MD, Suris-Regueiro JC, Varela-Lafuente MM (2006) Assessment of economic damages from the Prestige oil spill. Marine Policy 30:544–551

Gren I-M (2000) Winners and losers from Baltic Sea nitrogen reductions. In: Gren I-G, Turner K, Wulff F (eds) Managing a sea: the ecological economics of the Baltic. Earthscan Publications, London, UK, pp 95–105

Grieve C, Sporrong N, Coffey C, Moretti S, Martini N (2003) Review and gap analysis of environmental indicators for fisheries and aquaculture. Institute for European Environmental Policy, London, UK

Grigalunas TA, Anderson RC, Brown GM Jr, Congar R, Meade NF, Sorensen PE (1986) Estimating the cost of oil spills: lessons from the Amoco Cadiz incident. Marine Resource Economics 2:239–262

Hannesson R (2003) Aquaculture and fisheries. Marine Policy 27:169–178

Holmer M, Perez M, Duarte CM (2003) Benthic primary producers – a neglected environmental problem in Mediterranean maricultures? Marine Pollution Bulletin 46:1372–1376

Huang C-H (1990) Economic valuation of underground water and man-induced land subsidence in aquaculture. Applied Economics 22:31–43

Katranidis S, Nitsi E, Vakrou A (2003) Social acceptability of aquaculture development in coastal areas: the case of two Greek Islands. Coastal Management 31:37–53

Ledoux L, Turner RK (2002) Valuing ocean and coastal resources: a review of practical examples and issues for further action. Ocean and Coastal Management 45:583–616

Liu WH, Ou CH, Ting KH (2005) Sustainable coastal fishery development indicator system: a case of Gungliau, Taiwan. Marine Policy 29:199–210

Machias A, Karakassis I, Labropoulou M, Somarakis S, Papadopoulou KN, Papaconstantinou C (2004) Change in wild fish assemblages after the establishment of a fish farming zone in an oligotrophic marine ecosystem. Estuarine, Coastal and Shelf Science 60:771–779

Muir J (2005) Managing to harvest? Perspectives on the potential of aquaculture. Phil. Trans. Royal Society B. 360:191–218

Muir J, Bruger C, Young J, Stewart J (1999) The solution to pollution? The value and limitations of environmental economics in guiding aquaculture development. Aquaculture Economics and Management 3:43–57

PACEC and Stirling Aquaculture (1999) The Economic Impact of Scottish Salmon Farming. Report for HIE and SOAEFD. Economic Research Paper No. 7

Pearce DW, Secombe-Hett T (2000) Economic valuation and environmental decision-making in Europe. Environmental Science and Technology 34:1419–1425

Perman R, Ma Y, McGilvray J, Common M (2003) Natural Resource and Environmental Economics. Longman, Edinburgh

Pongthanapanich T, Roth E (2006) Toward Environmental Responsibility of Thai Shrimp Farming through a Voluntary Management Scheme. IME Working Paper 70/06, Department of Environmental and Business Economics, University of Southern Denmark

Read P, Fernandes T (2003) Management of environmental impacts of marine aquaculture in Europe. Aquaculture 226:139–163

Santillo D, Johnston P, Langston WJ (2001) Tributyltin (TBT) antifoulants: a tale of ships, snails and imposex. In: Harremoës P, Gee D, MacGarvin M, Stirling A, Keys J, Wynne B, Vaz SG (eds) Late Lessons from Early Warnings: The Precautionary Principle 1896-2000, European Environment Agency Environmental Issue Report No 22, Office for Official Publications of the European Communities, Luxembourg, ISBN 92-9167-323-4, 2001, pp 135–148

Sathirathai S (1998) Economic valuation of mangroves and the roles of local communities in the conservation of resources: case study of Surat Thani, South of Thailand. Final Report Submitted to the Economy and Environment Program for Southeast Asia (EEP-SEA), Singapore

Sathirathai S, Barbier EB (2001) Valuing mangrove conservation in Southern Thailand. Contemporary Economic Policy 19:109–122

Scottish Parliament Information Centre (1999) Salmon farming and the impact of the ISA virus. Research Note 99/20

Secretan PAD, Nash T (1989) Aquaculture and Risk Management. ADCP/REP/89/41. FAO, Rome
Stanley D (2000) The economics of the adoption of BMPs: the case of mariculture water management. Ecological Economics 35:145–155
Tran Thanh Be, Le Canh Dung, Brenan D (1999) Environmental costs of shrimp farming in the rice-growing regions of the Mekong Delta. Aquaculture Economics and Management 3:31–42
Tveteras S (2002) Norwegian salmon aquaculture and sustainability: the relationship between environmental quality and industry growth. Marine Resource Economics 17:121–132
Van Anrooy R, Secretan PAD, Lou Y, Roberts R, Upare M (2006) Review of the current state of world aquaculture insurance. FAO Technical paper 493. Food and Agriculture Organisation of the United Nations, Rome
Whitmarsh D, Wattage P (2006) Public attitudes towards the environmental impact of salmon aquaculture in Scotland. European Environment 16(2):108–121
Ye Y, Beddington JR (1996) Bioeconomic interactions between the capture fishery and aquaculture. Marine Resource Economics 11:105–123
Younger A, Kershaw S (2004) Sewage discharge improvements in the Solent and their effect on the shellfishery. CEFAS Shellfish News No. 17 (May)

Appendix: The Economic Appraisal of Aquaculture Development Projects

1 Introduction

The assessment of the desirability of investment projects in aquaculture as in any other industry should be based on a clear framework for the valuation of both the private and social costs and benefits involved. A way to achieve this is through economic analysis. The following section clarifies the concept of economic analysis as a social appraisal of investment projects as opposed to financial analysis based on private costs and benefits, while Section 3 defines the procedure to be followed for the assessment of the value of an investment project. Section 4 introduces the concept of the economic value of environmental resources. The last section clarifies the ideas introduced earlier through an illustrative example of economic appraisal of an investment project in a shrimp farm.

2 Financial Versus Economic Analysis of Investment Projects

Often incorrectly used as interchangeable terms, financial analysis and economic analysis are based on different accounting systems (Burbridge et al. 2001). These are illustrated in Table 8.1. As Burbridge et al. (2001) explain, financial analysis deals with the input costs and the market value of production of a private investor (e.g., fish farmer, processor, retailer, etc.), for whom the main objective is likely to be the maximisation of profits. On the other hand, the accounting system of

Table 8.1 Accounting systems for financial versus economic analysis

Financial analysis	Economic analysis
Private revenues (= production value at market price)	Social benefits (= internal and external benefits)
Minus	*Minus*
Private costs (= fixed and variable costs)	Social costs (= internal and external costs)
Equals	*Equals*
Private profit or loss	Welfare gain or loss to society

economic analysis is much broader and encompasses elements that may be ignored by the private investor but which can imply significant costs to society. The need for economic analysis stems from the fact that projects intended to yield benefits in the form of the provision of goods and services may negatively affect society as a whole. This may be the case with the environmental impacts of aquaculture or other production activities. When these impacts are not compensated we are in the presence of a market failure, defined in economics as an externality. For instance, this may be the case where the presence of sea cages reduces the enjoyment of visitors to the coast, but this goes uncompensated by the fish farmer responsible. Another example is the lack of compensation offered to fish farmers and fishermen for the decrease in production due to an oil spill. Externalities do not appear in financial appraisals. Economic analysis attempts to appraise investment projects in ways that correct for market failures such as environmental externalities. In order to take into account the impact on social welfare of changes in the quantity or quality of environmental assets, economists assign a monetary value to them so that they are considered along with the ordinary inputs (labour, capital, raw materials) and outputs (goods and/or services) of the project being appraised.

3 The Value of an Investment Project

In both financial and economic appraisals, the procedure for assessing the value of a project is to convert the stream of future costs and benefits into 'present' values. In this way, costs and benefits that occur at different times become comparable. This is done through discounting. The need to discount future values stems from the fact that costs and benefits in the future are not valued as highly as equivalent costs and benefits occurring in the present. For example, given the choice of receiving € 100 today and € 100 in one-year time, most individuals if not all would prefer the first option. However, if the amount offered in the future was greater than € 100, for instance € 105, many of those individuals would prefer the future amount. This implies that their rate of time preference is at most 5% per year.

The difference between the discounted total benefits and costs is the net present value (NPV). NPV can be expressed as:

$$NPV = \frac{(B_0 - C_0)}{(1+r)^0} + \frac{(B_1 - C_1)}{(1+r)^1} + \cdots + \frac{(B_n - C_n)}{(1+r)^n} \quad (1)$$

where:

B = Benefits

C = Costs

r = discount rate

n = number of years

A project should be accepted only if NPV is positive, that is if the discounted benefits of the project are higher than the discounted costs.

4 The Economic Value of Environmental Assets

The valuation of environmental assets and services for inclusion in economic appraisals is based on the concept of total economic value (TEV), which recognizes the fact that the environment provides not simply direct use values but also a range of indirect and 'passive' use values to society (Perman et al. 2003). Direct use values arise from the direct consumption of a resource, for example the coast for recreation, or fish as food. Indirect use values are benefits that are derived from the environment without human intervention, as in the case of life support services (e.g., gas regulation function of the open ocean) or ecological services that are inputs into a process of production (e.g., mangroves as breeding grounds for fisheries). Passive use values are assigned to a resource for its mere existence (existence value) or for its availability to future generations (bequest value), as in the case of endangered animal species such as whales. A substantial literature has grown up in recent years on how such values can be quantified, and a useful introduction to the role of economic valuation in environmental decision-making can be found in Pearce and Secombe-Hett (2000). Our concern here is not with the technical methodology of valuation, but rather in demonstrating how the estimated values can make a difference in the appraisal of projects that impact on the environment.

5 Shrimp Farming: An Illustrative Example

The concepts introduced in the previous sections are illustrated in the following example based on shrimp farming. The aim is to show that financial appraisal of a project may produce quite a different result from an economic appraisal where, as in the case of shrimp farming, there are significant environmental costs which need to be accounted for. Though the data are hypothetical, the main benefit and cost items are representative of the situation that applies to many shrimp exporting countries. As such they can be regarded as a characterisation of the externalities problem in shrimp farming and an illustration of the commercial pressures to convert mangroves where such externalities can be ignored. Evidence from Thailand, where the major cause of mangrove conversion has been shrimp farming, demonstrates that such pressures are real (Sathirathai and Barbier 2001, Pongthanapanich and Roth 2006).

5.1 *Financial Analysis*

Table 8.2 presents the financial benefits and costs per hectare for a proposed investment project involving the conversion of mangroves to shrimp farming. The time horizon is 5 years, which is the typical life of a shrimp farm in countries such as Thailand. Experience suggests that after this period, productivity declines and sites are abandoned for new locations. It is assumed that set-up costs of $5,000 per ha are incurred initially (year 0), with net financial benefits of $3,000 per ha per year from the sale of shrimp being earned over the years 1–5. These net benefits have to be discounted to obtain the net present values. This is done by multiplying the net benefits by the discount factor, $1/(1 + r)^t$, where t is the number of years the net benefits have to be discounted for. Given a discount rate of 5%, we can see that the net present value of the project for the private investor is positive ($7,988 per ha). The investment is thus financially viable.

5.2 *Economic Analysis*

Table 8.3 looks at the same investment from the perspective of society. For simplicity we assume that the costs of production reflect the real alternative use value (i.e., the true opportunity cost) of the capital and labour employed. The main difference in

Table 8.2 Financial appraisal of a hypothetical shrimp farming project ($/ha)

Year	Benefits	Production costs	Net benefits	Discount factor	Net discounted benefits
0	–	5,000	–5,000	1.000	–5,000
1	18,000	15,000	3,000	0.952	2,857
2	18,000	15,000	3,000	0.907	2,721
3	18,000	15,000	3,000	0.864	2,592
4	18,000	15,000	3,000	0.823	2,468
5	18,000	15,000	3,000	0.784	2,351
				NPV =	7,988

Table 8.3 Economic appraisal of a hypothetical shrimp farming project ($/ha)

Year	Benefits	Production costs	Foregone mangrove benefits	Pollution costs	Restoration costs	Net benefits	Discount factor	Net discounted benefits
0	–	5,000	–	–	–	–5,000	1.000	–5,000
1	18,000	15,000	1000	200	–	1,800	0.952	1,714
2	18,000	15,000	1000	200	–	1,800	0.907	1,633
3	18,000	15,000	1000	200	–	1,800	0.864	1,555
4	18,000	15,000	1000	200	–	1,800	0.823	1,481
5	18,000	15,000	1000	200	–	1,800	0.784	1,410
6	–	–	–	–	8,000	–8,000	0.746	–5,970
							NPV =	–3,177

this case, however, is that other cost items now have to be included in the calculations. Given what we know of the environmental impact of shrimp farming and measures taken to remediate it, the following have been included: (i) The opportunity cost of mangrove, which is equivalent to the benefits that this resource would otherwise have provided but which have now been foregone. These include the direct use value of forest products and the indirect use values associated with offshore fishery linkages and coastal protection. (ii) External costs of water pollution from shrimp ponds, mainly caused by saline intrusion into freshwater supplies and the run-off of agricultural chemicals. (iii) Costs of rehabilitating the ponds after their abandonment (e.g., mangrove replanting). This is factored into the calculations as a single year cost (year 6), the assumption being that any longer-lasting environmental effects after this date have no further significance as externalities. Assuming the same discount rate as before (5%), the NPV of the investment is negative ($–3,177), and hence not worthwhile.

A number of lessons can be learned from this comparison. Firstly, a project that is adjudged to be a 'good investment' from the standpoint of the private investor may not be so from the perspective of society. This is born out from our simple example, which contrasted the results of the financial analysis (the project will make the investor better off) with that of the economic analysis (the project will make society worse off). Of course, in a real situation the verdict we come to about a project will depend on the magnitude of the estimated benefits and costs, which is why the accurate assessment of externalities is so crucial. Secondly, it draws attention to the potentially powerful incentives to go ahead with a project in situations where private investors can avoid incurring the full costs of the natural resources they acquire. This applies *a fortiori* to shrimp farming, where mangrove swamps are often de facto open-access and can be obtained at a price that is far less than their true economic value.

List of abbreviations DPSIR Driving forces-Pressure-State-Impact-Response, HABs Harmful algal blooms, ISA Infectious salmon anaemia, NPV Net Present Value, TBT Tributyltin, TEV - Total Economic Value, WTP Willingness to Pay

Chapter 9
Future Trends in Aquaculture: Productivity Growth and Increased Production

Frank Asche,[1,*] Kristin H. Roll[2], and Sigbjørn Tveterås[2]

Abstract The introduction of semi-intensive and intensive farming practice, where producers actively influence the growing condition of the fish, has been the main engine for growth in aquaculture production. The control of the biological production process has enabled a number of productivity enhancing innovations. These advances have reduced the production costs, increased the product range and reduced prices to the consumer. This has made aquaculture products competitive compared with, e.g., meat and wild-caught fish products. There is little doubt that aquaculture production will continue to grow. However, with a competitive marketplace not every country, region and species can succeed. Changes in relative productivity will determine where production takes place and the need for low unit costs will likely limit the number of high volume aquaculture species.

Keywords Aquaculture, productivity growth, future trends

9.1 Introduction

Worldwide demand for seafood will increase in the future (Delgado et al. 2003). This is partly due to population growth and partly due to economic growth. As seafood supplies from wild sources mostly are fully exploited, this provides a substantial opportunity for aquaculture provided that aquaculture production can be competitive. Recent development indicates that this is the case as production has increased from about 3.5 million tonnes in 1970 to about 59 million tonnes in 2004 (FAO 2006). We will look closer at economic drivers for growth in intensive aquaculture production, and based on that make some predictions with respect to future trends in aquaculture production.

[1]Department of Industrial Economics, University of Stavanger, N-4036 Stavanger, Norway, Tel: +47 51 83 22 86; E-mail: Frank.Asche@uis.no

[2]University of Stavanger, N-4036 Stavanger, Norway

* Corresponding author

M. Holmer et al. (eds.), *Aquaculture in the Ecosystem.*

Aquaculture is distinguished from other aquatic production by the degree of human intervention and control that is possible (Anderson 2002). Hence, aquaculture can be defined as the human cultivation of organisms in water. As such, it is in principle more similar to forestry and animal husbandry than to traditional capture fisheries. In other words, aquaculture is stock raising rather than hunting. The production process in aquaculture is determined by biological, technological, economic and environmental factors. Many aspects of the production process can be brought under human control. This control makes innovation possible, and is accordingly essential for the rapid technology development that has fuelled the production growth which has taken place since the early 1970s.

Although aquaculture is a very old food-producing technology, it was not very important in terms of quantity produced until the 1970s. Then a revolution occurred as humanity's accumulated knowledge in the area then allowed the introduction of semi-intensive and intensive farming practices, where the producers actively influenced the growing conditions of the fish with feeding, breeding, etc. The control of the production process that was achieved enabled a number of productivity enhancing innovations to take place. Improved productivity results in a reduction in production costs, and with a given price, this makes the production more profitable. High profits are the market's signal to increase production, and this will happen both because existing producers produce more and because new producers enter the industry. To sell the increased production, one needs to give the consumers a reason to buy the product, and in general the most important incentive used is a reduction of the price. A substantial part of cost savings due to productivity increase is accordingly passed on to the consumers. This can clearly be seen in the price development for most successful aquaculture species. Hence, one can sum up the most important drivers in the development of modern aquaculture as follows: Control with the production process allows technological innovations that reduce production cost. This makes the product more competitive, the industry profitable and leads to increased production and lower prices to the consumers.

Several criteria can be used to classify an aquaculture system (Bjørndal 1990). From an economic point of view the most significant criterion is intensity, i.e., the division into intensive, semi-intensive or extensive forms of culture. Measures of intensity include stocking density, production by area, feeding regimes and input costs, while the most interesting feature is the degree of control within the production process. In intensive salmon farming, fish are reared in pens and the farmer controls factors of production such as farm size, stocking and feeding of fish. For other species (e.g., turbot, shrimp) the pens can be replaced with land-based tanks, raceways or ponds. Traditional aquaculture varies between semi-intensive and extensive. Mussel farming is an example of an extensive method used around the globe, where the farmer primarily provides a rope or a stake for the mussel fry to fasten onto, but otherwise leaves the mussel to grow. The small ponds used in Chinese aquaculture were traditionally operated on an extensive basis, as the farmer did little to control growth and biomass. While this system is still common, many farmers have become semi-intensive as they actively feed their fish and maintain higher densities as well as adapting other production-enhancing technologies.

A relatively intensive production technology is necessary for aquaculture to become industrialised. While most of the world's aquaculture production cannot be characterised as intensive, this seems to be the direction in which it is heading. The higher degree of control over the production process allows technological innovation to a much larger extent than other operation modes. This allows large-scale production that can benefit from cost-saving economies of scale, which is necessary if aquaculture is to fulfil its promise as a major food-producing method with global benefits. It also allows market-oriented production and logistics, so that the fish can be sold in the markets that provide the producer with most added value.

Since price in most cases is the most important argument with respect to which product in a group of products a retailer will stock, total production cost will be the main factor explaining which aquaculture products will be produced. By total production cost one means the total cost of bringing the product to the consumer, which then includes transportation and processing costs. What products and species will be produced will then depend on which species gives the lowest production costs. Choice of production location will depend on which area offers the most competitive advantages in terms of access to suitable land/sea localities, good market access, favourable regulations, etc.

If one looks at agriculture and meat production where there are four main species (beef, pork, chicken and lamb), it is likely that there will not be many aquaculture species that are produced in large volumes. Exactly how many species there will be will partly depend on how many species that can have a similar productivity growth and how distinct consumers think large fish is relatively to portion sized fish, fresh to frozen, or finfish to crustaceans and shellfish. There will also be a smaller high-end market where a substantial part of the wild fish is sold (similar to quail, rabbit, turkey, deer, boar, duck, geese, etc.), as well as a low end/feed production segment.

In this paper we will discuss these issues further. For a historical background we will focus particularly on salmon and shrimp, since these are among the most successful and valuable intensively farmed species, and the species where most development has taken place so far. The discussion will focus mostly on salmon, because of data availability, but can be generalised to other species. The chapter is organised as follows. Section 2 gives a brief overview of global aquaculture production. Section 3 discusses productivity growth and how this leads to lower production costs. This is the key to understanding why aquaculture production will continue to increase. Section 4 discusses some implications for the aquaculture industry in the future before some concluding remarks are offered in Section 5.

9.2 Production of Aquaculture

Following the introduction of semi-intensive and intensive production technologies, there has been a substantial increase in aquaculture production. Figure 9.1 shows the total global seafood production from 1970, together with wild and aquaculture production. As can be seen, aquaculture was relatively insignificant in 1970, making

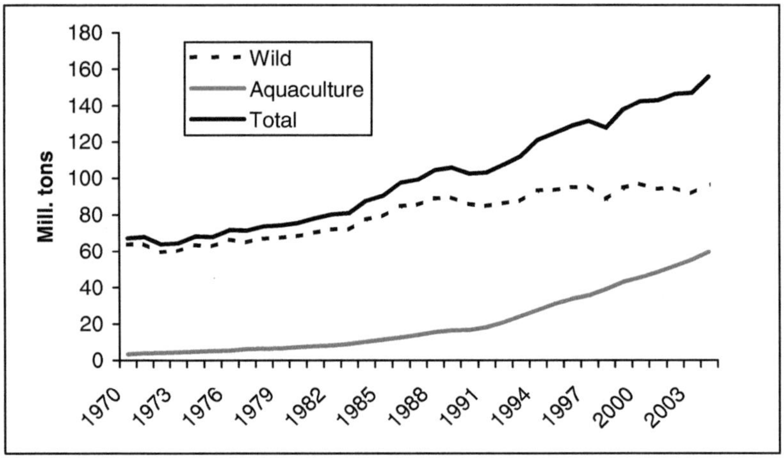

Fig. 9.1 Global production of seafood

Table 9.1 Aquaculture production in 1000 t by species in 2004 (FAO 2006)

Species	Volume	Percent
Carps, barbels and other cyprinids	18,304	40
Oysters	4,604	10
Clams, cockles, arkshells	4,117	9
Freshwater fishes	3,740	8
Shrimps, prawns	2,476	5
Salmons, trouts, smelts	1,978	4
Mussels	1,860	4
Tilapias and other cichlids	1,823	4
Scallops, pectens	1,167	3
Marine molluscs	1,065	2
Total	45,481	

up only about 5% of total seafood production with a production volume of about 3.5 million tonnes. In 2004 aquaculture production had increased to about 59 million tonnes and constituted 38% of the total seafood supply. While landings of wild fish have been stagnant since the late 1980s, aquaculture production has grown so much that it has maintained a rate of increase in seafood supply that exceeds global population growth. As a result, the global per capita supply of seafood has increased in all of the three previous decades. Hence, it is clear that aquaculture already plays a very important role in the global supply of food.

There are a number of species being farmed, using many different production techniques. Table 9.1 shows production where the species has been aggregated according to ISSCAP groupings (ISSCAP = International Standard Statistical Classification of Aquatic Animals and Plants), in which aquatic plants have been excluded. As can be seen, herbivore species like carps, barbels and other cyprinids account for a major part of global aquaculture production in terms of volume, making up 40% of the total volume. This is followed by production of oyster, clams and other molluscs. With a 5% and 4% of total production volume respectively, shrimp and prawns, on the one hand, and salmon and trout, on the other, account for a modest share of aquaculture production. It is clear that aquaculture produces large quantities of a substantial variety of species with carps, shellfish, shrimps as well as finfish species on the top ten list.

Quite a different picture emerges when we look at the ranking of species in value terms, in Table 9.2. The group including carps is still the largest, but with 26% of the total value it accounts for a considerable smaller share for than it did for volume. Although eight of the groups on the 'volume' list are still on the 'value' list, shrimp and prawns has moved from fifth place to second, and salmon and trouts from sixth place to third. Jointly the two groups accounts for 25% of total value of aquaculture production. Hence, the most intensively produced species are also among the most valuable. These are also some of the species with the highest export shares, with their major trade flows from Southeast Asia, Chile and Norway to the EU, Japan and the US. The production of these species is not increasing significantly faster than other species, however, suggesting that production costs associated with these species are declining at a similar rate as other species.

Aquaculture is a truly global production technology, with close to 180 countries reporting some level of aquaculture production. However, as shown in Table 9.3, there are substantial regional differences in production volume. Asia makes up about 92% of the production measured by volume and 81% by value. All the other regions have a higher value share than volume share, as they produce a higher-value product. This is particularly true for South America. China is by far the largest production country with a value share of 53% and a volume share of 70%. Measured by value, Japan, India, Chile, Vietnam, Thailand, Indonesia, Norway, Bangladesh,

Table 9.2 Aquaculture production in mill. US$ by species in 2004 (FAO 2006)

Species	Value	Percent
Carps, barbels and other cyprinids	16,422	26
Shrimps, prawns	9,735	15
Salmons, trouts, smelts	6,637	10
Freshwater fishes	6,000	9
Freshwater crustaceans	4,017	6
Clams, cockles, arkshells	3,312	5
Oysters	2,818	4
Coastal fishes	2,536	4
Tilapias and other cichlids	2,202	3
Scallops, pectens	1,747	3
Total	63,493	

Table 9.3 Aquaculture production by region in 2004 (FAO 2006)

Region	Volume	Value	Percent (volume)	Percent (value)
Asia	54,367.4	56,821.1	91.5	80.8
Europe	2,238.7	5,583.5	3.8	7.9
South America	1,137.8	4,563.2	1.9	6.5
North America	955.2	1,994.2	1.6	2.8
Africa	570.1	893.2	1.0	1.3
Oceania	139.3	447.2	0.2	0.6

South Korea and Brazil are the other top-ten producing countries. Egypt is the largest producer in Africa and is number 16 on the list. Hence, aquaculture is clearly strongest in Southeast Asia, and is primarily conducted in developing countries.

9.3 Productivity Growth and Lower Production Costs

The success of aquaculture has led to substantial increases in production for several species. A substantial increase in the production of a particular aquaculture species usually results in a significant drop in the price of that species. For the production to be profitable, technological innovations must take place to increase productivity and reduce production cost. We will examine this process of change in this section, focussing particularly on salmon since this is the species on which most research has been conducted and for which most data are available. It is also the large-volume species with the most intensive production practice, giving it the largest potential for productivity improvements and the biggest challenges with respect to environmental sustainability. While conditions will vary in the production of other species, most of the trends and relationships described in this section can be generalised to other species.

9.3.1 Quantity Increase and Price Reduction

Shrimp and salmon are good examples of species where production increases have been accompanied by price drops. Figure 9.2 shows the global production of farmed shrimp and the real price for the period 1984 to 2004. Production in this period increased from about 170,000 to 1.8 million tonnes. Prices were at their highest in the late 1980s, at more than 10 US S/kg and then fell consistently, to about US $5/kg in 2004. A similar trend is seen for Atlantic salmon over the period 1981 to 2004 (Fig. 9.3). Production of Atlantic salmon increased from about 20,000 tonnes in 1981 to about 1.6 million tonnes in 2004 and prices declined from a high of

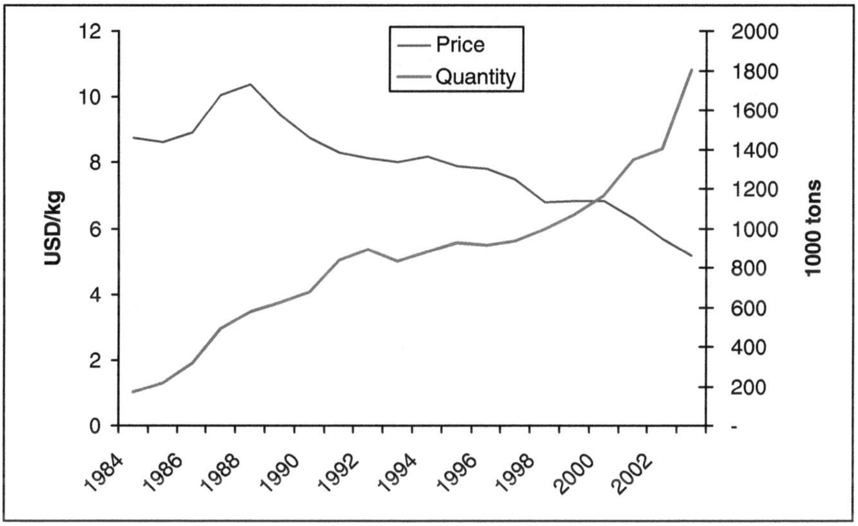

Fig. 9.2 Global aquaculture production of shrimp and real US import price (2003=1) (FAO 2006, Jim Anderson and Ass., personal communication)

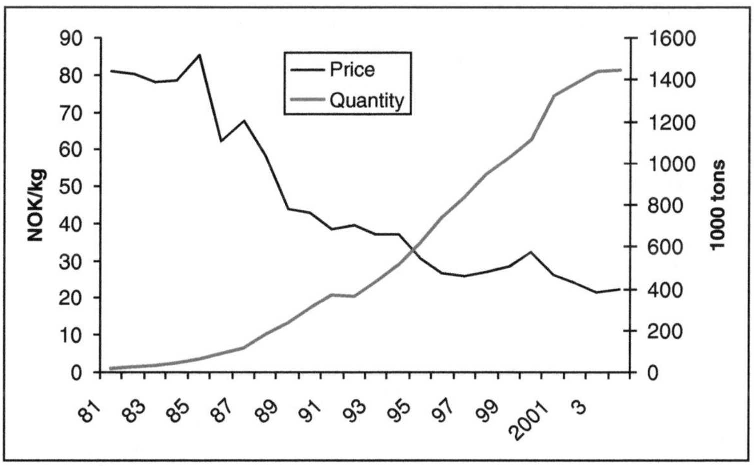

Fig. 9.3 Global aquaculture production of Atlantic salmon and real Norwegian export price 1981–2004 (2004 = 1) (FAO 2006 and the Norwegian Directorate of Fisheries)

almost 90 NOK/kg in the mid-1980s to about 22 NOK/kg in 2004. The story is the same, on a more limited scale, for other salmonides like coho and salmon trout. It is also similar for sea bass, sea bream, catfish and tilapia, although the strength of the price decline varies (Asche et al. 2001).

It is worth noting that the price reductions are not necessarily immediate. When the aquaculture species is first introduced, there often seems to be an early period when demand is increasing faster than supply and prices are actually increasing.

This can be explained by the fact that a stable supply of high-quality fish presents market opportunities that have not existed for the wild supply of similar fish. For instance, there is not going to be any price pressure if the farmed fish is sold in periods when there had previously been no supply of similar wild-caught fish, due to seasonality. Moreover, demand can increase when the logistical systems can operate with a stable and relatively predictable supply.

Somewhat simplified, one can say that there are two main market structures that an aquaculture producer or country can face, following an increase in their production. If the market size is limited and there are few other species or products from which one can win market share, prices will decline rapidly. If, on the other hand, there is a large market where the producer or country in question only produces an insignificant share, there may be no or only a weak price effect. There is of course a continuum between these two structures, and the main reason for shrimp prices declining at a lower rate than salmon is that the global production of shrimp is substantially larger. If one looks closer at the shrimp producers, one will also observe that there have been substantial changes in the top 10 list of producing countries within short time periods (Anderson 2003), illustrating how little effect each of the large producer countries has on the price. The larger the market, the weaker the effect of any single country's production on the price and the more exposed that production will be to the impacts of changes in other parts of the world.

The production and price of Egyptian tilapia presents another interesting case. Egypt is the world's second largest producer of tilapia after China, but imports and exports very little. Hence, one can say that tilapia producers in Egypt serve a market of limited size – the domestic Egyptian market. As shown in Fig. 9.4, the period 1997–2002 saw an increase in production from about 40,000 tonnes to about 160,000 tonnes, and a halving of the nominal price of Egyptian tilapia. The

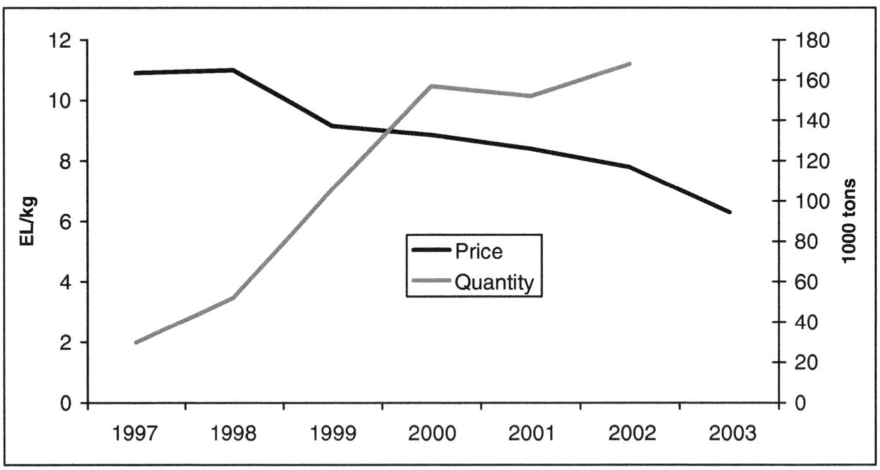

Fig. 9.4 Egyptian production and nominal wholesale price in Egyptian pound (EL) for tilapia (FAO 2006 and Ana Norman, personal communication)

observed price decline would be even stronger, if adjusted for inflation. Hence, the same economic forces that influence the global market for salmon and shrimp, also work in the domestic market for tilapia in Egypt.

9.3.2 *Productivity Growth*

So what causes this observed relationship between production levels and price? We will try to clarify this issue by looking more closely at salmon. For any product, its profitability determines the development of its production volume, with production tending to increase if it is very profitable. On the other hand, production will decrease if other uses of capital and labour are more profitable and if producers are losing money. The decline in the price of salmon has been necessary to induce greater consumption of the product. For this to be profitable, production costs must also have been substantially reduced. The main factors behind reduced production costs are productivity growth and technological change. In this section we will discuss the reduction in production costs for salmon aquaculture, focusing on Norway since this is the country for which data are most widely available. As the largest producer of farmed salmon, Norway can be considered fairly representative of other producers. However, at the end of the section we will also relate these results to other salmon-producing countries.

9.3.2.1 Determinants of the Production-Price Relationship

Figure 9.5 shows real production cost and export price for salmon in Norway. Both variables have a clear downward trend and the gap between them is consistently small. The average price in 2003 was about a quarter of the price in 1985 and the reduction in production cost is of the same magnitude. The important message here is that there is a close relationship between the development of productivity and the falling export prices. Productivity gains are therefore able to explain a great deal of the decline in farmed salmon prices, as the price has been moving down with the production cost, keeping the profit margin relatively constant. This is also as expected in a competitive industry, since high profitability is the market's signal to increase production. As the cost reduction has been translated into lower prices, it is also clear that the productivity gains have been passed on to consumers. The main effect for the producers is that they become larger and hence earns a higher profit because of larger quantities produced.

The reduction in production costs has been due to two main factors. First, fish farmers have become more efficient so that they produce more salmon with the same inputs. This is what is normally referred to as the fish farmers' productivity growth. Second, improved input factors (such as better feed and feeding technology and improved genetic attributes due to salmon breeding) make the production process less costly. This is due to technological change for the fish farmers, and productivity

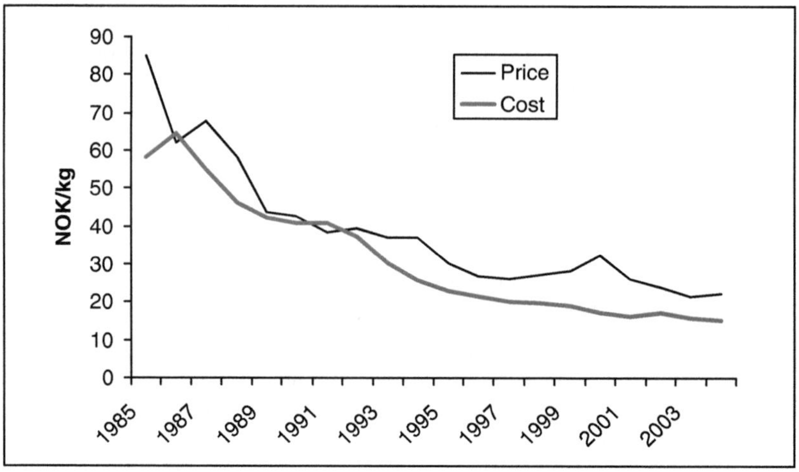

Fig. 9.5 Real production cost and producer price per kg in NOK 1985–2004 (2004 = 1) (Norwegian Directorate of Fisheries)

growth for the fish farm suppliers. This distinction is often missed and the productivity growth for the farmers as well as for their suppliers is somewhat imprecisely referred to as productivity growth for the whole industry. In addition, while the focus is on the production process, productivity gains in the distribution chain to the retail outlet are equally important. In the end, consumers are primarily interested in the final price for a product of any quality, and whether a price reduction is due to better feed or better logistics is of little importance.

The most important input in salmon farming is the salmon feed, which represented around 52% of operating costs during the period 1985–2004. Other inputs are smolts (15% cost share), capital (5%), labour (9%), insurance (2%) and materials (17%). The share of feed has been increasing (from about 25% in the mid-1980s) making the production process more feed intensive. As feed is the factor most closely related to the production volume, this development indicates better exploitation of the capital and labour employed at each farm. This can be explained to a large extent by increased production on each farm. Several studies using data from the 1980s found that substitution was possible between feed, capital and labour. For instance, hand feeding was at the time more efficient than machine feeding. However, with the increased cost share of feed these substitution possibilities have been reduced. Guttormsen (2002) suggests that they have largely disappeared in the 1990s. This implies that salmon production now, after investments in capital equipment have been made, can be characterised as a technology with a close to fixed relative factor share in the production process. The production process then becomes one of converting a cheaper feed into a more desirable product for the consumers. So, even if the substitution possibilities between capital, labour and feed are limited, the farmers can substitute between different types of feed. Currently, about 35% of the feed is fishmeal which has been partly substituted with vegetable meals. About 40% of the feed is oil, of which fish oil currently makes up about two-thirds.

A cost share of one factor, feed, at over 50% may seem high, but not when compared to other comparable industries such as pork and poultry production. For example, the cost share for feed for the most efficient poultry producers is over 80%. This suggests that there is still a substantial efficiency potential for salmon, and production costs can be further reduced if other factors are exploited even more efficiently.

That the composition of the input use varies over time suggests that the production technology has been changing over time, and this is certainly an important factor in explaining the productivity growth. Tveterås and Heshmati (2002) found that technical progress at the farm level explains only about one-third of the reduction in production costs, with the remainder accounted for by reduced prices for input factors, or technological innovations amongst the suppliers of input factors. Tveterås and Heshmati (2002) also found that productivity growth was anything but smooth, indicating that technological progress at the farm level and among the suppliers comes in leaps and is unpredictable. With the long production time in salmon farming, this can create cycles in profitability as production costs decline, since lower production costs initially give higher profits, which induce farmers to expand production. The expanded production then drives the prices down, reducing profits.

9.3.2.2 Cycles in Profitability

Cost and price do not move in complete synchronicity (see Fig. 9.5). In particular, the margins between price and cost were narrow in 1986, 1991, 1997 and 2001, and especially wide in the intervening years. In other words, some years were much more profitable than others. This structure is commonly seen in biological industries and other industries with a substantial time lag between the decision to increase production and the entry of the increased production into the market. A high profit margin gives a signal from the market to increase the supply, but due to the time lag in increasing the production, the signal can be quite persistent. This often leads to over-investment and excess production with the result that prices may fall to production cost levels, or even lower, for a period. The low margins will then be a signal to reduce production, which again takes time, and production will often be reduced too much, giving rise to a new period with very good margins. In a stable world, one would expect producers to work out the production level that gives normal margins. Unfortunately, the world is anything but stable and the production volume that gives a normal margin is a moving target, because of productivity growth and other supply shocks as well as exchange rate movements, demand shocks and market growth. The delay in responses from the producers will therefore produce boom-and-bust cycles at irregular intervals and with different strengths in industries like salmon production.

Cycles in profitability are not a problem in themselves, as one usually retains a substantial portion of profits at the top of the cycle to cushion the bottom of the cycle. However, many owners do not retain earnings, with the result that more firms get into trouble at the bottom of every cycle than necessary – a feature the salmon industry shares with other primary industries. The cycles also make salmon and other aquaculture industries very susceptible to trade conflict, as a number of producers will lose money at the bottom of the cycles.

9.3.2.3 Productivity Development in Norway Relative to Other Producers

Norwegian salmon producers have been the main object of several studies on productivity. Bjørndal (2002) compares cost data for Norwegian and Chilean salmon farms and concludes that the cost level in Chile is similar or lower than that in Norway, although the cost composition is different since Chilean processing costs are lower, but transportation costs are higher. Industry sources normally indicate that average production costs in Scotland are between 0.1 and 0.3 euros/kg higher than in Norway.

While a lack of data prevents us from reporting on the specific productivity development among salmon producers, it is possible to make assumptions by investigating the development in production shares. In a free market, changes in production shares exist due to differences in productivity development or production costs. On the other hand, in markets with trade restrictions and regulations, the development in production shares will show the combined effect of trade restrictions, regulations, and the relative productivity growth.

In Fig. 9.6, we see production shares of the four largest producers of salmon – Norway, Chile, the UK and Canada – which combined represent about 90% of the global production of farmed salmon. The Chilean figures are somewhat uncertain, as some sources report higher production than the figures used here, but that does not change the main picture.

The dominating trend in Fig. 9.6 is the development of Norwegian and Chilean shares, which decreased and increased respectively throughout the period. Norway's market share fell from 70% in 1981 to 40% in 1992. To some extent, this decline was probably bound to happen, as a result of the diffusion of best-practice production technologies from Norway to other countries. However, there is no doubt that

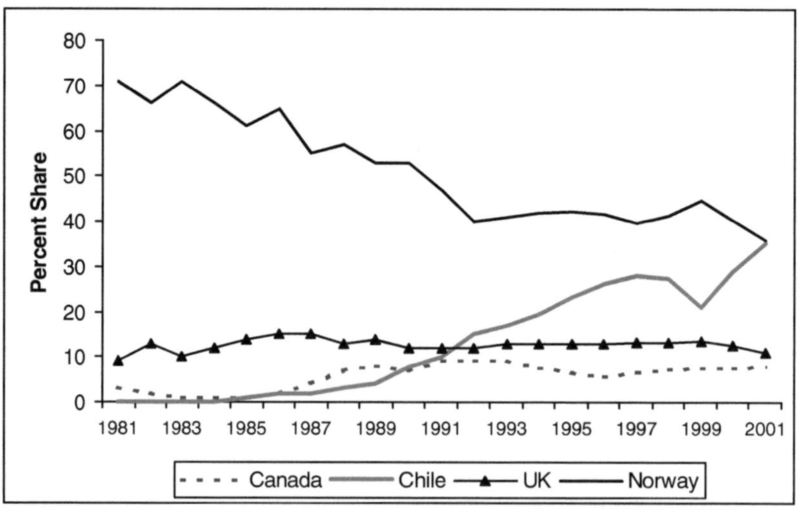

Fig. 9.6 Production shares for the four main salmon-producing countries, 1981–2001 (FAO 2006 and Norwegian Seafood Exports Council 2005)

it was accentuated by Norwegian entry and ownership regulations, as they represented incentives to invest in other countries. Since the second half of the 1980s, Norwegian capital has been involved in salmon farms in virtually all salmon-producing countries. The salmon market crisis around 1990 led to the abandoning of Norwegian ownership regulations, which has prevented consolidation in the industry. A restructuring process then started in Norway, as firms then were allowed to merge and larger firms were created, actually increasing Norwegian market share from 1992 to 1995. Then, following anti-dumping allegations from the EU in 1996, new regulations were introduced, including feed quotas per farm that effectively limit production. Ever since, with the exception of 1999, Norway has been losing market share and ended at 36% in 2001.

In the 1990s, Chile became a major producer of farmed salmon. Currently, Chile is the second largest salmon producer, with about 34% of total production, and the country is expected to surpass Norway as the largest producer relatively soon. The large increase in the Chilean production has been possible due to few restrictions on salmon farming, a low cost level, and many foreign firms in the industry providing for the same knowledge base as the competitors. However, Chile also has some major disadvantages, including a lack of infrastructure in Region XI, the southernmost region in Chile, where much of the future industry expansion may take place, and the long distance to the markets that causes high transportation costs. Furthermore, its position as one of the major producers has led to anti-dumping complaints. The only setback in Chile's production share was in 1999, which can be attributed partly to the Asian crisis in 1997/98 influencing demand in key markets, and partly to the uncertainty following the US dumping complaint.

Canada and the UK both have access to major salmon markets: the US (NAFTA), and the EU. Norwegian regulations and trade problems and Chilean trade problems were expected to benefit the Canadian and British salmon producers, but both countries' production stagnated in the 1990s, and the British salmon production reached an historic low of about 11% in 2001. Both the Canadian and UK industries seem to have experienced a productivity growth close to industry average over the period, but neither producer has been able to benefit from the trade restrictions and regulations faced by Norway and Chile. The UK industry has also been hit by disease problems and a high value of the pound sterling that have reduced profitability levels for Scottish farmers. This is a concern for Chilean and Norwegian farmers, as it provides an incentive for anti-dumping complaints by UK producers.

The four main producers have increased their combined share of production during the last decade. The only smaller producer growing at a similar pace to the four major ones is the Faeroe Islands. Japan, however, the second largest producer in the world in the early 1980s, as well as the US, Australia, Ireland, and Iceland have fallen behind. It seems that regulations and problems with suitable locations have hindered growth to a large extent, even though production in most of these countries has been growing in absolute terms. It may also be that these industries, because of their small size, never realised the external scale effects associated with agglomeration and cluster effects that can be associated with a larger industry.

Agglomeration effects have been revealed for Norway (Tveterås 2002), and are most likely present for the other three main producing countries as well.

9.3.2.4 Cost Reductions in the Supply Chain

Productivity growth is most easily observed in the production process and in the main input factors but also stems from improvements in distribution and supply chain logistics. When looking at the growth of the salmon industry, it is important to keep in mind that improved logistics account for a substantial part of the productivity growth, as economies of scale and transportation methods that have not been used for other types of fish have reduced the cost of bringing the product to the consumer. To illustrate this, one can look at Norwegian and Icelandic exports of fresh cod to the UK (Asche et al. 2007). The fishermens' share of the retail value is about 10–15%, which is in the range observed for wild fish all over the world, and also for many farmed species. In contrast, salmon farmers receive about 50% of the retail value. If cod had the same efficient logistics as salmon, its price could be reduced by about 70%. We find a similar example in France, where the price of salmon in supermarkets is about 2 euros/kg lower than in fish markets and fishmongers' shops.

Salmon currently has the most efficient distribution and logistics system, and it is not obvious that all other species and producers will be able to achieve the same level of efficiency. This is largely down to the need for a high degree of organisation. Small scale aquaculture producers in many developing countries will face supply chains with market clearing at each level, similar to what traditional fishermen face, and have a competitive disadvantage because of this. This is an issue that is difficult to overcome, as it is often related to how the society around the farm is organised. However, there are some examples, such as in Vietnam, where larger-scale operators invest not only in the production, but also in the whole supply chain to obtain cost savings and competitive advantage where they can be found.

9.3.2.5 Environmental Factors as a Limitation to Growth

A number of authors have claimed that aquaculture growth will be limited by environmental factors (e.g., Naylor et al. 2000). These possible limitations appear as two different issues; local environmental carrying capacity and limited availability of food (the fishmeal trap).

Local Environmental Issues

Any production process that interacts with the environment has the potential to damage the environment around the production site. This includes destruction of natural habitat and pollution from the production process that influences habitat and wildlife around the site. For salmon farming the main issues have been pollution from organic waste and the interaction between wild and farmed salmon. Farmed salmon may transmit diseases and parasites to wild salmon. Increased number of the sea lice parasite on wild salmon has been associated with escaped farmed salmon. Farmed salmon may also attempt to spawn in rivers and may impact the genetic pool. These experiences are not unique for salmon farming. Shrimp farming has received even more negative publicity than salmon farming in relation to detrimental environmental effects, such as destruction of mangroves, salinization of agricultural areas, eutrophication, and disruptive socio-economic impacts.

The environmental issues that arose in intensive salmon and shrimp farming during the 1980s and through the 1990s, must be seen in relation to the introduction of a new technology that uses the environment as an input. The larger the production at any site and the more intensive the process, the larger the potential for environmental damage. However, the greater degree of control with the production process in intensive aquaculture also makes it easier to address these issues. With all new technologies there will be unexpected side effects, and there will be a time lag from an issue arises until it can be addressed. First the impact and the causes must be properly identified. Second, the solution to the problems will require modifications of existing technology or maybe entirely new technology. In both cases pollution reduction implies some form of induced innovation. In this relation, Tveterås (2002) argues that industry growth has a positive effect on pollution, in line with the Environmental Kuznets Curve (EKC). The EKC hypothesis refers to an empirical observation that pollution tends to increase with economic growth up to a certain point, after which growth will reduce pollution. This gives the pollution profile over time the shape of an inverted U. Use of antibiotics in Norwegian aquaculture is a good example, as shown in Fig. 9.7.

There are two main causes for the industry to address environmental effects; (1) the effects reduce productivity and therefore profits, and/or (2) government regulations force the industry to do so. Industry size contributes in the sense that a large industry allows larger investments and thereby more efficient innovation of abatement technologies. Detrimental environmental effects of aquaculture not accounted for in market prices are by definition external effects, i.e., negative externalities. Asche et al. (1999) argued that internalisation of the externalities explain why some of the major environmental issues have been resolved in aquaculture. The arguments go along the following lines: Productivity in aquaculture depends on an environment where farmed fish thrive. Fish farms with environmental practices that deteriorate the local environment will experience negative feedback effects, where poor water quality reduces on-farm productivity. These negative environmental feedback effects are well known, and can be amply exemplified by reference to salmon and shrimp farming. The results are reduced biomass growth through deteriorating fish health and, in the worst case, disease outbreaks that wipe out entire on-farm fish stocks. Consequently, one is concerned with cultivating management practices that avoid such negative repercussions on productivity.

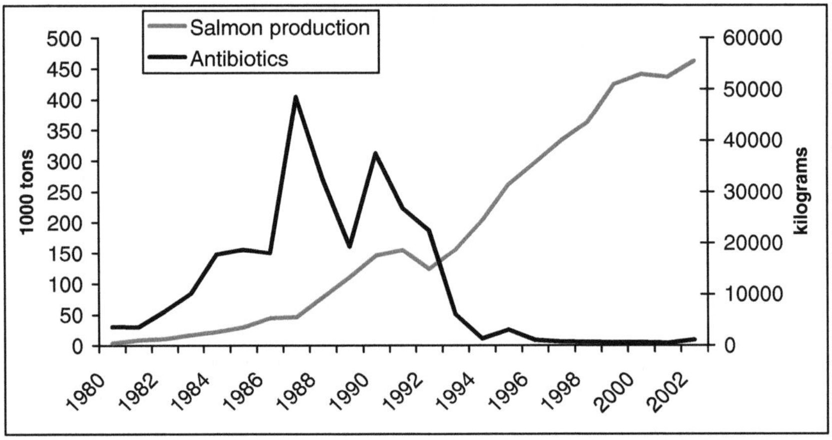

Fig. 9.7 Norwegian Salmon Production and Antibiotics Use in Norwegian Aquaculture (Norwegian Directorate of Fisheries and Norwegian Medicinal Department, personal communication)

If there is no negative feedback on expected profitability, however, it is unlikely that the industry will internalise detrimental environmental effects. In this case the government has to regulate the industry if the effects are to be avoided. The rapid growth of global aquaculture has represented an environmental challenge for authorities. First, knowledge about the environmental effects of aquaculture has been limited, or at worst lacking. This has called for extensive research to identify causes and effects. Second, in many places local governments do not have the resources to implement and enforce regulations. Finally, it is desirable with regulations that, on the one hand, are efficient in addressing the externalities and, on the other, allow the aquaculture industry to be economically sustainable.

The Fishmeal Trap

The 'fishmeal trap' is the name of a hypothesis that claims that aquaculture is environmental degrading because it leads to increased fishing effort to satisfy increased demand for feed (Naylor et al. 2000). Moreover, it follows that the availability of marine feed will put a limit to how much the aquaculture sector can produce. While the fishmeal trap is mentioned in relation to aquaculture in general, it is clear that it is an issue only in some forms of finfish farming, and does not apply to farming of sea weeds and shellfish. Furthermore, it will only apply to species that are fed with feed using marine inputs. This is a substantial part of the sector, as this is the case not only for carnivorous species like salmon and sea bass, but also for omni- or herbivorous species because the use of feed increases the growth rate. There are, however, some conditions that must be fulfilled for the fishmeal trap to occur (Asche and Tveterås 2004).

To what extent the fishmeal trap associated with aquaculture growth represents an environmental problem can be decomposed into two key issues, one pertaining to the regulation of capture fisheries and one pertaining to the market for protein meals. To what extent increasing demand for fishmeal increases fishing effort is related to the management regime in operation for the fishery in question. Hence, whether growth of aquaculture production can lead to unsustainable capture fisheries is primarily a fisheries management problem. However, as the track record of many fisheries management systems is not too good, this can be a real problem. It does, however, require that aquaculture growth increase total demand for fishmeal.

Even if fishmeal production has not increased during the last thirty years, in which industrialised aquaculture has expanded, evidence indicates that the fishmeal trap has not been an issue. However, from early 2005 to mid-2006 fishmeal prices have more than doubled to a record high level. While aquaculture expansion may have influenced recent price development, the main cause of the dramatic price increase is the economic growth in China. The Chinese income growth has led to an increased demand for animal proteins and in China fishmeal is widely used in animal feeds such as for poultry, in addition to aquaculture.

Since productivity growth is the main engine of growth in aquaculture, the increased fishmeal prices would prevent further growth for species that is too dependent on marine sources for food. The commercial breakthrough of cod aquaculture, for example, will probably be constrained if fishmeal prices are to remain at current high levels. Hence, scarcity may constrain growth of high-priced carnivore aquaculture species, in particular in the short run when feed technologies are given. Most aquaculture species, however, are herbivore, and even salmon is these days turning into a semi-vegetarian, so, in terms of volume, fishmeal availability should not limit aquaculture growth.

In addition to innovations that reduce the dependencies of marine proteins in aquaculture feeds, it should also be mentioned that recent developments in technologies for krill fisheries represent a potential alleviation to limited fishmeal supply, as it taps into an abundant resource of marine proteins.

9.4 Lessons for the Future Development of the Aquaculture Sector

Although aquaculture is an age-old food-producing technology, its development only really picked up pace in the 1970s. A revolution then occurred with the introduction of semi-intensive and intensive farming practices, as producers started to actively influence the growing conditions of the fish with feeding, breeding, etc. The control of the production process that was obtained allowed a number of productivity enhancing innovations to take place. These led to semi-intensive practices as formula based feeding and intensive and industrial practices as in sea bass farming.

The substantial increase in productivity has reduced production costs substantially in intensive and semi-intensive aquaculture production. This gives strong incentives for existing producers to increase production and for new producers to enter the industry, because it is profitable. Substantial increases in the production of species such as salmon and shrimp, but also tilapia, sea bass, etc., have led to large reductions in price. The speed with which price reductions follow the increased production depends on market growth and the extent to which the species in question can win market shares in existing markets rather than having to create new markets. Typically, prices will decrease faster in isolated markets, whether they are domestic or export-oriented. This is the case also for other species and for developing countries, like tilapia in Egypt, as well as developed countries, like sea bass and sea bream in southern Europe.

With this structure, the only way for companies to survive and remain profitable in the face of decreasing prices is to reduce production costs through productivity growth. The fact that shrimp and other species continue to be farmed in increasing volumes despite reduced prices is evidence that they follow the same pattern as salmon, even though the specific elements contributing to the productivity growth can differ. For species, such as turbot, where the productivity growth is less rapid, the production increase is also substantially smaller.

For large-volume species such as salmon and shrimp, production takes place in different regions of the world. Relative productivity development (including the negative effects of diseases, trade issues and environmental costs) will determine where production takes place, both between and within regions. In the future, we are likely to see this kind of competition appear also between species; such inter-species competition already exists to some extent between small whitefish species such as tilapia and catfish and similar species in the same market segments. Hence, although technical progress is likely to contribute to an increase in the global aquaculture production, the production of specific countries, regions or species may be reduced if they are not competitive.

Cost considerations are likely to limit the number of high volume species. If one looks at the world's meat production very few aquaculture species are likely to be of high volume. However, for these high-volume species there will certainly be quality variations, giving some producers a higher margin even though the price movements are determined in the larger market. Moreover, there will certainly be a number of niche species that are produced in relatively low volumes (which still may be counted in tens of thousands of tonnes), that will target the most valuable market segments where variation and uniqueness gives a higher price. In temperate high cost regions like Europe, such species are likely to be the most important.

It is impossible to say which regions will be the leading producers in anything but a short run perspective, as this will depend on which species that take the lead and which environment they prefer. China is currently the world's largest aquaculture producer. However, China and South-East Asia's leading role is likely to diminish in a longer term, as economic progress will make land and labour more expensive. Cost considerations will also indicate that Africa and South America are the regions where one is most likely to see the largest growth.

This development is an opportunity as well as a challenge for many developing countries. Increased substitution gives more opportunities to gain market access and to win market shares. However, it also increases the potential competition. Different regions have different potential advantages. Seafood species grow faster in warmer waters, and tropical and subtropical regions therefore have a clear advantage. However, these environments suit different species than those found in colder climates, and it may be more difficult to gain consumer acceptance for these unfamiliar species. Infrastructure and production structures with many small producers can also present a challenge in gaining access to and fair treatment in global markets.

The control of the production process in aquaculture in many ways makes aquaculture similar to any other growing industry. Accordingly, the growth in other industries should hold a number of lessons and perspectives for the future growth of aquaculture. Although it is not perceived as equally dynamic in many parts of the world today, agriculture is in many ways the industry that is closest to aquaculture. By becoming increasingly intensive, agriculture has enabled humanity to increase the global food producing capacity tremendously. Certainly it is not equally intensive everywhere, and hunting, gathering or very extensive farming practices are still used as food producing technologies many places in the world today. But because such production techniques are not very efficient, the share they represent of the world's food supply is relatively less important.

Agriculture certainly has had its problems, and still has. It has a huge environmental impact as landscapes are transformed, something that still continues with a substantial deforestation. Erosion or overuse of the soil can make the land unproductive, and there are accordingly numerous examples indicating that agriculture can be conducted in an unsustainable manner. These issues are a set of challenges that has to be mastered for agriculture to be sustainable, and although one does not succeed everywhere, the general experience for the two previous millennia is positive. Aquaculture faces many of the same opportunities and challenges as agriculture have faced, and in general solved well.

9.5 Conclusions

There is no doubt that aquaculture production will continue to grow, and substantially. As shown, e.g., by Delgado et al. (2003), demand for seafood will grow both because of increased economic growth and an increased global population. This provides an environment where growth is possible provided that aquaculture products are competitive. It is clear that a lower production cost due to productivity growth is the main engine for growth in aquaculture production. Lower production cost makes aquaculture production of different species profitable in a large number of countries. This is also what makes, and will continue to make aquaculture products competitive in the markets where they are sold, independently of whether these are export or domestic markets. This productivity growth is possible because of the higher degree of control over the production that is present in aquaculture relative to traditional fisheries.

To obtain this control with the production process, one needs to move towards relatively intensive production techniques. Unfortunately, while this is what has made it possible for aquaculture to become an important source of food, it has also been responsible for creating some major environmental challenges. At this stage, it is impossible to claim that most aquaculture will follow environmentally sustainable practices, but the evidence from other industries indicates that this is likely as the environmental issues can generally be addressed. Moreover, environmentally detrimental practices will in general hamper productivity growth or increase production cost, and make such farmers uncompetitive. This is true for shrimp farmers that mine locations, as they do not undertake the investments necessary for adapting new technologies, although such practices will potentially always be profitable in poor countries with poor regulations. It will also be true for species that requires substantial quantities of fishmeal if marine inputs for feed become scarce. This is because these species will be uncompetitive because of higher production costs if they cannot substitute this feed for feed with a lower or no content of marine inputs.

In many ways, aquaculture development is still in its infancy. For many species one has not even closed the production cycle yet, i.e., one still depends on the harvesting of wild fingerlings rather then producing them from a domesticated stock. Hence, there is a substantial potential for further productivity growth, and for aquaculture production to become less costly. There will certainly be boom-and-bust cycles as production at times increases faster then the productivity growth, but the underlying trend is clearly one of sustained growth. In a worst-case scenario, there may be import bans from the EU and the US because of environmental concerns. However, continued productivity growth will ensure that aquaculture becomes an increasingly important food supplier locally because it is profitable and produces an easily traded commodity. It is also unlikely that there will be import bans on most aquaculture species, and any trade limitations stemming from environmental concerns are likely to affect only a few species. On the other hand, trade restrictions, if they are not targeted at achieving sustainable practices, can limit economic development and local food supply. Many dumping cases indicate that this is a problem.

The evidence so far indicate that within a group of species there can be strong competition (Asche et al. 2001), but so far there has been little competition between different aquaculture species. As the industry matures, this is likely to change profoundly. If one looks at agriculture, only a few species are likely to be large volume species. These are likely to be segmented into groups like meat and vegetables, but there are not likely to be many such groups with large volumes. Within each group, there are likely to be a few important parameters that distinguish the species, like size, flesh texture and colour. The species that will be the large volume species are those where the potential for productivity growth is largest and where production cost can be the lowest. As higher water temperature in general leads to higher growth, producers in tropical and subtropical areas have an advantage. This should indicate that most of the aquaculture production will continue to take place in developing countries, provided that suitable sites are available with respect to water quality etc. However, as the EU, Japan and USA are the main markets, infrastructure also is a main concern, as the consumers do not care where in the supply chain costs occur, but only about the final price. This gives producers in other regions an opportunity, in particular since fresh product for most species are the most valuable product form.

There are not likely to be many wild species at the seafood markets in the future, and again the parallel to agriculture is strong. However, in contrast to agriculture, a few sectors are likely to remain. In particular, a whitefish market, mainly frozen, is likely to exist. There may be one or more similar markets too, but the remaining species from the wild as well as the aquaculture sector is likely to go into two very different types of markets. There exists a high-end market for food where exclusivity and exoticness are important attributes and prices in general is high. In these markets there are no high volume species, but a substantially variety. A large number of seafood products, wild as well as farmed, are likely to be sold in such markets. Even though prices are relatively high, margins are likely to be low also in this market segment for most products as competition to be a part of this market will be tough. For the species that cannot get into this market, only low price markets remains. This will partly be low value products to the poorer countries in the world, who do not have the ability to pay for other foodstuff, and partly be as inputs in the feed for other food products like fishmeal.

Acknowledgement Financial support from the Norwegian Research Council is acknowledged.

References

Anderson JL (2002) Aquaculture and the future. Marine Resource Economics 17:133–152

Anderson JL (2003) The International Seafood Trade. Woodhead Publishing: Cambridge

Norman A (2005) Personally Collected figures from the Obour market

Asche F, Bjørndal T, Young JA (2001) Market interactions for aquaculture products. Aquaculture Economics and Management 5:303–318

Asche F, Guttormsen AG, Tveterås R (1999) Environmental problems, Productivity and innovations in Norwegian salmon aquaculture. Aquaculture Economics and Management 3:19–30

Asche F, Tveterås S (2004) On the relationship between aquaculture and reduction fisheries. Journal of Agricultural Economics 55(2):245–265

Asche F, Roll KH, Tveterås R (2007) Productivity growth in the supply chain – another source of competitiveness for aquaculture. Marine Resource Economics 22:329–334

Bjørndal T (1990) The Economics of Salmon Aquaculture. Blackwell: Oxford

Bjørndal T (2002) The competitiveness of the Chilean salmon aquaculture industry. Aquaculture Economics and Management 6:97–116

Delgado CL, Wada N, Rosegrant MW, Meijer J, Ahmed M (2003) Fish to 2020, Supply and Demand in Changing Global Markets. International Food Policy Research Institute, Washington D.C., USA

FAO (2006). FAO FISH STAT Plus 2004 Database, Rome, Italy

Guttormsen AG (2002) Input Factor Substitutability in Salmon Aquaculture. Marine Resource Economics 17:91–102

Naylor RL, Goldburg RJ, Primavera JH, Kautsky N, Beveridge MCM, Clay J, Folke, Lubchenco J, Mooney H, Troell M (2000) Effects of aquaculture on world fish supplies. Nature 405:1017–1024

Norwegian Directorate of Fisheries (Various Years) Økonomiske Analyser av Fiskeopdrett (In Norwegian, Economic analysis of fish farming), Bergen

Norwegian Seafood Export Council (2005) Salmon Database, Tromsø
Tveterås S (2002) Norwegian salmon aquaculture and sustainability: The relationship between environmental quality and industry growth. Marine Resource Economics 17:121–132
Tveterås R, Heshmati A (2002) Patterns of productivity growth in the Norwegian salmon farming industry. International Review of Economics and Business 49:367–393

Chapter 10
Status and Future Perspectives of Marine Aquaculture

Yngvar Olsen[1,*], Oddmund Otterstad[2], and Carlos M. Duarte[3]

Abstract The predicted growth in aquaculture production to 80–90 million tonnes year^{-1} in 2050 is constrained by important drivers, which can be divided into three main clusters: (1) a resource cluster (availability of resources such as space, feed and energy); (2) an attitudinal cluster (public and consumer attitudes, legislation etc.) and (3) an innovation cluster (new technology and market developments). This chapter discuss solutions to these bottlenecks based on the current status of aquaculture and possible developments in e.g. feed technology, off-shore and land based farms. A model for the possible future interactions between the three clusters is discussed. From this analysis, it is concluded that a major challenge for aquaculture is to achieve better control of the feed availability in the future. Only if this can be realised, aquaculture may grow in a similar way as agriculture. Space for the industry and public environmental concern are other main driving factors of the development, but these constraints can most likely be mitigated through technological improvements.

Keywords Bottlenecks, environment, technology, attitudes, policy

10.1 Introduction

In this chapter, we shall examine some important bottle-necks for further development of aquaculture during the next decades. By identifying present trends in marine aquaculture development, we shall point out a few relevant scenarios towards year 2050.

[1]Trondhjem Biological Station, Department of Biology, Norwegian University of Science and Technology, Trondheim, Norway. Tel: +47 73591592; Fax: +47 73591597; E-mail: yngvar.olsen@bio.ntnu.no

[2]Department of Sociology and Political Sciences, Norwegian University of Science and Technology, Trondheim, Norway

[3]IMEDEA, CSIC-UIB, Miquel Marqués 21, 07190 Esporles (Mallorca), Spain

* Corrosponding author

The growth of the human population, forecasted to reach about 9 billion inhabitants along the 21st century, along with the increasing per capita consumption of food and water is likely to be conducive to increasing difficulties to feed humanity. Indeed the projected population is within present estimates of the carrying capacity imposed by the availability of freshwater, estimated at 10–16 billion people using conservative assumptions (Cohen 1995). Indeed, the average per capita water consumption already exceeds by 50% that used in deriving the estimated carrying capacity above (Cohen 1995), suggesting a smaller carrying capacity than anticipated. These estimates suggest that the availability of water to produce food will become a bottleneck to the human population along the 21st century; a situation that has already been reached in many densely populated regions of the world (e.g. Asia and Africa) (Vörösmarty et al. 2000). The only source of food that does not consume freshwater is the ocean as provided through wild harvesting or, increasingly, aquaculture. Enhancing food production from the ocean may, therefore, help feed humanity by releasing pressures on freshwater resources.

The total fish supply and hence consumption has been growing at a rate of 3.6% per year since 1961, while the world's population has been expanding at 1.8% per year (WHO 2003). The average apparent per capita consumption of fish products has increased from about 9 kg per year in the early 1960s to 16 kg in 1997 (WHO 2003), close to the recommended intake of 50 g day^{-1}, or about 2–3 servings per week, of fish and seafood products for a healthy diet. Provided the rise in human population to near 9 billion in the 21st century, the required fish production to satisfy market demands and dietary requirements will be about 160 million tonnes $year^{-1}$, which is twice as high as the current fisheries catch, which has reached a plateau at about 82 million tons (Fig. 10.1). Wild fish captures are unlikely to increase, as most fish stocks are depleted (Myers and Worm 2003). Accordingly, aquaculture, which already supplies one third of the fish production, must develop further to produce the recommended dietary components.

These two considerations: (1) the likely exhaustion of the capacity of agriculture to feed a growing human population, and (2) the need to increase fish production beyond the ceiling of exploitation of wild fish stocks to meet market needs, are likely to act as motors driving a continued expansion of aquaculture. Aquaculture technologies and practices must greatly improve relative to current practices to meet these challenges. Here we examine present trends and driving forces of marine aquaculture development and discuss possible projections and scenarios of aquaculture along the next decades on the basis of possible scenarios and bottlenecks.

10.2 Current Trends of Development

A shift in paradigm for the development in global marine fisheries took place around 1990. The change in trend towards a stagnation of capture from global marine living resources was verified around one decade later. FAO statistics

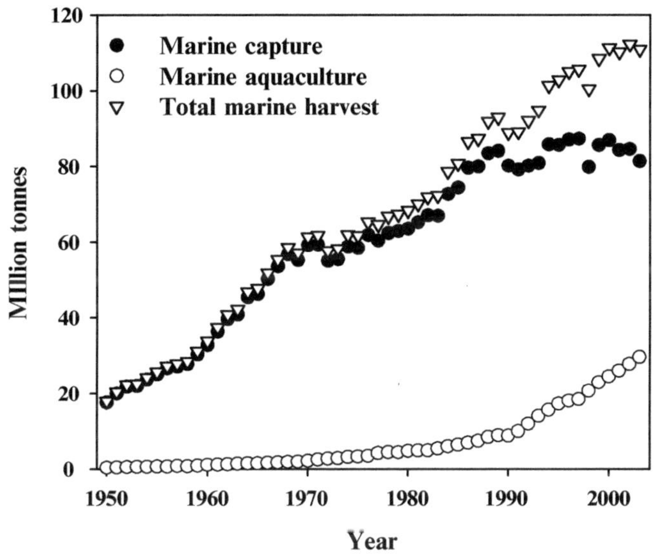

Fig. 10.1 Historical development of global marine capture, marine aquaculture, and the total marine harvesting of seafood (FIGIS, FAO statistics)

(FIGIS[1]) clearly reveal that the total harvest of marine biological resources, defined as the sum of resources captured through fisheries (termed fisheries or captured) and biomass produced and harvested from mariculture (termed cultured), has with minor fluctuations showed a relatively steady increase of 1.5–2.0 million tonnes year^{-1} in the period 1950–1990 (Fig. 10.1). This increased harvesting mainly reflects an increased yield from fisheries, driven by improved technology and higher requests in the market (Caddy and Garibaldi 2000). It is now a general view that the fisheries is beyond the level of exploitation that secures an optimal multispecies yield of marine biological resources (Myers and Worm 2003), although there may be minor stocks of fish that are still under-exploited.

A major uncertainty for the estimation of maximal sustainable yields of capture is the lack of information on quantities that are lost or discarded during fishing. It is a general assumption that a significant fraction of the catches are lost or discarded, mostly as a result of economic and political conditions (27 million tonnes in late 1980s, Alverson et al. 1994; 20 million tonnes in 1998, FAO 1999, Hall and Mainprize 2005). It is, of course, a major challenge to reduce this problem in the future, both from resource availability and management perspectives.

Figure 10.1 illustrates the marked increase of marine aquaculture production from around 1990, the time when fisheries yields levelled off. Mariculture has a

[1](FIGIS, http://www.fao.org/figis/ servlet/static?dom=root&xml=index.xml)

long tradition in some regions of the world, particularly in South-East Asia. The historical and economic importance of mariculture has, however, been low compared with freshwater aquaculture. It is remarkable that the increase in mariculture with time has quantitatively replaced the earlier increases in captured resources yielding a pattern of steady increase in total harvesting at 1.5–2.0 million tonnes per year in the period 1990–2003 despite the stagnation of wild-stock catches. This clear trend in captured and cultured marine organisms clearly indicates that a future increase in seafood production must come from culture.

The production yields originating from freshwater aquaculture are still higher than those from mariculture, and the rate of increase in the period 1990–2003 is also higher (Fig. 10.2). A major structural difference between freshwater and marine aquaculture is that fish is a major product in freshwater and only a minor in seawater. The Chinese production of carps is by far the dominant component of freshwater aquaculture, and the increased freshwater production after 1990 took place mainly in China. The trend illustrated in Fig. 10.2 may suggest that the potentials of freshwater culture should be more thoroughly considered.

A future expansion of freshwater aquaculture may be feasible in some parts of the world, with excess freshwater available, but freshwater is a limiting resource in most densely populated regions, for example in China, and its availability is forecasted to decline in the future (Cohen 1995). The increased yields of freshwater fish is to some extent a result of intensification of pond cultures of carps, and history has shown that intensification of low technology systems like ponds are relatively risky, as experienced in shrimp aquaculture. Recent experiences have revealed that

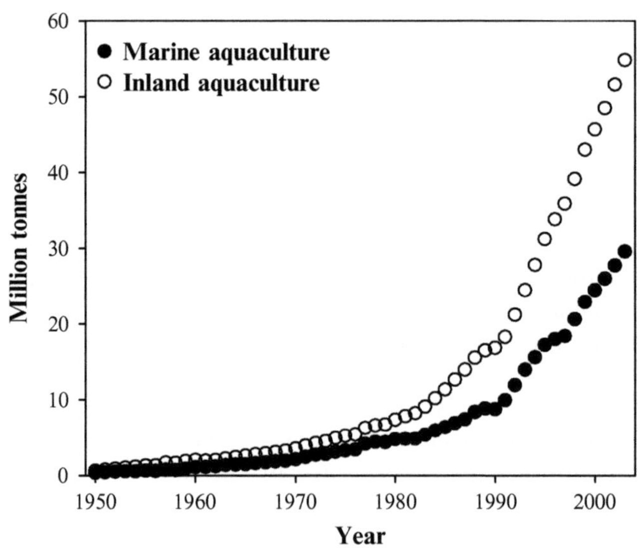

Fig. 10.2 Historical development of global freshwater aquaculture and marine aquaculture (FIGIS, FAO statistics)

sustainable ponds yields for shrimp production are lower than expected (Fast and Leung 2003), and negative effects caused by extensive phytoplankton blooms (Alonso-Rodriguez and Paez-Osuna 2003) and chemicals added to mitigate ecological problems (Graslund and Bengtsson 2001) are well documented problems during intensification. The general prospect for freshwater aquaculture is promising, but the global potentials seem all together to be more restricted than for mariculture because of the limited supply of freshwater.

A closer view of FAO statistics reveals that marine plants (macroalgae) and shellfish (crustaceans and molluscs) are the major group of organisms produced in mariculture on a global scale (Fig. 10.3A). This production is mainly taking place in SE Asia,

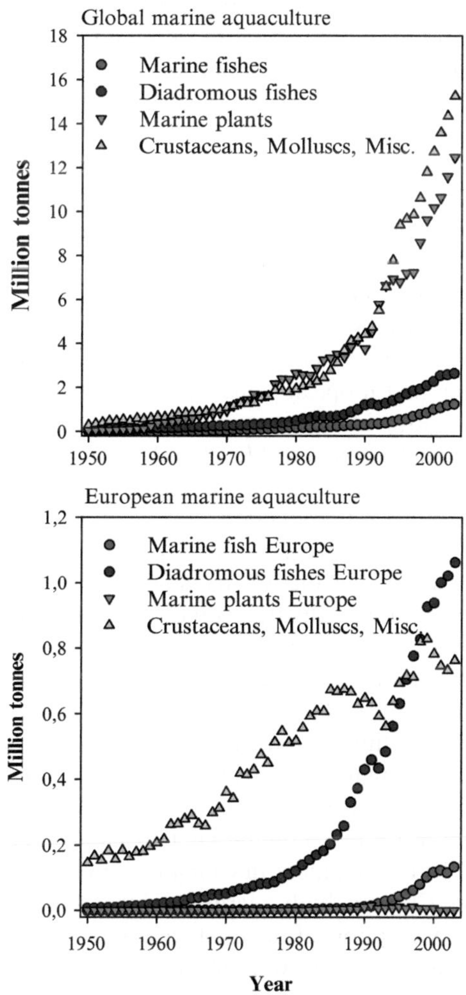

Fig. 10.3 Historical development of some main groups cultured in the sea. A: Global production values and B: European production values (FIGIS, FAO statistics)

and these groups represent the majority of the increased production after 1990. The global increase in production of diadromous fish, mainly comprised by Atlantic salmon, and marine fish species (many different species) is quantitatively less, but these groups have exhibited a steady and significant rate of increase since 1990, especially Atlantic salmon. The cultivation technology of marine fish is throughout more complicated than that for most other groups of organisms (Moksness et al. 2004). It is in particular the technology for producing fish juveniles that has inhibited developments of marine finfish culture, and this is most likely why cultivation of marine fish has grown the slowest on a global scale.

European marine aquaculture is structurally different from the general global picture. There is no significant marine plant production in Europe whereas there is a historically lasting production of shellfish like oysters and blue mussels. On the other hand, the European production of salmonids constitutes almost half of the world production. The most important species is Atlantic salmon, mainly produced in Norway and the UK. Europe is also responsible for more than 15% of the global marine fish production, and the most important species are sea bass and sea bream, which are mainly produced in the Mediterranean countries. Fish is among the most highly valued products of seafood, and the high finfish production in Europe is quite successful and also competitive. Salmon is exported to Southern Europe and also worldwide. The pattern is different for the Mediterranean species, which are mostly consumed in the region where they are produced (Asche et al. 2007, Chapter 9, this book), except for red tuna fattened in Mediterranean farms, which is entirely exported to Japan. The high share of highly valued finfish in European aquaculture, organisms that are relatively knowledge- and technology-intensive as compared to many other cultured groups, suggests a relatively high technological level of the European aquaculture industry. Other countries, like Chile and Canada, are rapidly emerging as significant competitors to Europe in salmonid production.

Although global statistics reveal a very rapid growth in global mariculture from 1990, a closer view shows that China is responsible for the majority of the global growth in mariculture production (FIGIS, FAO statistics), with total aquaculture increasing by 34% for the period 1990–2000 (Brugere and Ridler 2004). The increasing production in China started at the same time as captures levelled off. There is, therefore, most likely no other cryptic market related mechanism behind the coincidence in time of reduced fisheries and increased culture. China is not the only country showing a growing mariculture trend, this is a relatively consistent trend also in other SE Asian countries like Vietnam and Thailand, and in other countries with a major commitment to aquaculture, like for example Norway and Chile. The development in China is, however, important for the rest of the world, because China is both a large producer and committed to mariculture. The average consumption of seafood in China has increased from 10 to above 25 kg per capita from 1990 to 2002 (Vannuccini 2004), exceeding global average values, but the country has an overall objective to increase the export of high-priced seafood products.

Table 10.1 Predictions made for future developments in global aquaculture

Reference and forecast period	Assumption: price	Assumption: per capita consumption (kg $year^{-1}$)	Assumption: development of capture (% $year^{-1}$)	Predicted growth rate (% $year^{-1}$) (range)
IFPRI 2020	Flexible	17.1	0–0.7	1.8–3.5
Wijkström (2003)	Constant	14.2–19.0 17.8–30.4	0–0.7	0.4–4.6
2010				3.4–5.3
2050				3.2–3.6
Ye (1999), 2030	Constant	15.6–22.5	0–0.7	0.6–4.2
Trend analysis of mariculture (Fig. 10.1)			0	
1.5 million tonnes $year^{-1}$				3.6
2 million tonnes $year^{-1}$				4.5

There are a number of predictions made for the future needs for seafood and the production yields from aquaculture (Ye 1999, Wijkström 2003, IFPRI-model 2020[2]). The assumptions made for these predictions, compiled by Brugere and Ridler (2004), are different, and it is not always clear if the predictions are made for total or marine aquaculture (Table 10.1). Among the major uncertainties of the predictions of the future human demands for seafood, are the development of natural captures, the development of aquaculture in China, and the future trends in developments of price and per capita consumption (see Brugere and Ridler 2004). All estimates are therefore advanced guesses. The predictions of growth in aquaculture production vary between 0.4% and 5.3% $year^{-1}$ for a period between present and 2020–2050 (Table 10.1).

A simple prediction which we base on recent steady trends suggesting that total marine harvesting will continue to increase by 1.5–2 million tonnes $year^{-1}$ up to 2020, like through the period before (Fig. 10.1), yields annual, exponential increase rates of 3.6–4.5% $year^{-1}$ in mariculture production, with marine plants included. This implies that mariculture production, with plants included, will increase to 55–64 million tonnes by 2020. Other predictions, published by other authors, compiled in Table 10.1 are based on different assumptions; the published values include total aquaculture with plants excluded. All predictions suggest, however, that mariculture production of seafood will continue to increase during the first decades of our century. If the trends up to 2020 become similar to that in earlier decades, the mariculture production will grow with an average exponential rate of about 4% $year^{-1}$ during the next two decades. The possible constraints for the increasing mariculture production that are predicted are discussed below.

[2] International Food Policy Research Institute

10.3 Main Driving Factors of Future Development of Mariculture

Despite their many uncertainties, predictions on the future demands and production of mariculture are of strategic importance, because they stimulate thinking on the social, economic, and technological factors that may affect the realization of those predictions. Such evaluations, often termed as foresight studies, may be more important than the predictions themselves. The identification of the main driving factors for the development in mariculture and the consideration of how the aquaculture and fishery industries and the supporting science should prepare to meet the future challenges are, therefore, of crucial importance.

In a foresight study carried out by a multidisciplinary group at the Norwegian University of Science and Technology (see methods in Box 10.1), the optional driv-

Box 10.1 Aquaculture Foresight Study, brief explanation of methods

The objective of the foresight study within aquaculture technology at the Norwegian University of Science and Technology (NTNU) was to establish a basis for future long-term research and education strategies of the university within the sector of aquaculture, including the complete value chain from fisheries to market (*"Aquaculture from sea to table"*).

Key individuals connected with the seafood industry, researchers from NTNU and SINTEF, resource management officials and prominent international researchers and experts within aquaculture were invited to contribute through the writing of more than thirty short feature articles on the future of aquaculture. Most of the feature article writers also knew fisheries and aquaculture intimately, but a number of typical all-rounders were also invited. The authors were particularly asked to bring up their main future concerns for the industry in a personal paper without references.

The response from the professions was surprisingly positive. These feature articles formed the starting point for a seminar where key researchers and representatives from the research council, resource management and the industry contributed further to clarifying the challenges and possibilities in the years to come. A project group of six people and a steering council from key communities within NTNU further refined the material.

A conclusion from the analysis of the incoming material was that the main driving factors for the development of aquaculture and the seafood industry seem tied to three main clusters of "external driving factors". These were: a) The availability of resources, e.g., water, feed, energy, and human resources b) human attitudes tied to ethics, the environment and health, and c) our ability to develop knowledge and innovate.

The model generated (Fig. B1) was influenced by an environmental feedback model known as the DPSIR model (Turner et al 1998), but more important was the inspiration from the actual synthesizing of the empirical material collected through the foresight process at NTNU. Each of the axes indicates a crises index – from no crises (or balance) in origo (A), through a normal adaptation to problems, passing over to the stress level and ending up with a situation out of control. We have placed the points (B) and (C) to symbolize a situation with high crises. The dotted line symbolizes the threshold value where the crises turns really "wild" and we get cumulative effects where everything rapidly gets out of control. Every position inside of the dotted arc is more or less "sustainable". The model describes the trajectory of an object of analysis (here mariculture) where the path is composed of chronological snapshot. In some cases the trend is moving toward crises, in other cases it is brought back to comfortable balance and in other cases it ends up in chaos. The model might be used for comparison between different nations as well as it might be applied at different structural levels.

The resource axis indicates when a shortage of one or more important resources is registered (resources that are crucial for humans to survive; like food, water, fresh air, oil, etc.) This shortage might be revealed gradually, or it happens without any warning. The governance or attitude axis indicates the level of political activities. Usually in our days there is a close relationship between the resource and the attitude axes. For instance, a resource shortage is naturally followed by compensatory actions. The third cluster of drivers, the innovation cluster, might be seen as a third axis (z) of the model (see 3D figure). Innovations and new knowledge (or communication of exciting knowledge to stakeholders) are the only means to bring the situation back to balance.

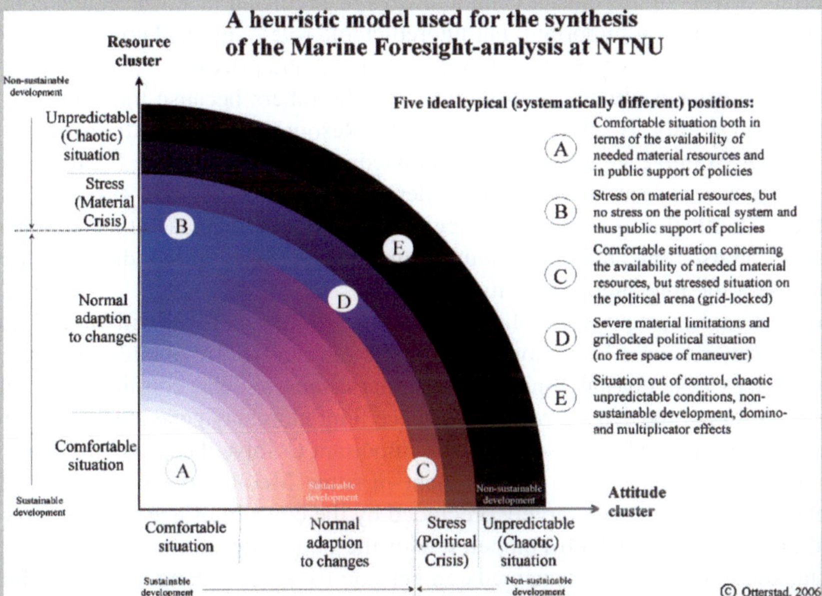

Fig. B1 Conceptual model for analyzing the consequences of driving factors and perspectives of future marine aquaculture

ers of the future developments in fisheries and aquaculture were clustered in three main groups as follows:

Resource cluster – Availability of resources; e.g., space, feed, and energy

Attitude cluster – Public and Consumer Attitudes and legislation, related to e.g., ethics, environment and health

Innovation cluster – The ability to develop new knowledge and implement the required innovation of technologies and markets

The Resource cluster involves factors related to resources, which may for example be space for operation in coastal and open ocean locations, environmental quality for safe food production, adequate feed resources, energy at an acceptable cost, acceptable infrastructure, investments, and human resources (competence). All of these optional drivers, ranging from natural resources to social issues, can be interpreted as resource related factors which may affect the development and potentials of mariculture.

The Attitude cluster involves factors more related to emotions, culture, and legislation, with many groups involved from the public, management authorities, industries, and political stakeholders. Such driving factors may for example be public environmental concern, legislation that affects the mariculture industry, other requirements for sustainable development, sudden unpredictable events coming up, and adverse public relation campaigns from competing sectors and other stakeholders opposing aquaculture developments. This group of drivers is to a great extent related to society with its great variety of stakeholders. Many groups define themselves as producer of premises for mariculture development, among them the competing industries, for example the chicken industry. The factors that become important drivers in this cluster cannot always be predicted in advance, and contingencies often play an important, but unpredictable role (e.g., mad cow disease and bird flue, and their positive impact on fish demand and prices).

The Innovation cluster is different from the others because these factors all relate to our ability to solve problems of the Resources and the Attitude clusters through for example generating new knowledge, establishing new break-through technologies, consensus among stakeholders, new markets, and new applications of products. Political commitment, strategic decisions of the aquaculture sector and the states, legislation, and the economic power of the aquaculture industry will certainly affect the innovation capacity.

In most cases, factors within the Resource and Attitude clusters will interact quite strongly. Any resource limitation, for example the availability of sites for mariculture activity in the coastal zone, will generate discussion and conflicts with other potential users of the coastal zone. This space limitation may generate a wide range of public concerns, which indeed may be important to solve the space problem.

For instance, aquaculture development in some Mediterranean locations (e.g. Balearic Islands) has been largely precluded by the concern of negative impacts on the tourism industry, which competes with the aquaculture sector for use of the coastal zone. In cases of a sustained competition for space among many stakeholders, it may be difficult to solve the resource problem for mariculture through time-consuming consensus processes and industrial improvements. In the ultimate situation, there may be a need for a technological break-through to release the resource constraint in a way that will also satisfy important stakeholders. Technology for offshore mariculture may, for example, release the pressure on coastal sites and may therefore be a solution that may satisfy both the industry and the stakeholders involved. This solution may, however, involve new stakeholders and conflicts and will depend strongly on technological innovations.

The above case is just one example of possible issues affecting aquaculture development, and the exercise to examine interactions can be illustrated in a 3D coordinate system with the main clusters of drivers as coordinates (Fig. 10.4). The most favourable situation is close to the origin; with an acceptable recourse situation and few social and political conflicts. Any upcoming problems of resource availability or attitudes will displace the situation away from the origin, signalling potential bottlenecks affecting the development of aquaculture. If the problems become too severe, the situation may be out of control entering an unstable state, where the situation can be quite unpredictable (Fig. 10.4, e.g., mad-cow decease in agriculture, situations of war, or unstable governments). The situation may improve by itself, time is then important (e.g., consensus process), or it may be solved through major innovations (*z*-axis) that brings the situation back to origin, conducive to aquaculture development.

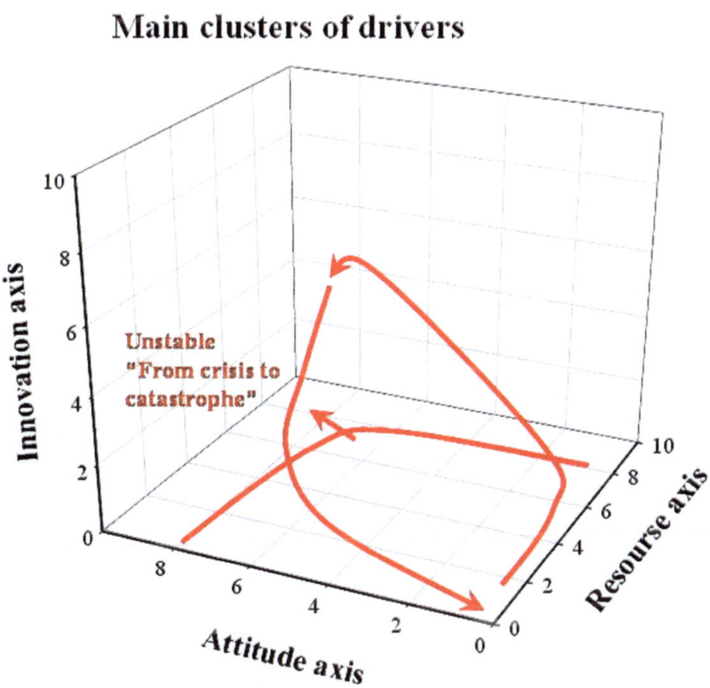

Fig. 10.4 Main clusters of drivers, describing a resource – attitude – innovation space, for analysing potential effects of different driving factors on future aquaculture development

It is likely that industrial developments and innovations will require some resistance and challenges during periods; an everlasting origin-situation may lead to stagnation and losses of competitive force. If so, we can conclude that there is a very positive long-term effect on sustainability of continuously challenging a developing industry even if the short time effect may seem to be negative for the industry ("competition makes strong"). The driving factors discussed below have been evaluated broadly using the methodology represented in Box 10.1 and Fig. 10.4.

10.4 Future Perspectives on Mariculture

The future development of mariculture is dependent on many factors and is not easily predictable. The above section treated some driving factors that may, under given situations, affect developments quite pronouncedly. A primary factor not mentioned specifically, which is relatively predictable, is the political, social and demographic effects of the growing global human population. The growing population, and an increasing market demand, will most likely become a powerful driving force for a further increase in mariculture activity during the 21st century. For our prediction of the future development from 2000 to 2050, we assume the following scenario:

- The global population will develop as predicted by UN, reaching 9 billion people by 2050.
- There will be a parallel, steady increase in the demands for seafood in the world market.
- Capture from fisheries levelled off at today's value of 80 – 90 million tonnes year^{-1}.
- There will be no major world crisis that will reduce the purchasing power of the global population during 2006 – 2050, which will continue to rise steadily.

10.4.1 Bio-resources For Fish Feed (Resource Type Driver)

There is a fundamental difference in the developmental stage of agriculture and aquaculture. The provision mode for seafood is now gradually moving from capturing to culturing. A comparable development took place on land at the beginning of the Holocene, when agriculture gradually took over hunting in the ancient urban human communities. The driving force for this process was most likely the increasing needs for food of the growing human population, an increase which could not be achieved by harvesting nature. The domesticated animals were controllable herbivores that exploited plants from nature. The comparable activity in the sea is termed sea ranching; it involves release of juveniles, and some times also habitat improvement and control of predators, but this activity is far more complicated and economically questionable than animal farming on land. The concept of sea ranching was intensively studied from the early 20th century in Europe and the USA, but is commercially used only in SE Asia where this type of mariculture has a stronger tradition (Svåsand and Moksness 2004).

In traditional mariculture, the cultured species are maintained under strict control in relatively intensive cultures (high biomass). One consequence is the need to provide fish food. The availability of feed resources for aquaculture is believed to become a main driver of mariculture development. Food resources are already becoming limiting, particularly during El Niño events, when the major fisheries of some of the upwelling regions in eastern Pacific and Atlantic waters can be greatly reduced. The major bottleneck is, in fact, the availability of marine lipids (Opsahl-Ferstad et al. 2003). The feed companies are already adding more and more lipids and proteins from land agriculture into fish feed, and a major part of their research budgets over the past decade has been related to this specific task.

Marine species require food of marine origin for growth, or more specifically they need to be supplied with long chain ω3 fatty acids such as DHA (22:6 ω3, docosahexaenoic acid) and EPA (20:5 ω3, eicosapentaenoic acid). Such highly unsaturated ω3 fatty acids (ω3HUFA) are only present in high amounts in marine or aquatic organisms. Mariculture of marine organisms can therefore not easily be based on lipids from agriculture, because the essential fatty acid composition is different. Oils from higher plants are typically rich in ω6 fatty acids and short chain ω3 fatty acids (e.g., 18:3 ω3, α-linoleic acid). This fact represents a major potential constraint for developing mariculture during the next century, because the feed resources in mariculture are not as easily available as the feed resources in agriculture (plants). Mariculture feed for most type of organisms is based on herbivore animals that are harvested from wild stocks. It is therefore not obvious that mariculture will become as successful as agriculture, because such wild stocks have a limited availability (e.g., Myers and Worm 2003).

There are some potential ways to provide new marine fatty acids rich resources for fish feed. An ultimate challenge is to reduce discards from fisheries; some 20 million tonnes year^{-1} might then become available (Hall and Mainprize 2005). Other options are:

- Selected suitable oils (and protein) from agriculture – these resources can be used to dilute the marine resources, but cannot replace them entirely.
- "New" marine resources – a significant increased harvesting of resources can only be achieved by harvesting at low trophic levels of the marine food web, emphasizing the use of marine plants and herbivore invertebrates.
- Production of "single cell biomass" using microorganisms with marine-type lipids, ω3 HUFA rich.
- Production of genetically modified organism (GMO) with marine-type lipids – e.g., a transgenic microorganism ("single cell biomass") or a higher plant with ω3 HUFA genes transferred from algae

Use of resources produced in agriculture for mariculture will not contribute essentially to the world supply of food, and these conflicts are more easily seen from a densely populated region as SE Asia than from western countries, where the general attitude is that food is produced in too high amounts. The use of agricultural products for fish feed tends already to be a more difficult issue in, for example, Vietnam and China.

Some regions of the world's oceans have large stocks of herbivore copepods and krill that are abundant and potentially exploitable, depending on their distribution in time and space. The standing stock biomass of Antarctic krill (*Euphausia superba*) is estimated at 500 million tonnes (range 125–750) (Nicol and Endo 1997), meaning that the annual production must be over 100 million tonnes year^{-1} (life span 6 years). The annual production of red feed (*Calanus finmarchicus*) in Nordic Seas has been estimated to 74 million tonnes (Aksnes and Blindheim 1996). Some 5–10% of the annual herbivore production is of comparable magnitude as the entire production of zooplanktivore fish (1st carnivore), and 1% is still a significant resource as compared with traditional regional fisheries. If zooplankton production in the sea tends to be food limited, we may expect that moderate harvesting will primarily result in less mortality of zooplankton and not necessarily in reduced standing stocks and availability for planktivore fish. If herbivore zooplankton is used as a source of feed for carnivore fish, for example salmon which feeds on this prey in nature, this would imply that cultured salmon would move one trophic level down in the seafood chain (Fig. 10.5).

The fact that zooplankton is food for important fish stocks brings up management and political dimensions, and the potential interaction with fisheries must therefore be thoroughly examined. But, considering that fish stocks have been decimated in the ocean (Myers and Worm 2003), the logical consequence is that their prey, zooplankton, must have been relieved from predator control, increasing in abundance and switching to food limitation instead. An increased availability of marine biomass from low trophic levels has, however, a limit because these potential supplies will become limiting at a later stage.

Marine macroalgae are other unexploited resources for marine lipids (Radwan 1991). Such algae are currently used as resources for an industrial production of

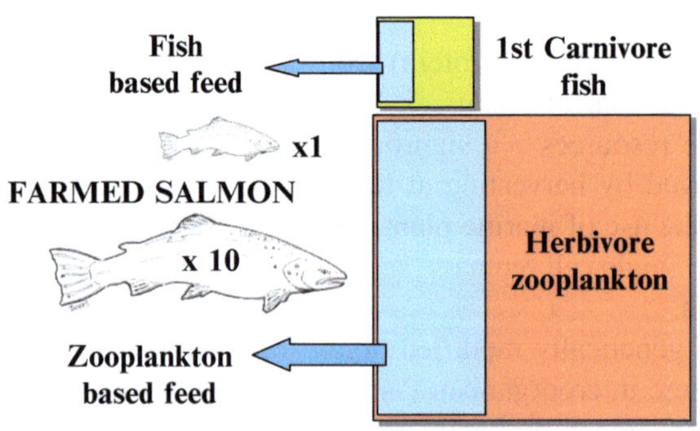

Fig. 10.5 Theoretical scheme illustrating the potential benefits of harvesting resources for salmon feed on herbivore zooplankton level. Feed for farmed salmon is currently based on first carnivore fish (arbitrary amount for fish farming, illustrated by blue square). If the zooplankton produced on the trophic level below is harvested instead – this zooplankton is the food of first carnivore fishes in nature – the salmon production can be increased by a factor of 10

biopolymers, like alginate, carrageenans, and agars, which are widely used, for example as additives for food and for pharmacological and medical purposes. It will be highly beneficial to also exploit the lipids from the harvested macroalgae that are processed to isolate polysaccharides, which may require a modified industrial extraction process. This, together with the fact that macroalgae contain only low contents of lipids, some 2% of dry matter, may explain why this resource is still not exploited as a source of lipids.

If a cost efficient controlled production of the food used in mariculture is established, mariculture is likely to generate changes of a scale equivalent to a new paradigm. It was first in the 20th century that domesticated animals were mainly fed cultured plants, alone or as a supplement to harvested plants. This was probably made possible by the introduction of cheap artificial fertilisers and improved technology, and it represented a new step towards full control with the agriculture food cycle, which thereafter has developed very rapidly to the western type of highly industrial and controlled agriculture of today.

To produce animal biomass for fish feed may seem unlikely, but mass production of a "single cell biomass" of marine quality seems more feasible and closer. A microorganism with marine lipids that can be grown based on for example cheap carbohydrates or methane has not yet been identified, but it has indeed already been searched for (Hinzpeter et al. 2006). It is perhaps most likely that such an organism will be produced using modern genetic methods (GMO). An alternative is that genetically modified higher plants will be used to produce marine lipids (Napier et al. 2004).

Based on the above premises and discussion, we can infer that mariculture development will be heavily dependent on the availability of resources for fish feed. The situations that may develop as a result of our ability to derive new bio-resources for fish feed can be illustrated by the following three scenarios for 2050, which indeed are both subjective and speculative:

Scenario 1 – No new marine feed resource for mariculture is made available
Scenario 2 – Large herbivore zooplankton stocks are utilised for fish feed in addition to agriculture products
Scenario 3 – Equal to Scenario 2, but "single cell biomass" becomes gradually available

The first scenario assumes that no new marine-type feed resources become available in the period (2000–2050), only new resources from agriculture become available. The second scenario assumes that harvested herbivore zooplankton develops into an additional major new resource. In the third scenario, zooplankton is still a new resource, but a marine type "single cell biomass" produced based on methane becomes gradually available by ca 2020 and is the major component of fish feed in 2050. The last scenario implies that feed for fish is no longer produced based on resources that could be used directly for human food. The use of methane as a substrate for single cell biomass production, primarily derived from human waste degradation, represents a potential sustainable, ecological, and everlasting resource for human food production, and aquaculture would have, like agriculture, full control of its food chain.

Scenario 1 – No New Marine Feed Resource for Mariculture is Made Available

Global fishery activities have continued along relatively traditional lines. The global harvest of fisheries is around 120 million tonnes, thanks to a better understanding of the marine ecosystem, a reduction in losses, improved harvesting technology, a strategic use of nutrient emission from agriculture to enhance fish production in the coastal zone, and new efficient international regulations and management principles of fisheries. Carnivorous species are mainly made available from fisheries, and have high prices in the market. The large catches of zooplanktivore fish, earlier used primarily for fish feed, are almost entirely used for direct human consumption, a result of an increased global economic welfare. The losses of traditional feed resources has at the end affected the conditions for mariculture developments negatively as no other ω3 HUFA rich bio-resources has been made available for fish feed. It is in particular the production of high-valued marine fish and squid that have been most impacted. Mariculture development has been most successful for macro-algae, herbivore animals, e.g. mussels and tilapia grown in brackish waters, and other fish species that can be produced with a high fraction of oils and protein from agriculture (e.g., ω3 rich oils from rapeseed) and only minor amounts of marine feed. Salmon cultivation continued to expand, because salmon is relatively flexible with regard to lipid nutrition, but its production levelled around 2020. Mariculture in Europe produces mainly shellfish. The culture of carnivore fishes dominating European aquaculture in the early part of the period is strongly reduced. Cultured seafood is to a great extent imported from other regions that have been more successful with their herbivore fish species. There is, however, an increasing competition also for agriculture products in the world market, and the global perspectives are relatively negative.

The global mariculture production by 2050 is 60 million tonnes. The total global marine harvest (i.e., the sum of fisheries and mariculture) is 180 million tonnes, corresponding to an increase of 1.4 million tonnes per year from 2000 to 2050. The perspectives for the future are negative; over-fishing is still a severe problem, developments in mariculture is slow and the sea supplies less food per capita to human nutrition than in the year 2000.

Scenario 2 – Large Herbivore Zooplankton Stocks Is Utilised for Fish Feed in Addition to Agriculture Products

Methods for harvesting, refining, and managing stocks of large herbivore zooplankton became available and commercial by 2015, and the global harvest of fisheries in 2050 reached 190 million tonnes, of which 80 million tonnes were herbivore zooplankton. This implies a major change in harvesting strategies facilitated by a better understanding of the marine ecosystem, an efficient harvesting technology for large herbivore zooplankton, and new efficient international regulations and management principles of fisheries. Harvesting of fish is to a great extent directed toward the large zooplanktivore stocks that are well priced, because their use for direct human consumption has increased. Piscivore species are less abundant in catches, they are at an acceptable price, but most of these species are mainly produced in mariculture. Zooplankton is the main feed resource, and cultured carnivore marine fish, diadromous salmonids, and squid are now produced at a lower trophic level in culture than in nature, because their feed is partly produced from herbivore zooplankton biomass and agricultural products. The use of zooplankton for fish feed has released the fishing pressure on most zooplanktivore stocks, and fishery management has improved. This positive effect on the stocks outweighs the potential negative consequences of harvesting part of their food.

A great variety of species is produced in mariculture by 2050, including macroalgae, herbivore animal species (e.g., other mussel species), crustaceans, and fish. A few major species are responsible for the main volume produced, and large international companies are responsible for the majority of this production. The diversity of cultured "herb species" is still high thanks to local priorities, and smaller producers undertake most of this production. Feed of agricultural origin is mainly used for species that can be grown on such feeds, but it is used as a feed supplement for all species.

European mariculture is doing quite well, thanks to the early investments in R&D during the beginning of the century that made the industry highly competitive. The Mediterranean production of sea bass and sea bream is now considerably increased, thanks to the success of developing an export industry of fresh products to continental Asia and Latin America. Salmon production has continued to increase, although at a lower rate than in the last century. The competition is gradually becoming stronger towards 2050, and again the future prospects for feed supply are questionable.

The ecological efficiency is far higher in culture than in nature, and this contributes to a total global mariculture production by 2050 of 80 million tonnes, with a high component of high-priced species. The total global marine harvest (i.e., the sum of fisheries and mariculture) with zooplankton included is 190 million tonnes, corresponding to an increase of 1.6 million tonnes per year from 2000 to 2050. Fisheries and aquaculture contribute more to the per capita food supply of humans than in the year 2000.

Scenario 3 – Equal to Scenario 2, But "Single Cell Biomass" Becomes Gradually Available

The development is identical to that of Scenario 2 up to late 2020s. Already from around 2020, a new cheap source of marine type "single cell biomass" of a genetically modified bacteria produced based on methane was commercially tested as a resource for fish feed a microorganism that is a nutritionally optimal food for marine animals has been genetically modified to contain ω3 HUFA and is now produced in more or less the same way yeast is produced today. It took some time, but the genetically modified organism was at last accepted world wide for use as animal food. This happened after a strong pressure created by market demands, competitor decisions, and serious human health considerations. The "single cell biomass" was initially used as a supplement in feed for marine and agricultural animals, but it became gradually cheaper and dominant.

European mariculture is doing very well, thanks to the early investments in R&D during the beginning of the century and the fact that Europe took the lead in developing the technology for producing the marine type "single cell biomass" for feed. Both the European feed and the aquaculture industry are highly competitive, and the industries are involved worldwide. Sea bass, sea bream, salmon, trout, and squid are major export products, and mariculture has a significant impact on the economies of all coastal nations. The prospects in 2050 are still very positive. European aquaculture has taken full control with its food cycle. There are no major resource constraints on production anymore, the problems with environmental impact and space for production was solved when the technology of safe offshore mariculture became established, after major R&D efforts, made in cooperation and competition with other nations.

The total global marine harvest (i.e., the sum of fisheries and mariculture) is slightly higher compared to Scenario 2, but the future perspectives are far more positive. Mariculture has a great potential for further growth. The oceans may become more important than land in producing human food already in the early 22nd century.

The above scenarios are highly speculative, but there is considerable confidence in that the availability of feed for cultured fish will become a major driver of the future developments of mariculture. We have not very seriously considered how globalization of production and trade will interact with feed availability during the next half century. There are, moreover, other factors that will affect developments and make predictions more complicated. The effects that feed resources will have on the quantitative yields of mariculture is uncertain, but the effect on the structure of the production and the species that will become successful, is relative predictable.

10.4.2 Environmental Concern and Space (Attitude and Resource Type Drivers)

There is a growing competition for space in the coastal zone in many densely populated regions of the world, and mariculture is one of the relatively space requiring industries that competes for space in many regions (Whitmarsh and Palmieri 2007, Chapter 8 in this book; Dempster and Sanchez-Jerez 2007, Chapter 3 in this book). In many regions, as illustrated by the Balearic Islands in the Mediterranean, there is a main conflict between tourism-recreation and mariculture, and in some cases, mariculture activity is not at all accepted, as it is believed to impact on environmental quality besides its modest visual impacts. This is particularly the case in many western countries. In other regions, characterized by a complete different social situation, mariculture may have a positive perception in the public even though it already utilizes absolutely all the space available in coastal waters. The human perception is different, but both situations imply problems for further growth of the industry and for the provision of seafood to people.

There are, however, environmental limits to the density and intensification of aquaculture, which presently generates considerable environmental impacts (Tett 2007, Chapter 1 this book). There is a growing concern on environmental issues in most countries of the world. Aquaculture is a young industry, and the environmental impacts of the industry have been continuously challenged through its early stage, probably more intensely than for the more mature agriculture industry. Although this has been at least temporarily problematic for the mariculture industry, the industry may on a longer time scale have taken advantage of such pressures, as mentioned above.

One example is the problem of the rapidly increasing use of antibiotics for salmonids, starting in the late 1980s. This practice was opposed by many stakeholder groups, including the educated public, scientists, journalists, and many NGOs. The situation was dramatic for the salmon industry for some years, but the pressure resulted in a drive for innovation, in this case, the development of vaccines which could replace the prophylactic use of antibiotics added to the feed. The use of antibiotics, in fact, dropped to a very low level within a few years of the introduction of vaccines in the market. The situation has remained stable thereafter, with very low use of antibiotics for prophylactic purposes. Public concern was the main driving factor (attitude-type driver). This concern resulted in research activity and innovation (new vaccines), and a more sustainable situation of optimism in the industry was achieved at the end, and is still there. There is little doubt that the industry today takes advantage of the fact that the use of antibiotics for prophylactic purposes have been phased down, but the process that caused the changes was indeed very painful.

Mariculture may affect coastal waters, but it is also important to recognize that mariculture is an industry which more than most others require a clean and unpolluted environment, because the industry produces food for a concerned and critical market. The issues of food safety has for a decade been on the political agenda, particularly in industrialized countries. Food production is completely incompatible with industrial pollution, and also with pollution that originates from mariculture activity itself. It is particularly the emissions of toxins that are threats against human health and fish welfare which are problematic, and some coastal regions and marginal seas are already affected by, for example, relatively high contents of PCBs and dioxins. The scale of this problem is already beyond the stage where captured fish resources cannot uncritically be used for fish feed. This might itself have represented a further amplifying factor for a reduced availability of bio-resources for feed, but there is already a low-cost technology established to remove pollutants from fish oils that are used for fish feed (Bell and Waagbø 2007, Chapter 6 this book). The exposure to environmental toxins is, however, still a potential problem for mariculture development, a problem which to great extent is created by other industries and practices.

An increased loading rate of organic matter and mineral nutrients for plants do not represent the same threat against food safety, and to a certain extent mariculture activity itself produces this type of pollution (Islam 2005, Mente et al. 2006). It has, however, become apparent that the production efficiency of mariculture, and thus its economic potential, is quite sensitive to organic and inorganic loading. This problem is particularly experienced in densely populated regions with very intensive mariculture activity. For instance, disorganised proliferation of milkfish pens in Bolinao (Pangasina, Luzon, The Philippines) lead to a major red tide event that killed all of the fish in the pens and much of the wild fish in the adjacent reefs. There is, therefore, a limit to the density of fish farms, which places limits to intensification where the sediment and water quality become unacceptable for economically feasible mariculture.

It is still a challenge for most cultured species to improve the efficiency of fish feed use in mariculture, and to greatly reduce the environmental impacts per unit fish production. Indeed, mariculture obligations to the environment must comply with emerging legislation related to use and quality of coastal waters. Available space, environmental concern, and mariculture requirements to the environment, or environmental quality, are presumably main drivers for the development of environmental-friendly production technologies that will be used by tomorrow's mariculture activities (see below).

10.4.3 Unpredictable Emotional Events and Trade Restrictions (Attitude Type Drivers)

The problem of unpredictable or emotional events that may affect the development of mariculture on a global scale is probably minor, but the different countries may be relatively seriously affected (Whitmarsh and Palmieri 2007, Chapter 8, this book). Indeed, public opinion is characterized by sudden, unpredictable opinion shifts (Richard and Bouchaud 2005), which often affect consumer attitudes and can have catastrophic market consequences. The events that most likely can trigger this type of response on a large or even global scale may be issues of food security, such as news that question the healthiness of eating fish. A similar event as the mad cow disease for cultured fish is one example. This event harmed in particular the UK beef industry, but even this severe problem was handled and brought under control by the European agriculture industry and the state agencies involved. There is no reason why a similar or equally serious event should not happen for cultured fish, but cow mad disease was a demonstration to science, authorities and the public that the impossible was possible; a disease was transferred by a protein. Recently, information telling that caged salmon contained higher contaminant loads than wild catches raised sudden consumer alarm (Hites et al. 2004). It was subsequently shown that the conclusions of the analyses were biased by comparisons based on wild stocks from pollution-free waters with aquaculture products grown in waters with relatively high pollutant loads, where wild fish showed similarly pollutant loads (Bell and Waagbø 2007, Chapter 6, this book).

Another possible scenario is that of a major environmental accident in the sea, such as a dramatic coastal nuclear power plant or an atomic submarine accident, which may result in high contamination levels of major fish stocks on a regional scale. This may result in both reduced fisheries for human consumption, enhanced prices for seafood, and reduced resources for feed. This is a problem that cannot easily be managed and counteracted; it is presumably only time that can mitigate the negative effects of this problem. A severe contamination problem can seriously affect the perception of seafood in the public, including that in regions that are not directly, only emotionally, affected. In all events, some countries, and perhaps also larger regions, will suffer for long a time. The effect will, however, hardly become global for an extended time.

There are many other unpredicted events that may harm a single group of producers or a country, but never have a global scale. This may typically include trade embargoes, legislation to protect own markets, and negative public relations on a product or a practice. The latter type of problems can easily be amplified by competing industries throwing so-called adverse public relations campaigns or a package of negative rumours brought to and distributed by the press. This may be done to push the negative perception of a product even further, and such activities can be initiated without any foregoing event of negative public concern. It is also important to mention that all industries appear to use these questionable methods, including the mariculture industry, if they see an immediate benefit. It is important, however, that these types of driving factors do not affect mariculture development on a global scale.

10.5 Technological State and Developments of Mariculture – A Basis for Innovation

The history of aquaculture goes back more than 2,500 years, when common carp was produced in freshwater pond systems, which remained the main technology for rearing of aquatic organisms up to our times. Pond systems are still being used worldwide both for marine and freshwater aquaculture. The main technological concepts used for mariculture after 1960 are quite diverse. Some main concepts are as follows:

- Net-cage systems – floating structures suspending nets, maintained in open waters, used primarily for fish
- Rope structures – long line, floating structures of different design, maintained in open waters, used for mussels and seaweed
- Land- or seashore-based systems with flow through water or water recycling (RAS; Recycling Aquaculture Systems) and optional control of waste emissions, mainly used for high valued fish, but applicable for most types of organisms
- Ponds or concrete dams, sometimes with reuse of water, used for crustaceans and fish, but also other types of organisms
- Sea-ranching with controlled production and stocking of juveniles, used for all types of organisms

According to our view, some of these methods have greater future potentials than other, but cultural factors as well as availability of technology will affect our choices of technology also in the future. Sea-ranching and pond systems have limited cultural tradition in the western world, but it is important in Asia. Pond cultures do generally not comply with the requirements of intensive cultivation (see above). The available technology of mariculture represents a constraint for the innovation response that may follow a specific problem.

10.5.1 Ponds and Sea Ranching

Ponds were, as mentioned above, the traditional low-technological systems, and sea ranching also has a relatively long history. These methods were developed over time to great extent through trials and errors. Sea ranching represents low-intensive mariculture, with production within the carrying capacities of nature. Such methods are often mentioned as extensive mariculture, with low production as compared to most other methods. The methods of intervening with nature during sea ranching are selective stocking of juveniles, predator control, and a careful selection or improvements of the habitat. Other means of controlling this type of production are virtually absent. Juveniles of fish and most other groups that are restocked are normally produced in land-based farms, but the seed can also be collected from natural spawning habitats. Restocking of populations in coastal waters is a widely used technique in East and South-East Asia, but such techniques were introduced also in Western countries already in the first part of the 20th century (Svåsand and Moksness 2004). These techniques are currently used for salmonids, for example in Western Europe (Aprahamian et al. 2003), whereas Asian countries may restock most of their important stocks of fish and shellfish in coastal waters.

Ponds are still widely used in most parts of the world, particularly for shrimps, fish and a variety of other organisms. The techniques used are currently highly diverse depending on the type of organism cultured, climate, and the environmental and cultural conditions. Pond and concrete culture systems, which can be characterized as closely related, are most suitable for semi-extensive mariculture, because the means of maintaining control of cultivation is relatively poor. Pond culture is not suitable for maintaining high exchange rates of water, which is perhaps the strongest means available for controlling environmental and biological conditions in cultures. Pond systems are also relatively shallow, with strong benthic-pelagic interactions. Even a moderate accumulation of wastes in sediments will therefore easily affect living conditions of the cultured animals. There are major efforts made to increase feeding of pond culture in order to increase production, for example of high valued shrimps, but these attempts have some times failed (Fast and Leung 2003). Means to enhance the carrying capacity of pond culture are for example strong aeration or oxygenation, water circulation, regular draining/cleaning procedures with removal of sediments, and introduction of scavenger organisms that may utilize wastes from the original organism (multi-trophic culture). The history of the last decades has shown that slightly intensified pond cultures can be controlled adequately, but it is even clearer that such systems are not adequate for intensive production. Phytoplankton blooms, anoxia, frequent outbreaks of diseases with collapses of standing stocks have generated great losses to the shrimp industry, and appear to still remain a problem (Graslund and Bengtsson 2001, Alonso-Rodriguez and Paez-Osuna 2003, Avnimelech and Ritvo 2003, Fast and Leung 2003).

10.5.2 Cages and Rope Cultures

Cage fish farms and land based fish farms using flow through or recycling technologies represent the high-technological solutions for fish cultivation. The majority of the world's cage fish farms are, however, still relatively primitive wooden cage systems only suitable for protected locations. On the other hand, the most cost-efficient production of fish today is undertaken in large cage farms situated in relatively open and exposed locations. There are two main types of modern cage fish farms, one system with cages suspended in a rigid framework of a steel construction and one with flexible, wave resistant cage units made by plastic tubes, suspended and moored independently to form an overall flexible main structure. The latter structure is apparently becoming the most feasible solution.

Feeding and maintenance is highly mechanized and automated in the most modern cage farms. The European production of salmon, sea bass and sea bream is mainly undertaken in modern cage fish farms, and salmon farming has been the main driver of these technological developments. The feed conversion efficiency (feed invested per fish produced) was traditionally a potential main weak point of cage culture, because of severe feed losses. Modern salmon farms are constructed and operated in a way that this is no longer a problem, the feed conversion efficiency is normally around 1 kg feed used per kg fish produced.

The long-line cultures represent cage culture like system that can be used for benthic plants and animals, and these are primarily used for mussels and seaweeds. The long-line structures used around the world are highly diverse, adapted to cover the local requirement, species and production costs. Shallow waters have required special adaptations, for example for oyster farming. The structures carrying the oysters are net systems suspended from the bottom to the surface. The technological state of the systems for shellfish and seaweed cultivation is also highly variable, the most advanced and industrialized technology used for shellfish appears to be that used for production of green mussels in New Zealand.

10.5.3 Land-Based Farms

Intensive farming of fish in flow-through and recycling aquaculture systems (RAS) has made major progress during the last decades, and there are now developed concepts which allow inland production of marine fish. In RAS, the water drained from the production tanks is carefully and successively treated with biofilters, skimmers, UV, ozone and other optional treatment methods to remove organic and inorganic waste from the water before it is reused for aquaculture (Blancheton 2000). It will still take some time before such systems are widely used, because they require competence and technical skills. Flow-through fish farms is still most common. Land based farms can be seen as a development of pond cultures that are organized and operated to allow intensification under adequate production control. One of the main differences between recycling systems and cage cultures is that the physical environmental conditions can be fully controlled in RAS. For instance, temperature and light cannot be directly controlled with cages. Manipulation of the growth environment can help reduce production time in RAS systems, through, for instance, the control of temperature around optimal levels. These improvements can compensate for the higher capital and maintenance costs of land-based farms as compared to cage cultures.

It is also important to note that RAS, contrary to cage systems, allows a better control of waste emissions. The particulate sludge can be used to produce other organisms or it can be collected and deposited on land. Dissolved nutrients can be retained in appropriate biofilters. It is in principle possible to reuse waste products from cage cultures as well, but this cannot be done in the same controlled way. Combining the cultivation of species of different trophic levels, through the development of integrated multi-trophic aquaculture (IMTA) systems, allows waste recycling for cage cultures. Mussels capture particulate wastes, fish faeces and feed particles downstream of the farm. Seaweeds and phytoplankton absorb inorganic nutrients, and the enhanced phytoplankton growth downstream of the cage system supports mussel production (Neori et al. 2004).

10.5.4 Main Lines of Technological Development

Environmental concern and legislation, along with the reduced available space in coastal zone, will most likely be the main drivers for future innovations and developments of mariculture technology. In the western world, these drivers are anticipated to steer mariculture technology along two main lines of developments:

- Cage culture for fish and long-line culture for mussels and seaweeds will move gradually away from the shoreline, towards the open ocean (ultimately offshore mariculture)
- Land-coastal based fish farms, with reuse of water and control of waste emission.

The capacity to produce marine fish in open-ocean conditions is very high, although it can be risky, under strong hydrodynamic conditions, where wastes from the fish farms will be diluted very efficiently (Olsen et al. 2005). It is very likely that dynamic water conditions are beneficial for fish welfare; at least it is associated with high feed conversion efficiency for salmon. Last, but perhaps most important, is the fact that the big modern cage farms are very cost efficient. Mussels and seaweeds can grow fast in open ocean sites, but the availability of wastes and excess nutrients released from the cage fish farm will be low. More closed and less dynamic locations are therefore better suited for IMTA.

Land based RAS is another compromise to mitigate environmental concern and space requirements, but land based farms will normally have higher production costs than cage cultures. It is currently quite common that the larval and juvenile stages of fish are produced in land based systems (Moksness et al. 2004), including Atlantic salmon and sea bream, and RAS are becoming beneficial because of low water use, reduced energy costs, and high production capacities as a result of the stable environmental conditions that can be maintained in these systems. RAS allow complete environmental control, and will certainly represent a main line of development in western countries in the decades to come. Growth of adult salmon, sea bass and sea bream will most likely continue to take place in cages, but there are other species that most likely will be produced in land-based farms, for example species of flatfish like turbot.

10.6 Concluding Remarks

The present composition of cultured species will to some extent also reflect the future development of species on a short time scale, because markets and the efficiency of the cultivation technology are already responsible for the current situation. The few most dominant cultured species are Pacific cupped oyster, Japanese carpet shell, and Yesso scallop among the molluscs, Atlantic salmon among fishes, and Whiteleg shrimp and Giant tiger prawn among the crustaceans. Beside these species, there are a high numbers of species which are produced in lower, although variable, quantities. The situation may, however, change quite significantly over a longer time perspective. We will suggest that the availability of feed resources will become a main driver for species composition of future marine aquaculture. If the further increase in feed must be derived from use of agricultural sources, and not from new marine or other marine type of sources, there will most likely be a gradual change towards a higher proportion of herbivorous and omnivorous species with lower ω3 HUFA requirements than marine and diadromous fishes. The global production of molluscs and crustaceans, characterised by lower n-3 HUFA requirements, is already increasing faster than fish production, but it cannot be clearly related to the current feed situation. It is a major challenge of aquaculture to achieve better control of the feed availability in the future. Only if this can be realised, aquaculture may grow in a similar way as agriculture. Space for the industry and public environmental concern are other main driving factors of the development, but these constraints can most likely be mitigated through technological improvements.

Acknowledgements We express our thanks to all persons that have participated in the foresight study at NTNU, including employees, SINTEF personnel, external contributors from management and industry, and the international experts.

References

Aksnes DL, Blindheim J (1996) Circulation patterns in the North Atlantic and possible impact on population dynamics of *Calanus finmarchicus*. Ophelia 44(1–3):7–28

Alonso-Rodriguez R, Paez-Osuna F (2003) Nutrients, phytoplankton and harmful algal blooms in shrimp ponds: a review with special reference to the situation in the Gulf of California. Aquaculture 219(1–4):317–336
Alverson DL, Freeberg MH, Murawski SA, Pope JG (1994) A global assessment of bycatch and discards. FAO Fisheries Technical Paper No. 339, FAO, Rome, 233 pp
Aprahamian MW, Smith KM, McGinnity P, McKelvey S, Taylor J (2003) Restocking of salmonids – opportunities and limitations. Fisheries Research 62(2):211–227
Asche F, Roll KH, Tveterås S (2007) Future trends in aquaculture: Productivity growth and increased production. Chapter 9 of this book
Avnimelech Y, Ritvo G (2003) Shrimp and fish pond soils: processes and management. Aquaculture 220(1–4):549–567
Bell JG, Waagbø R (2007) Safe and nutritious aquaculture produce: Benefits and risks of alternative sustainable aquafeeds. Chapter 6 of this book
Blancheton JP (2000) Developments in recirculation systems for Mediterranean fish species. Aquacultural Engineering 22(1–2):17–31
Brugere C, Ridler N (2004) Global aquaculture outlook in the next decades: An analysis of national aquaculture production forecasts 2030. FAO Fisheries Circular No. 1001, Rome, FAO, 49 pp
Caddy JF, Garibaldi L (2000) Apparent changes in the trophic composition of world marine harvests: the perspective from the FAO capture database. Ocean and Coastal Management 43:615–655
Cohen JE (1995) How Many People Can the Earth Support, W.W. Norton, New York, USA
Dempster T, Sanchez-Jerez P (2007) Aquaculture and coastal space management in Europe: an ecological perspective. Chapter 3 of this book.
FAO (1999) The state of world fisheries and aquaculture 1998. FAO, Rome, Italy
Fast AW, Leung PS (2003) Rise and fall of freshwater prawn (*Macrobrachium rosenbergii*) culture in Hawaii: 1965–1999. Reviews in Fisheries Science 11(3):243–290
Graslund S, Bengtsson BE (2001) Chemicals and biological products used in south-east Asian shrimp farming, and their potential impact on the environment – a review. Science of the total Environment 280(1–3):93–131
Hall SJ, Mainprize BM (2005) Managing by-catch and discards: how much progress are we making and how can we do better? Fish and Fisheries 6:134–155
Hinzpeter I, Shene C, Masson L (2006) Biotechnological alternatives for omega-3 polyunsaturated fatty acids production. Grasas y Aceites 57(3):336–342
Hites RA, Foran JA, Carpenter DO, Hamilton MC, Knuth BA, Schwager SJ (2004) Global assessment of organic contaminants in farmed salmon. Science 303(5655):226–229
Islam MS (2005) Nitrogen and phosphorus budget in coastal and marine cage aquaculture and impacts of effluent loading on ecosystem: review and analysis towards model development. Marine Pollution Bulletin 50(1):48–61
Mente E, Pierce GJ, Maria Begoña Santos MB, Neofitou C (2006) Effect of feed and feeding in the culture of salmonids on the marine aquatic environment: a synthesis for European aquaculture. Aquaculture International 14(5):499–522
Moksness E, Kjørsvik E, Olsen Y (2004) Culture of cold-water marine fish. Blackwell, 528 pp
Myers RA, Worm B (2003) Rapid worldwide depletion of predatory fish communities. Nature 423:280–283
Napier JA, Sayanova O, Qi BX, Lazarus CM (2004) Progress toward the production of long-chain polyunsaturated fatty acids in transgenic plants. Lipids 39(11):1067–1075
Neori A, Chopin T, Troell M, Buschmann AH, Kraemer GP, Halling C, Shpigel M, Yarish C (2004) Integrated aquaculture: rationale, evolution and state of the art emphasizing seaweed biofiltration in modern mariculture. Aquaculture 231:361–391
Nicol S, Endo Y (1997) Krill Fisheries of the World, FAO Fisheries Technical Paper 367
Olsen Y, Vadstein O, Slagstad D (2005) Assimilative carrying capacity: Contribution and impacts on the pelagic system. Aqua Culture Europe 2005 "Optimising the Future", Trondheim, Norway, August 5–9, 2005. Plenary Keynote Presentation, Abstract

Opsahl-Ferstad HG, Rudi H, Ruyter B, Refstie S (2003) Biotechnological approaches to modify rapeseed oil composition for applications in aquaculture. Plant Science 165:349–357
Radwan SS (1991) Sources of C-20-polyunsaturated fatty-acids for biotechnological use. Applied Microbiology and Biotechnology 35(4):421–430
Richard Q, Bouchaud JP (2005) Theory of collective opinion shifts: from smooth trends to abrupt swings. European Physical Journal B – Condensed Matter 47:151–159
Svåsand T, Moksness E (2004) Marine stock enhancement and sea-ranching. In: Moksness E, Kjørsvik E, Olsen Y (eds) Culture of cold-water marine fish. Blackwell, 528 pp
Tett P (2007) Fishfarm wastes in the ecosystem. Chapter 1 of this book
Turner RK, Adger WN, Lorenzoni I (1998) Towards Integrated Modelling and Analysis in Coastal Zones: Principles and Practices. LOICZ Reports & Studies No. 11, iv + 122 pp. LOICZ IPO, Texel, The Netherlands
Vannuccini S (2004) Overview of fish production, utilization, consumption and trade, based on 2002 data. FAO Fishery Information, Data and Statistic Unit, Rome, Italy
Vörösmarty CJ, Green P, Salisbury J, Lammer RB (2000) Global Water Resources: Vulnerability from Climate Change and Population Growth. Science 14(289):284–288
Whitmarsh D, Palmieri MG (2007) Aquaculture in the coastal zone: pressures, interactions and externalities. Chapter 8 of this book
WHO (2003) Diet, nutrition and the prevention of chronic diseases. WHO Technical Report Series 916, 160 pp
Wijkström UN (2003) Short and long-term prospects for consumption of fish. Veterinary Research Communications. 27 Suppl. 1: 461–468
Ye Y (1999) Historical consumption and future demand for fish and fishery products: Exploratory calculations for the years 2015–2030. FAO Fisheries Circular No. 946, FAO, Rome, Italy

Epilogue

Modern, industrial marine aquaculture is an emerging industry and uses marine resources to produce food to the human population. It builds upon long human experience of fishing, culturing and handling of fish and marine organisms around the world, combined with experience derived from much longer tradition of agriculture. Seafood products are tasty and healthy, and society is willing to pay a high price to get the best products. Due to stagnant or declining fisheries, aquaculture products are becoming ever more important as supplementary products to wild catches. The ten chapters of this book are based on at least two decades of intensive research in aquaculture with contributions from researchers from all over the world. Aquaculture is a global industry, and with this book we hope that the reader has achieved an overview of the current trends in environmental, economic and social aspects of this growing industry.

During the preparation of this book, it became apparent that aquaculture has the potential to continue feeding the human population products from the sea, and even contribute significantly to global food production, and feeding the growing human population in the future. In contrast to the production of crops and meat on land, marine aquaculture production is almost independent of fresh water as a resource. Fresh water is likely to become a limiting resource at given a human population of 9 billion, which is expected to be reached by 2050 (Duarte et al. in preparation). In contrast to fisheries, which are stagnant and may even decline due to overfishing of particularly top-carnivore species, aquaculture production has the potential to increase, but only if certain bottlenecks of the industry can be overcome. In this book, the four most important bottlenecks for aquaculture are identified as: (1) the availability, suitability and cost of fish feed; (2) space availability for aquaculture purposes; (3) the environmental impacts of aquaculture and (4) technological and energetic requirements (Duarte et al. in preparation).

There are several ways to solve the "feed-trap". Instead of producing high volumes of top-carnivore species such as salmon and cod, a shift to lower trophic levels such as marine algae and bivalves will increase the feed efficiency significant. Marine algae are already the most important product by volume in aquaculture and could be used either for human consumption or in a polyculture system as feed for higher organisms. Bivalves have the advantage of filtering the water, are independent of artificial feed inputs and may even be beneficial to the environment through removing excess particulate nutrient from the fish production in a polyculture system. The present substitution of fish meal and fish oil by terrestrial products in salmon feed (Chapter 6) is only beneficial in the short term, as fresh water in the long term is predicted to limit terrestrial production. Other marine resources, such as the large stocks of zooplankton present in the oceans, could be possible solutions to the feed-trap, although there are some technological as well as ecological constraints, which have to be considered (Chapter 10).

Space is already a limiting factor along several coasts, primarily due to competition from other users, e.g., tourism and urban development, but aquaculture's need for space is relatively limited and the prospected aquaculture production will use <1% of the available shelf area. Proper planning and development of off-shore technology can provide solutions to this bottleneck.

Aquaculture has, as presently conducted, serious impacts on the environment, which are likely to multiply along with increasing production unless initiatives are taken to develop the industry in a sustainable way. Aquaculture impacts the environment through the detrimental effects of waste products and the chemicals, and impact the wild populations by depletion of natural seed stocks, genetic dilution by escapees, release of genetically modified organisms and harvest to produce fish flours and oils. Initial setting is a critical point for avoiding destruction of fragile habitats such as coral reefs, seagrass meadows, salt-marshes and mangroves (Chapter 2). The impacts of waste products can be minimized by use of polycultures, where marine algae utilize the dissolved nutrients and bivalves the suspended particulate waste. Closing the production cycle, including production of feed and seed stock, should be the ultimate goal to minimize the environmental impacts in future aquaculture production. Because of its requirements for relatively good environmental status, aquaculture has the potential to shift from being a negative agent in the marine environment at present to become, if a sustainable model is found, a positive force in the marine environment.

Production technology is not likely to limit future production, as the industry is based on low technology, which over the years will develop into more efficient systems minimizing the environmental impacts (Chapter 10). Energy is used at various stages in the production and a reduced distance between production units and markets may be necessary in a future with higher fuel prices.

Considering these bottlenecks, this book suggests that marine aquaculture has the potential to supply humanity with healthy products from the sea in the future. It is likely that the oceans will become a much more important contributor to global food production, as it is not constrained by the limitations in fresh water resources as observed on land. The expansion will, however, require significant social, scientific, technological and policy developments. Most important is that current practices become a positive force for the marine ecosystem rather than a threat, where polyculture and appropriate siting are two of many solutions to be considered. Not only the industry, but also society has to participate in this development of food production. Intelligent decisions and planning are required.

We hope that this book provide a useful departure point to face the future challenges in aquaculture research and development. There are numerous issues to be addressed extending from the environmental impacts to human health, to societal developments and to policy making. Aquaculture stands at the forefront of a revolution in food production, and can become an important player (and maybe the only solution) to feed the much larger humanity in the future, but only if the existing bottlenecks are solved quickly and in an environmental sustainable manners.

Reference

Duarte CM, Holmer M, Olsen Y, Soto D, Marbà N, Guiu J, Black K, Karakasis Y (in preparation) Will the oceans help feed humanity?

Index

MIX
Papier aus verantwortungsvollen Quellen
Paper from responsible sources
FSC® C105338

If you have any concerns about our products,
you can contact us on
ProductSafety@springernature.com

In case Publisher is established outside the EU,
the EU authorized representative is:
Springer Nature Customer Service Center GmbH
Europaplatz 3, 69115 Heidelberg, Germany

Printed by Libri Plureos GmbH
in Hamburg, Germany